AF371495

Engineering Materials

For further volumes:
http://www.springer.com/series/4288

James Njuguna
Editor

Structural Nanocomposites

Perspectives for Future Applications

 Springer

Editor
James Njuguna
School of Engineering
Institute for Innovation, Design & Sustainability
Robert Gordon University
Aberdeen
UK

ISSN 1612-1317   ISSN 1868-1212   (electronic)
ISBN 978-3-642-40321-7   ISBN 978-3-642-40322-4   (eBook)
DOI 10.1007/978-3-642-40322-4
Springer Heidelberg New York Dordrecht London

Library of Congress Control Number: 2013954843

Printed on acid-free paper

Springer is part of Springer Science+Business Media (www.springer.com)

Preface

This book acknowledges that the key to success of modern structural components is tailored behaviour of materials. A relatively inexpensive way of obtaining macroscopically desired responses is to enhance base material properties by addition of micro or nanoscopic matter, i.e. to manipulate the macrostructures. Accordingly, in many modern engineering designs, materials with high complex microstructures are now in use.

The macroscopic characteristics of modified base materials are the aggressive change of an assemblage of different 'pure' components especially at nanoscale. A key challenge to nano-enhancement is dispersion methods for the development of polymeric nanocomposites, which is covered in "Dispersion Methods of Carbon Nanotubes for the Development of Polymeric Nanocomposites: Characterization and Application". Nanostructured materials have great potential as reinforcement materials in polymers (polymer matrix) when used to modify and improve their physical and mechanical properties. Examples of unique properties achievable today are covered on nanotubes/thermoplastic nanocomposites on nanotubes/ thermoplastic nanocomposites (Thermoplastic Nanocomposites with Carbon Nanotubes) and thermal and mechanical properties of graphite-based nanocomposites (Graphite-based Nanocomposites to Enhance Mechanical Properties). With ease to manufacture in bulk, nanomaterials are finding increasingly more practical applications (e.g. in the manufacture of plastic auto-aero parts, structures, implants and ceramics to microelectronics). "Optimization and Scaling up of the Fabrication Process of Polymer Nanocomposites: Polyamide-6/Montmorillonite Case Study" therefore focuses on optimization and scaling up of the fabrication process of polymer nanocomposites.

Due to the molecular size of their reinforcement, polymer nanocomposites offer the possibility to develop new materials with unusual properties. These include photoactivity (Influences of Morphology and Doping on the Photoactivity of TiO_2 Nanostructures), foam-glass-crystal materials presented in "Foam-Glass-Crystal Materials" and high temperature polymer nanocomposites covered in "High Temperature Polymer Nanocomposites" is focused both on physical modification of traditional polymers as well as synthesis of new monomers from which new polymers are obtained with interesting new properties that can provide opportunities for novel applications. Impact and energy absorption performance (experiments, modelling and simulations) are presented in both "Energy

Absorption and Low-Velocity Impact Performance of Nanocomposites: Cones and Sandwich Structures" and "Predictions of Energy Absorption of Aligned Carbon Nanotube/Epoxy Composites". It is shown that nanotechnolgy and nanoscience indeed offers promising results and a unique level of mechanical properties enhancement and/or control by involving the use of nano-sized organic and inorganic particles. An important area of research on antibacterial applications is the central focus of "Silver Nanocluster/Silica Composite Coatings Obtained by Sputtering for Antibacterial Applications", Materials with antibacterial properties are more and more requested in several fields where the risk of microbial contamination is considered a relevant issue.

Foams play a key role in many technological applications. A huge amount of research was done in the last two decades to develop low-toxicity fire retardants on foams. The use of flame retardants to reduce the flammability of polymers and production of smoke or toxic products during their combustion is nowadays an important aspect of the research, development and application of new materials. As an example, "Recent Advances on the Utilization of Nanoclays and Organophosphorus Compounds in Polyurethane Foams for Increasing Flame Retardancy" looks into recent advances on the utilization of nanoclays and organophosphorus compounds in polyurethane foams for increasing flame retardancy. Finally, "Ecological Assessment of Nano Materials for the Production of Electrostatic/Electrochemical Energy Storage Systems" covers the ecological assessment of nanomaterials for the production of electrostatic/electrochemical energy storage systems. The ecological implications (regarding known environmental effects) of carbon-based nanomaterials are analysed using Life Cycle Assessment (LCA) approach.

This book covers important issues and topics that are attractive to the scientific community. It is a useful tool for scientists, academicians, research scholars, polymer engineers and industries as it is a unique set of valuable contributions from renowned experts. This book is also supportive for undergraduate and postgraduate students and hope fully an inspiration to many young scientists to devote their efforts in nanomaterials research.

The editor acknowledges the great effort put into this book by the contributing authors. He is very thankful to all the contributors and also to a great team at Springer for brilliant editorial support.

James Njuguna

Contents

Dispersion Methods of Carbon Nanotubes for the Development of Polymeric Nanocomposites: Characterization and Application

C. Manteca Martínez, A. Yedra Martínez and I. Gorrochategui Sánchez

Abstract Composites use is increased worldwide quickly because they present significant advantages over traditional materials, as metals. It is able to get improved and tailored properties to each application by means the formulation. Initially, composites were used in advanced technologically sectors as aeronautical, although the technological advance and manufacturing cost reduction have promoted their use in different industrial sectors as building, renewable energies, automotive, sports, nautic, etc. In all cases there are needs to reduce weight, to improve the resistance in aggressive environments and to keep high performances during the life-span in more demanding applications. In this chapter is shown the development of thermoset multiscale composites with improved properties, making special emphasis in the dispersion method of the nanoreinforcements. Thermoset multiscale composites consist of glass fibers as reinforcement and multiwall carbon nanotubes (MWCNTs) as nanoreinforcement. MWCNTs have a huge potential as nanoreinforcements polymeric matrices because they present unique and excellent mechanical, thermal and electrical properties with a competitive cost. Moreover, currently in the market there is enough availability of these nanomaterials for using in industrial scale. As polymeric matrix has been used polyester resin. These new low cost polymer composite/nanocomposite materials improved with nanoreinforcements opens new market opportunities in large-volume applications as structural composites: civil engineering works, transport sector, etc. The main challenge is to transfer the MWCNTs excellent properties to the polymeric matrix. It requires dispersing the nanoreinforcements as individual particles in the polymeric matrix, to avoid agglomerates. For this reason, it has studied and implemented an advanced dispersion techniques of high shear forces called Three Roll Mills. Following, it has been performed a novel and detailed quality and quantity characterisation of the dispersion rate combining different techniques: viscosity measurements and confocal Raman spectroscopy.

C. Manteca Martínez · A. Yedra Martínez (✉) · I. Gorrochategui Sánchez
Fundación Centro Tecnológico de Componentes (CTC), Santander, Spain
e-mail: ayedra@ctcomponentes.com

J. Njuguna (ed.), *Structural Nanocomposites*, Engineering Materials,
DOI: 10.1007/978-3-642-40322-4_1, © Springer-Verlag Berlin Heidelberg 2013

Later the samples was characterised from mechanical point of view. It has been obtained a high MWCNTs/CNFs dispersion rate within the matrix by using 3-roll-milling process. It has permitted to work with nanoreinforcements low content: from 0.1 wt % to 1 wt %, obtaining good results with 0.1 wt %. Moreover, it is a process easy to industrial scale up. MWCNT's distribution map based on intensity RAMAN spectra (at G band) has been obtained. All samples present a high dispersion rate of the nanoreinforcements. Furthermore, correlation between viscosity increasing and nanoreinforcements content and dispersion rate has been observed. From mechanical point of view, the samples present better behavior than original sample (without nanoreinforcements). Depending of nanoreinforcements content and parameters during the dispersion process, the mechanical improvements can be (in the tensile test) up to 13 % in tensile strength and 15 % in Young module, and (in flexural test) up to 26 % and 84 %. Finally is shown an example of structural application of these materials: panel sandwich.

1 Introduction

The use of advanced composite materials is becoming increasingly popular in the manufacturing of structural elements in recent years [1–3]. Thus, traditional materials such as steel, wood or aluminum are being replaced in some applications by advanced composite materials, with better specific properties [4].

Within the large variety of composite materials in the market, those formed by organic matrices (epoxy, vinylester, polyester, …) and high strength fibres (glass, carbon, aramid, …) are the most developed and used in an industrial level [5, 6].

These new materials have been applied almost exclusively in technologically leading industries such as aeronautics or aerospace, although at present, other industries as the civil works sector is starting to use them [2]. In particular, composite materials of glass fibre polyester resin are used in the transportation industry, chemistry, civil works and sports, due to its low density, good mechanical performance (its excellent relations stiffness/weight and resistance/weight), chemical stability, aggressive environment resistance, long life, low cost of manufacture, installation and maintenance [5, 6].

In recent years, the trend in the development of advanced reinforced composites with additives is the reduction of their size, from macro and micro scale to nanoscale [7]. The introduction of new nanoadditives can develop nanocomposites with properties tailored to the requirements of the end user application. Examples of such nanoadditives are: nanoclays, nanoparticles of aluminium oxide, titanium oxide, carbon nanotubes and nanofibres, etc. The use of these nanoadditives can improve mechanical, electrical, fire resistance composite properties [8, 9].

Nanostructured materials have great potential as reinforcement materials in polymers (polymer matrix), used to modify and improve their physical and mechanical properties. It is a consequence of the nanoscale dimensions of the fillers: increases the surface/volume ratio, reduces the distance between particles and increases the ratio form (s = l/d, where l:length and d:diameter). These new properties and/or improvements achieved will be influenced by the shape, size and volume fraction of the fillers, dispersion and interfacial bonding of nanoparticles in the polymer matrix [7, 10].

In particular, carbon nanotubes (CNTs) have one of the highest specific surface areas in nanostructured materials. This feature allows that a small quantity of nanostructured material is able to influence a large volume of the polymer matrix, because they can provide a contact area much higher than traditional particles [7, 10, 11]. This is a great advantage in comparison to traditional fillers used in composites, where it is necessary to introduce a higher quantity to get significant benefits that can worsen other aspects and features of the material [12]. To obtain an effective reinforcement of the polymeric matrix with CNTs, it is necessary to take into account the following requirements [7, 13, 14]:

- CNTs quality: length/diameter ratio, degree of purity and integrity.
- Dispersion: CNTs must be dispersed as individual particles into the polymer matrix.
- Functionalization of the nanomaterials: high interfacial bonding of the individual CNTs with the polymer matrix in order to achieve transfer of stresses from the matrix to the CNTs.

2 Materials

The most used fibre reinforced polymer (FRP) are thermoset resin reinforced with high strength fibres, because they present characteristics like low cost, easy fabrication, simplified installation, low weight and high variety of shape and size, that make them desirable for design in industrial applications [15].

The fibres are responsible to provide structural properties in composites like strength and stiffness. They also provide thermal stability and electrical conductivity or insulation (depending on the type of fibre). Reinforcing fibres show different forms: short chopped fibres, long continuous fibres, woven fabric or mat. The final properties of composites depend on the configuration and combination of the fibres [16].

The thermoset resin is the matrix material that joins the fibres, distributes and transfers the external loads to the fibres. The resin separates the fibres individually and prevents the propagation of cracks. It provides protection to fibres against external damage which may result from mechanical abrasion or chemical reactions with the environment. Finally the resin provides flexibility on the shape during the manufacturing and good surface finish quality [16, 17].

Glass fibres are the most widely reinforcement in FRP because of its availability, low cost (cheaper than other type of fibres like carbon, boron or aramid) and good mechanical properties. The typical diameter of glass fibres range from 5 to 25 μm, being flexible and easily conform to different kind of shapes [16].

Unsaturated polyesters (UP) form an important group of thermoset resins employed to manufacture polymer composites. This resin is used in different sectors and applications: roofing and building insulation, sport cars bodies, translucent roofing panel in lorries, public transport vehicles, aircraft radomes, ducting, spinners, chemical storage vessels, chemical plant components, swimming pools, stacking chairs and sports equipment [6].

Carbon nanotubes are an allotropic form of carbon. It consists of a number of graphene layers rolled to form a cylindrical tube. Its diameter ranges from one to ten nanometers and its length reaches several hundreds nanometers, even micrometers [18]. There are several types of carbon nanostructures: single-walled nanotubes (SWCNTs), multi-walled nanotubes (MWCNTs) and carbon nanofibres (CNFs) [10]. MWCNTs were observed in 1991 when Ijima imaged a product of an arc discharge experiment. CNTs show unique mechanical and electrical properties. They present high mechanical properties strength and stiffness) and high electrical and thermal conductivity, providing potential applications for a variety of areas such as reinforcements in composites, nanoprobes, displays, sensors, energy storage and electronic devices [10, 18, 19].

3 Nanomaterials Dispersion Within Thermosetting Resin

One of the most important aspects to transfer the excellent properties of the nanomaterials to the polymeric matrix is the dispersion rate. The main challenge is to disperse the nanoparticle as individual particles in the polymer matrix.

When the fillers size is reduced, the distance among particles in the matrix is reduced too. Some authors [8] have established a relationship between the distance between nanoparticles and their volume fraction (%) for single-wall nanotubes. It has been obtained based on the theoretical model of hexagonal dense packed [14]. For instance [8], at a volume fraction of 4 % the distance between two nanotubes is around 2.8 nm. If the polystyrene is assumed as polymeric matrix with a chain diameter of 0.8 nm, there is only a small gap for the polymer to penetrate between the nanotubes. So, in this example, the maximum volume fraction of SWCNT that could be used is 10 %. In general, it is very difficult to introduce high volume contents of nanoparticles in a polymeric matrix. When the nanoparticles volume content is high, they could be agglomerated.

On the other hand, nanomaterials present a high specific surface area because of its nanometric scale. In consequence they tend to aggregate due to adhesive forces (electrostatic and Van der Waals forces). For this reason, it is necessary to use effective dispersion techniques. These techniques have to be able to break up the agglomerates and distribute the individual nanoparticles in the polymer without producing damage in their nanostructure.

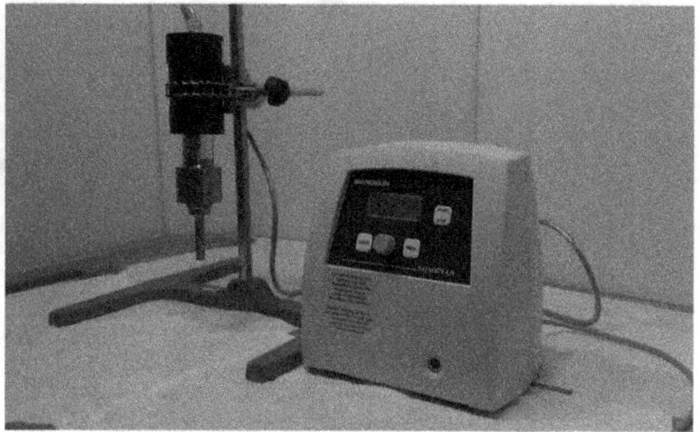

Fig. 1 Sonication equipment to disperse nanomaterials

Currently, different methods are being used to disperse nanomaterials into polymeric matrices. The main methods are:

- **Sonication**

  Sonication technique applies ultrasound energy to disperse particles in liquid media. It involves introducing an ultrasound probe into the mix of resin and nanofillers. It generates high local vibrational energy, separating the agglomerated nanoparticles. This method is suitable for small volumes and low viscous materials. However, this high local energy leads to breakage and damage of nanoparticles reducing the aspect ratio [14, 20] (Fig. 1).

- **Stirring**

  This method is suitable to disperse nanoparticles in medium liquid viscosity systems. This method applies high shear forces to disrupt the agglomerates. The high shear forces can be obtained by different principles: rotating dics, rotor–stator systems or ball mill [8, 14].

  Design of the impeller, mixing speed, size of balls, process time, among other parameters, control the dispersion result [14] (Fig. 2).

- **Calendering (Three roll milling)**

  This method is a suitable technique to disperse microparticles in different media. This technique breaks the agglomerates by the crushing forces of the rollers and the high shear forces resulting from the rollers. This method can be used to process from low viscosity products to thick materials and from small to large volumes [20, 21].

Following this method is explained in more detail because is the one applicable for CNTs in polymer matrices.

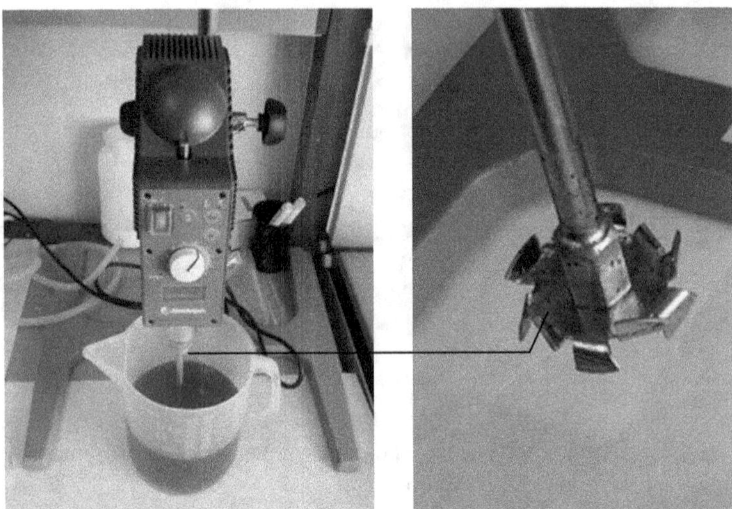

Fig. 2 Rotating disc to disperse particles in different matrices

4 Three-Roll Milling Process

4.1 Principles

One of the promising methods to disperse CNTs in thermoset resins is the three-roll milling technique. This technique is a standard method to disperse micro-particles in industry (e.g. colors, electronics, cosmetics, food, adhesives, etc.), but is becoming a promising technique to achieve good dispersion rate with nano-particles [14, 20]. The main advantages of this method include high particle distribution, temperature control, preventing contamination, small and large batches, easy process control, low material loss, etc. [21].

In this technique, mixing and dispersion of MWCNTs take place in the vortices which are located between the rollers by combining the crushing force of the rollers and the high shear forces resulting from rollers speeds and the distance between the rollers [21], as shown in Fig. 3.

The process to disperse the nanoparticles using this technique is described next. First, CNTs are manually mixed into the neat resin, and then the mix is fed into the feed roller and center roller by the hopper. In this gap occurs a first dispersion of the aggregates and the final dispersion is performed in the gap between the center roll and apron roller.

With this technique are achieved high shear forces able to obtain good results in dispersion of nanostructures within polymer matrices, besides being easily scalable to industrial level.

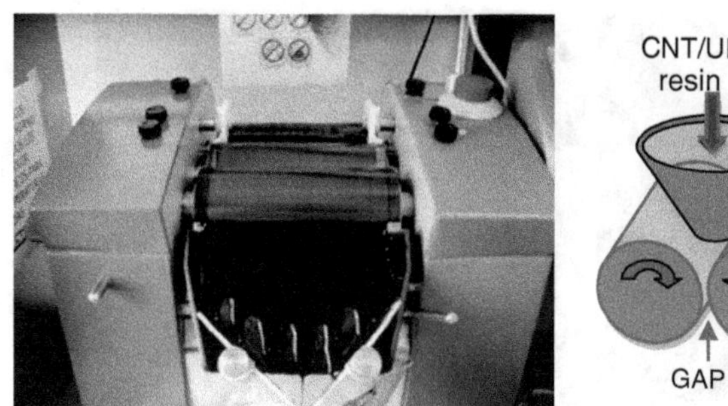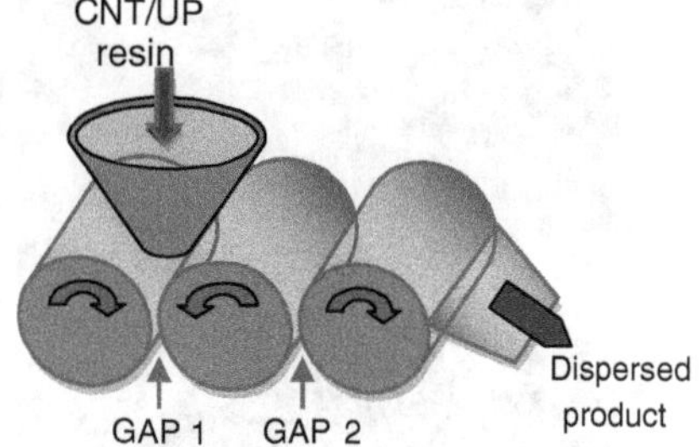

Fig. 3 Carbon nanotubes resin dispersion on the rollers and diagram of three-roll milling method

Other techniques that apply shear forces to distribute the CNTs into thermoset resins like rotating discs don't reach a degree of dispersion as high as three roll milling.

In both techniques, the dispersion of the nano-particles is due to the shear forces generated, responsible for breaking the aggregates of CNTs. In the case of rotating disc (mechanical stirring), these shear forces are created between the container wall and the stirring metal disc, however, in the three-roll milling the dispersion of the mixture takes place in the vortices which are between the rollers.

The strain rate and shear stress are calculated according to the following equations:

$$\dot{\gamma} = \frac{\Delta v}{Gap} \tag{1}$$

$$\tau = \eta \frac{\Delta v}{Gap} = \eta \cdot \dot{\gamma} \tag{2}$$

where Δv is the speed difference between metal disc and container wall for rotation disc technique and speed difference between rollers for the three-roll milling technique, Gap is the distance between the container wall and the metal disc or the distance between the rollers, and η is the viscosity of the sample.

According to this, assuming the parameters shown in the Table 1, for rotation disc technique the shear stress is constant because the distance between the container wall and the metal disc is fixed during the dispersion process, whereas for three-roll milling technique, decreasing the distance between the rollers increases the shear stress.

Table 1 Example of dispersion parameters

Dispersion process	Speed (rpm)	Gap
Rotation disc	2,000	$\approx$20 mm
Three-roll mill	230	40, 25, 15, 10, 5 μm

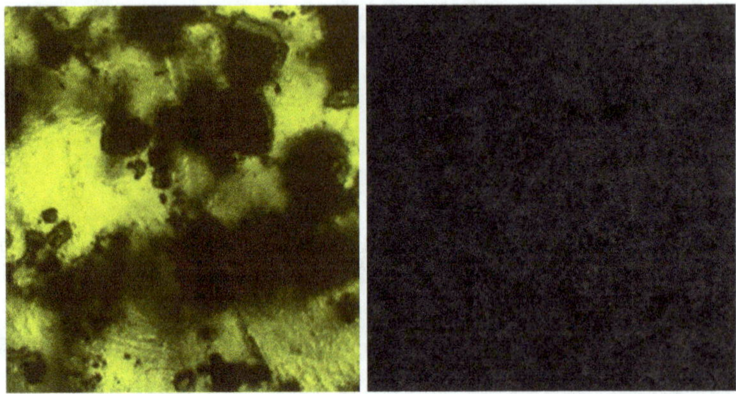

Fig. 4 Optical images of sample (resin + 0.1 wt%) produced by rotation disc technique (*left*) and three-roll milling technique (*right*)

The shear stresses generated are three orders of magnitude higher in the three-roll milling technique. These forces are able to break up the nanotubes aggregates and disperse them into the polyester resin, as shown in the pictures (see Fig. 4), without breaking or damaging the nanotubes, because the nanotubes have a much higher tensile strength that three-roll milling process generates.

4.2 Preparation of Samples

There are several processes to fabricate fibre reinforced thermoset polymers. For selecting a process, different factors influences the decision such as production rate, product shape and size, the type of thermoset polymer to be used, cost, properties required, quality, etc. [15, 16].

In general, the manufacturing process of fibre reinforced thermoset polymers consists of four steps [15, 16]:

- Impregnation: The target of this step is to achieve that the resin flows into the fibres to form a plate called laminate.
- Lay-up: Consists of placement the laminates at the desired form as dictated by the design.
- Consolidation: In this step the layers are in close contact and the air is removed.
- Solidification: This is the final step that achieves the final shape of the product. This stage depends on the type of polymer used. Usually, pressure or vacuum is needed during the step. The solidification rate depends on the type of thermoset polymer and the polymerization kinetics.

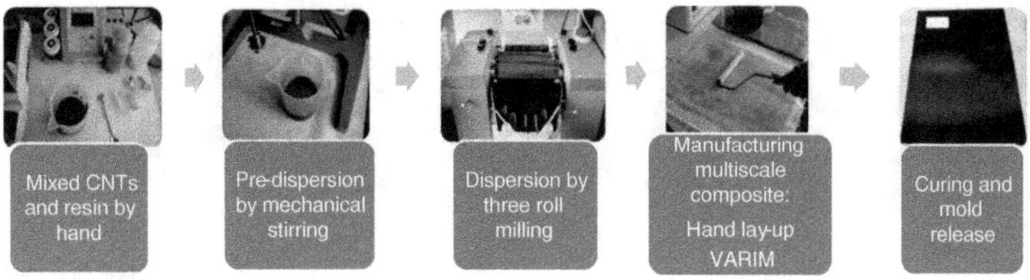

Fig. 5 Scheme of manufacturing nanocomposite

The introduction of nano-particles into thermoset polymer as additives involves changes in impregnation step. A suitable processing technique is necessary to disperse the nano-particles agglomerates into the thermoset polymer.

The manufacturing of fibre reinforced polyester resin nanocomposite will be carried out in three steps: (1) addition of the nanoreinforcement into the polyester resin and dispersion; (2) addition of the hardener and impregnation by hand lay-up or vacuum assisted infusion molding (VARIM); and (3) the curing and mold release stage. An example of a schematic illustration of the process is shown in Fig. 5.

Dispersion by three roll milling consist of several sequential stages, where in each one of them the distance between rollers (gap) was reduced up to 5 μm and rollers speed was kept constant (feed roller: 25 r.p.m., center roller: 76 r.p.m. and apron roller: 230 r.p.m.).

5 Samples Characterization and Potential Applications

5.1 Dispersion Characterization

There are several characterization techniques to determinate the structure and properties of the nanocomposite: x-ray diffraction, transmission electron microscopy, scanning electron microscopy, dynamic mechanical thermal analyses, mechanical properties, flammability, etc. [9].

These techniques could be combined with other methods as rheology, Raman spectroscopy and atomic force microscopy to provide a better knowledge of the dispersion of a polymer nanocomposite [9, 22]. An example of the use of these techniques is presented here.

The study of the degree of dispersion carried out by the measure of the viscosity in each stage of three-roll milling process and the determination of the stresses and shear strain rates that are generated during the dispersion process. Strain rates and shear stresses are calculated from expressions (1) and (2), discussed above.

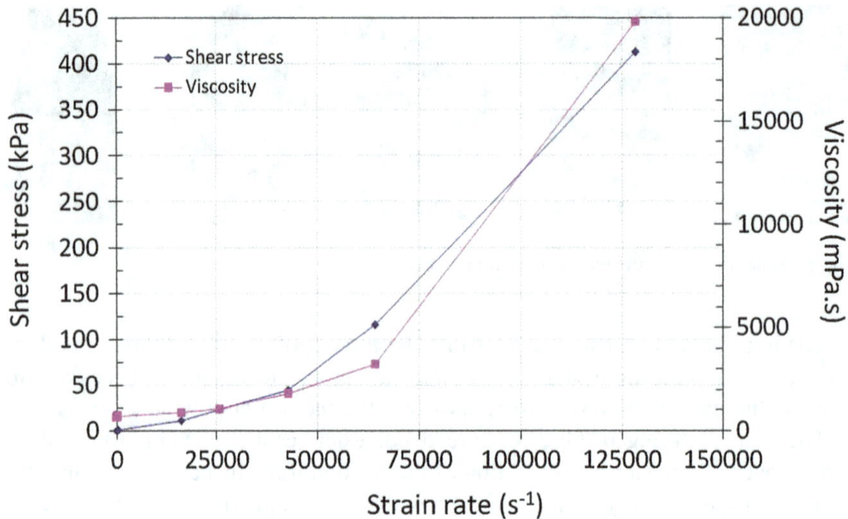

Fig. 6 Shear stress and viscosity of an UP suspension with 0.1 %wt MWCNTs as a function of strain rate

Figure 6 shows the shear stress and viscosity respect strain rate of each stage of three-roll milling dispersion process of an unsaturated resin containing MWCNTs (0.1 wt%).

Taking into account that a high dispersion rate of carbon nanotubes is related with high strain rate values during the dispersion process, it is observed that an improvement of the dispersion of carbon nanotubes involve an increasing of viscosity. Decreasing the distance between rollers, the shear stresses rise followed by an increase in viscosity of the sample. These shear forces are capable of disrupting the carbon nanotubes aggregates in the polymer matrix causing an increase of the sample viscosity two orders of magnitude higher respect the neat polymer matrix.

Viscosity measurements were performed in order to analyze the effect of increasing the carbon nanotubes content on the viscosity. Figure 7 shows the results of viscosity respect carbon nanotubes content of the samples after a three-roll milling dispersion.

Clearly, it is observed that as the MWCNT content increase, the viscosity of the sample rise too. Samples containing 1 wt% MWCNT (or more content) showed a high degree of viscosity. These high viscosity values become the mix in an unwieldy paste.

Raman spectroscopy is commonly used to characterize and identify materials, providing chemical and structural information of materials. It is a high resolution non destructive technique with an easy sample preparation and operated at room temperature [18]. It is a commonly technique for characterize carbon nanotubes. A general Raman spectrum of MWCNT is shown in Fig. 8. The Raman features of

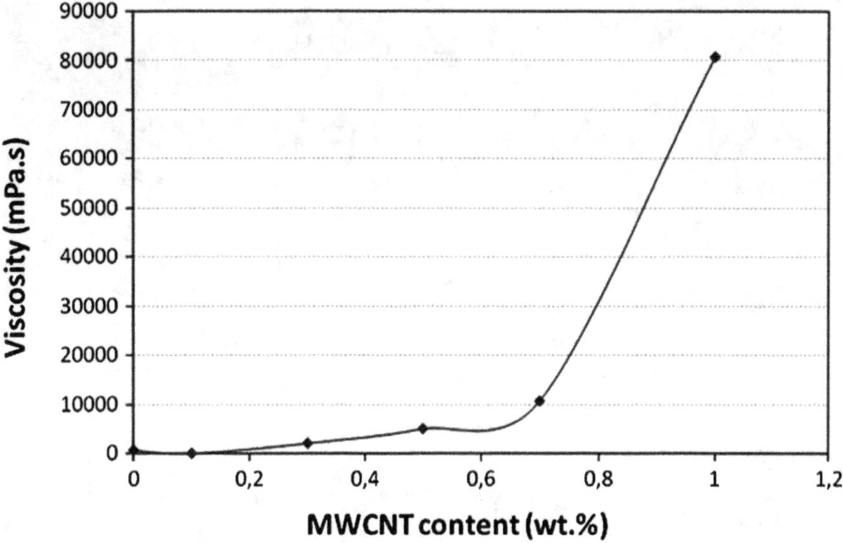

Fig. 7 Viscosity at various carbon nanotubes content for three-roll milling dispersion process

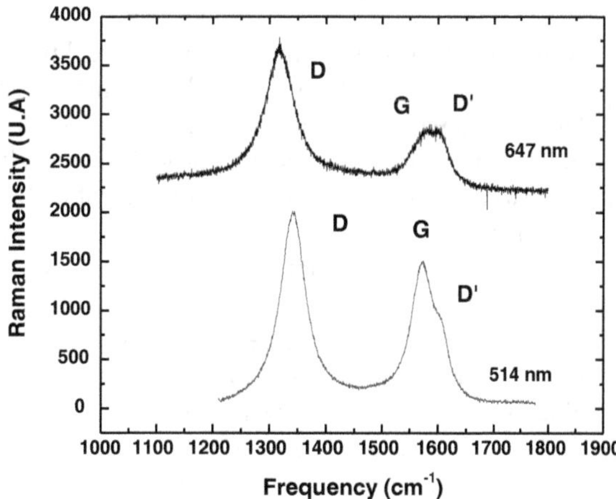

Fig. 8 Raman spectrum from MWCNTs bundles

MWCNT are the tangential (G band) mode at higher frequencies and the disorder-induced D band. The G band (at 1,582 cm^{-1}) has two peaks (G^+ and G^-) related to the circumferential (TO) and axial (LO) atomic vibrations. The D band appears at 1,350 cm^{-1} and its related to the presence of defects in the nanotubes walls [18, 23].

To know the carbon nanotubes dispersion within the resin were performed Raman tests in samples containing MWCNT in different percentages.

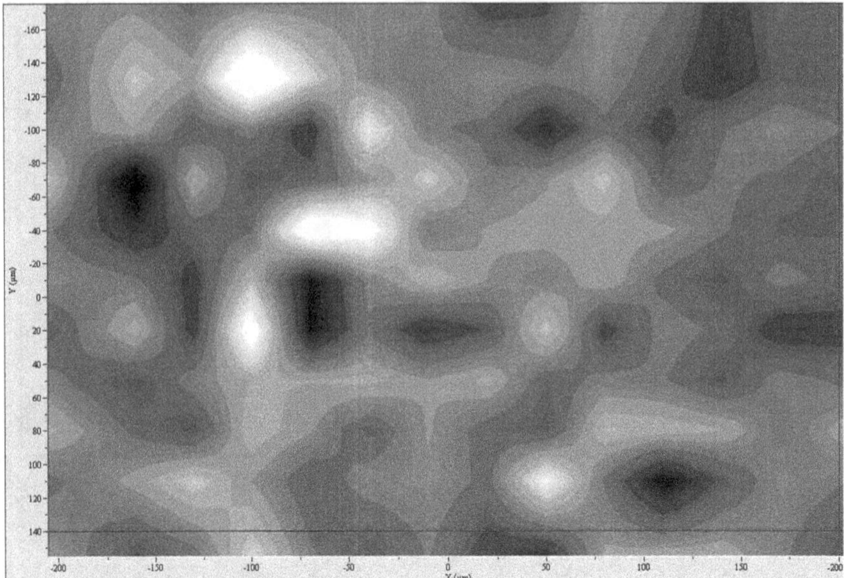

Fig. 9 Distribution map of UP/0.1 % MWCNT

From the intensities of the spectra (Band D) collected from the scans were obtained distribution maps of MWCNT within the matrix, as shown in Fig. 9.

The black areas of the image correspond to areas in the sample where there are not carbon nanotubes, corresponding to low intensities of the spectra. Gray and white areas are associated with the presence of carbon nanotubes, which results in higher intensities in the spectra obtained. The highest intensities correspond to high concentrations of nanotubes that suggest there are aggregates in the sample (white colour areas). Therefore, a correct dispersion will have greater percentage of gray colour areas and a lesser percentage of black and white colour areas.

The distribution maps of MWCNT's in the matrix were analyzed using a specific image analysis software. With this software it can be quantified the nanotubes content within the matrix in the scanned area. The results were shown in Table 2 and Fig. 10 shows some pictures of the samples analyzed.

According to the results, in general, a high degree of dispersion was achieved for all samples developed, as shown by the low percentage of black areas (% MWCNT absence) and high percentage of gray areas.

5.2 Mechanical Characterization

Carbon nanotubes present high strength and stiffness as well as high aspect ratio which make them candidate to improve the mechanical properties of polymers

Table 2 Results obtained after treatment of the images produced from the Raman spectra

Sample	% MWCNT filling		% MWCNT absence (% black area)	Scan size (μm)
	% white area	% gray area		
UP/0.1 %MWCNT	27.44	72.22	0.29	72 × 72
UP/0.5 %MWCNT	8.79	90.26	0.95	30 × 30
UP/0.7 %MWCNT	3.77	95.61	0.62	30 × 30
UP/1 %MWCNT	3.01	95.44	1.55	30 × 30

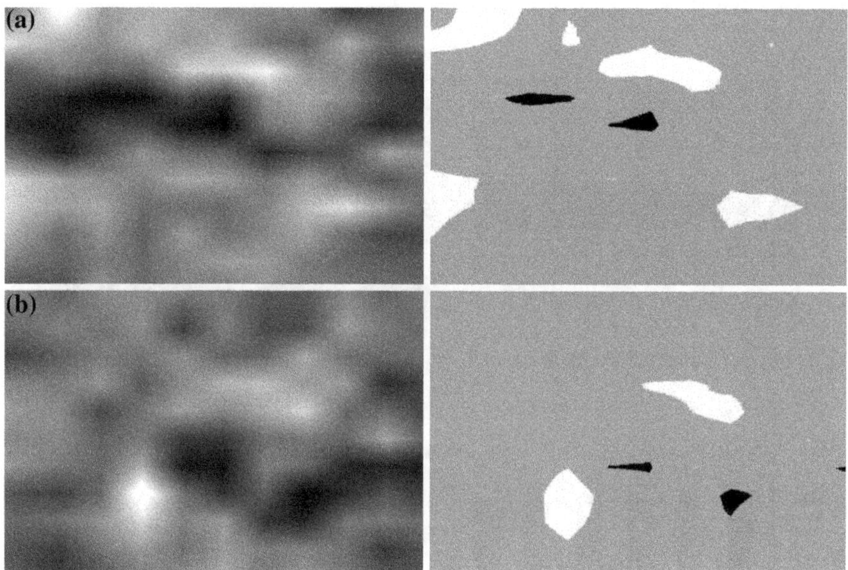

Fig. 10 Distribution maps of MWCNT/polyester composites: **a** MWCNT 0.5 wt%. **b** MWCNT 0.7 wt%

[14]. Figure 11 depicts the tensile behavior of nanocomposites under ambient conditions. Tensile characteristics were determined following UNE-EN ISO 52724 with 250 × 25 × 4 mm specimens [24]. It is observed that the addition of nanofillers improves the tensile properties. Depending of nanoreinforcements content the mechanical improvements can be up to 13 % in tensile strength and 15 % in tensile modulus [25]. The tensile strength decreases with rising nanotubes content but remains above that of the unfilled nanocomposite.

All the samples show a similar behavior. The failure mode was delamination (separate layers involving crack opening, sliding shear and scissoring shear) as a result of a reinforcement-matrix interaction weak [8, 14, 26, 27] (see Fig. 12).

Figure 13 shows the flexural behavior of nanocomposites. For bending properties UNE-EN ISO 14125 was the used standard with 800 × 15 × 4 mm specimens [28]. It is observed that the flexural strength and modulus improved with the

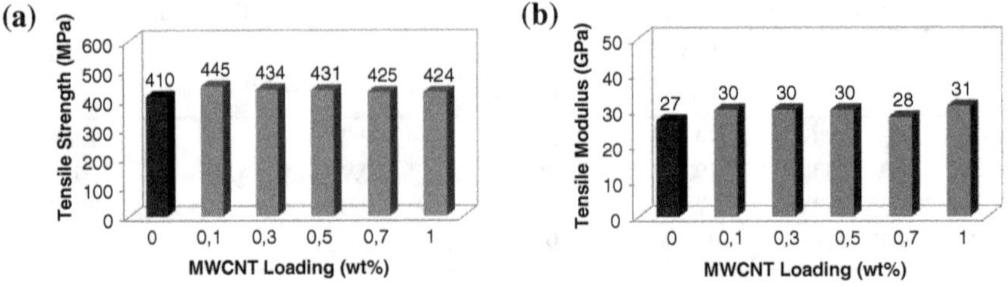

Fig. 11 Tensile strength (**a**) and tensile modulus (**b**) versus carbon nanotubes content for UP/Glass fibre/MWCNT nanocomposites

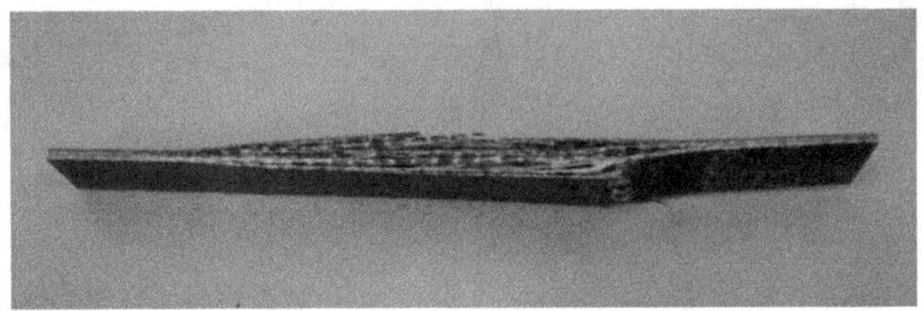

Fig. 12 Delamination failure mode of a tensile test specimen

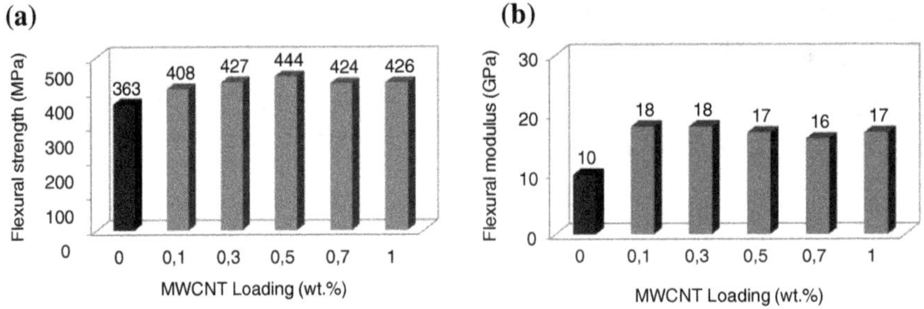

Fig. 13 Flexural strength (**a**) and flexural modulus (**b**) versus carbon nanotubes content for UP/Glass fibre/MWCNT nanocomposites

addition of carbon nanotubes. Depending of nanoreinforcements content the mechanical improvements can be up to 26 % in flexural strength and 84 % in flexural modulus [25].

Due to the material anisotropy two failure mechanisms were found: fibre tensile fracture and interphase shear failure [26, 27] (see Fig. 14).

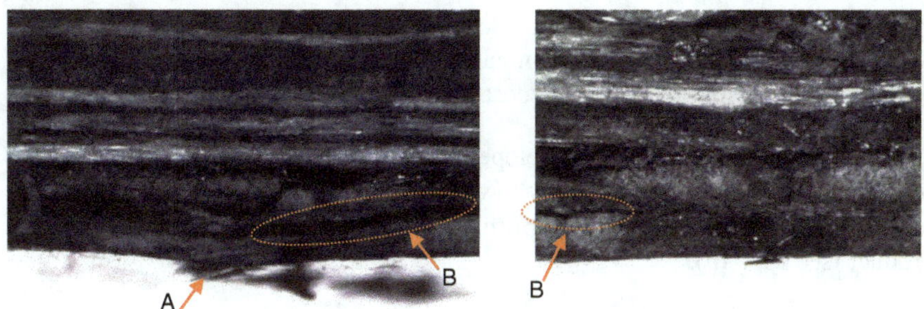

Fig. 14 Bending specimen. **a** Fibre fracture. **b** Interphase shear failure

5.3 Structural Application: Panel Sandwich

Structural sandwich panels are products used primarily in order to perform enclosures in building, refrigeration sector and transport sector (railway carriage). This widespread use is due to its physical and thermal properties, such as design, economic and safety [17].

Manufactured panels are formed by a skin made of a metallic material or reinforced polymeric matrix, and a core consisting of modified polyurethane rigid foam [17, 29].

Functions associated with structural sandwich panels are [17, 29, 30]:

- High bearing capacity with low weight.
- Fire protection in case of fire.
- High thermal-acoustic insulation.
- Good barrier to water and steam.
- Excellent air tightness.
- High resistance to weathering and harsh environments.

Fig. 15 Sandwich panel prototype

- Easy to install.
- Can be easily fixed or replaced in case of damage.
- Long life with low maintenance costs.

Sandwich panels have been developed with improved mechanical performance by incorporating MWCNT's in very low proportions (up to 1 wt%). Moreover, the raw materials used present high performances/cost rate, involving a low cost final product (see Fig. 15).

References

1. ACCIONA company: Web www.acciona.es
2. ARIES PULTRUSION S.L. company: Web www.ariespultrusion.com
3. STRONGWELL company: Web www.strongwell.com
4. Mayugo, J.A.: Estudio constitutivo de materiales compuestos laminados sometidos a cargas cíclicas. Doctoral Thesis, Universidad Politécnica de Cataluña (Spain), (2003)
5. García, S.K.: Análisis de laminados de materiales compuestos con precarga en su plano y sometidos a impacto. Doctoral Thesis, University Carlos III, Madrid (Spain), (2007)
6. Chanda, M., Roy S.K. (eds.): Industrial Polymers, Specialty Polymers and their Applications. CRC Press, US (2007)
7. Friedrich, K., Fakirov, S., Zhang, Z. (eds.): Polymer Composites: From Nano to Macro Scale. Springer, New York (2005)
8. Fiedler, B., Gojny, F.H., Wichmann, M.H.G., Nolte, M.C.M., Schulte, K.: Compos. Sci. Technol. **66**, 3115–3125 (2006)
9. Koo, J.H. (ed.): Polymer Nanocomposites: Processing, Characterization and Applications. McGraw-Hill, New York, (2006)
10. Ureña Fernández, A.: Materiales compuestos con refuerzo multiescalar. Summer School. University of Rey Juan Carlos I (Spain), (2007)
11. Tjong, S.C.: Structural and mechanical properties of polymer nanocomposites. Mater. Sci. Eng. R **53**, 73–197 (2006)
12. R&D Project: Desarrollo de paneles sándwich y perfiles tubulares mediante la tecnología de pultrusión. CENIT 2006–2009. ACCIONA—ARIES PULTRUSION S.L
13. Coleman, J.N., Khan, U., Blau, W.J., Gun'ko, Y.K.: Small but strong: a review of the mechanical properties of carbon nanotube-polymer composites. Carbon **44**, 1624–1652 (2006)
14. Fiedler, B., Gojny, F.H., Wichmann, M.H., Nolte, M.C., Schulte, K.: Fundamental aspects of nano-reinforced composites. Compos. Sci. Technol. **66**, 3115–3125 (2006)
15. Rosato, D.V. (ed.): Designing with Reinforced Composites. Hanser, Munich (1997)
16. Sanjay, K., Mazumdar, PhD.: Composites Manufacturing Materials, Product, and Process Engineering. CRC Press, US (2002)
17. Callister, W.D. (ed.): Introducción a la ciencia e ingeniería de los materiales. Reverté (1998)
18. Dresselhaus, M.S., Jorio, A. (eds.): Carbon Nanotubes: Advanced topics in the Synthesis, Structure, Properties and Applications. Springer Science, Berlin (2008)
19. Paradise, M., Goswami, T.: Carbon nanotubes-Production and industrial applications. Mater. Des. **28**, 1477–1489 (2007)
20. Peng-Cheng Ma et al.: Dispersion and functionalization of carbon nanotubes for polymer-based nanocomposites: a review. Composites: Part A **41**, 1345–1367 (2010)
21. EXAKT company. Web www.exkat.de

22. Huang, Y.Y., et al.: Dispersion rheology of carbon nanotubes in a polymer matrix. Phys. Rev. B **73**, 125422 (2006)
23. Dresselhaus, M.S., Dresselhaus, G., Saito, R., Jorio, A.: Raman spectroscopy of carbon nanotubes. Phys. Rep. **409**, 47–99 (2005)
24. Plastics: Determination of tensile properties: Standard EN ISO 527
25. Yedra A.: Development of low cost polymeric nanocomposite materials with high mechanical performances. In: Proceedings of Nanospain Conference (2012)
26. Manteca, C., Yedra, A., Gorrochategui, I., Miguel, R.: Estudio y análisis de fallo mecánico de paneles sándwich fabricados por Pultrusión, XXVI Encuentro del Grupo Español de Fractura (2009)
27. Gorrochategui, I., Manteca, C., Yedra, A., Miguel, R., del Valle, F. J.: Composite material pedestrian bridge for the Port of Bilbao. In: Proceedings of International Conference on Structural Nano Composites, UK (2012)
28. Fibre-reinforced plastic composites: Determination of flexural properties. Standard EN ISO 14125
29. Web www.panelsandwich.com
30. Web www.sandwichpanels.org

Thermoplastic Nanocomposites
with Carbon Nanotubes

Shyam Sathyanarayana and Christof Hübner

Abstract The potential impact of light-weight structures with multifunctional properties for engineering applications drives significant research and development activities on nanocomposites. Polymer nanocomposites especially those with carbon nanotubes (CNTs) are very attractive for conductive composites with good structural characteristics. The biggest challenge facing the commercial success of CNT based composites is the intrinsic strength of their agglomerates which prevents good filler dispersion in the matrix, a key attribute for any reinforcement. This chapter is a review of the CNT incorporated thermoplastic composites processed mainly containing economic multi-walled carbon nanotubes (MWCNTs). We present an overview of CNTs (their structure, production process, properties, surface modification, applications etc.) and the mechanism of their dispersion in thermoplastic matrices. The processing of thermoplastic matrices via the conventional twin-screw compounding approach is discussed along with the influence of process parameters on MWCNT dispersion. The role of secondary processing operation on composite properties is also highlighted. The mechanical, electrical and thermal properties of the thermoplastic-CNT composites are reviewed highlighting the key findings and shortcomings. We conclude throwing some light on the developments on multi-scale reinforcements in polymeric matrices.

An erratum to this chapter is available at 10.1007/978-3-642-40322-4_13

S. Sathyanarayana (✉)
Polymer Engineering, Fraunhofer Institute für Chemische Technologie ICT,
Joseph von Fraunhofer Strasse 7, 76327 Pfinztal, Germany
e-mail: Shyam.Sathyanarayana@basf.com

C. Hübner
Advanced Materials & Systems Research, BASF SE, B001, Carl-Bosch-Strsse 38,
67056 Ludwigshafen, Germany
e-mail: Christof.Huebner@ict.fraunhofer.de

J. Njuguna (ed.), *Structural Nanocomposites*, Engineering Materials,
DOI: 10.1007/978-3-642-40322-4_2, © Springer-Verlag Berlin Heidelberg 2013

1 Introduction

The ever increasing demand for improved properties of polymer matrices especially for light-weight structures and multifunctional characteristics has strongly shifted the focus of researchers on nanomaterials as reinforcements. The development of polymer nanocomposites has been a significant area of research and has evolved significantly over the last two decades owing to the ability of nanoscale reinforcements to create remarkable property enhancements at relatively low filler concentrations, compared to conventional composites. Nanocomposites based on polymeric matrices gained significant interest after the report of nanoclays filled nylon-6 by Toyota in 1993 [1], but the actual mention of the term was first found in the work on Lan and Pinnavaia [2]. Surprisingly automobile tires in which carbon black acts as a reinforcement is also an example of a nanocomposite. However, the term "nanocomposite" is not generally used to describe such composites. A nanocomposite is considered to be a multiphase solid material where one of the phases has one, two or three dimensions of less than 100 nm according to Ajayan et al. [3]. According to the ISO/TS27687 standards the nanofillers are classified based on their dimensions as 1-D (e.g. platelet, lamella; thickness <100 nm), 2-D (e.g. tube, fiber; diameter <100 nm) and 3-D (e.g. bead, sphere; all dimensions <100 nm) nano-objects [4].

The growth of different types of nanomaterials starting from nanoclays, cellulose nanowhiskers, carbon nanofibers, carbon nanotubes (CNTs), graphenes, nano-oxides like nanosilica, nanoalumina, titanium dioxide etc. has led to the development of composites with extremely attractive macroscopic properties—multifunctional in most circumstances depending on their inherent characteristics. Excellent electrical, thermal, mechanical, optical, fire-retardant, barrier, anti-bacterial and scratch resistant properties of these composites have been reported and the results are only getting better with time. The high surface to volume ratio of these fillers facilitates the attainment of desired macroscopic functionality at substantially lower filler loading fractions. The biggest challenge with these nanofillers however has been with dispersing them in the host matrix to take advantage of their theoretical potential. Table 1 gives an overview of the areas of application for the well known and widely employed nanofillers.

Table 1 Areas of possible applications for different nanofillers

Area of application	Type of nanofillers
Tribology	Fullerene like tungsten sulphide, zeolith, nanoclay, nanosilica
Antibacterial properties	Titanium dioxide, zinc oxide, copper oxide, silver
Electrical properties	Carbon nanotube, carbon nanofiber, graphene
Mechanical properties	Cellulose nanowhisker, nanoclay, carbon nanotube, graphene
Flame retardancy	Nanoclay, carbon nanotube
Barrier properties	Nanoclay
Self cleaning surfaces	Titanium dioxide, nanosilica
Surface modification	Nanosilica (AEROSIL®)
Coatings	Nanosilica, nanoalumina, zinc oxide

Carbon based nanomaterials are highly attractive due to their ability to transition an insulating polymer matrix to a conductive composite, in addition to the proven advantage in achieving excellent structural properties. CNTs stand out among the other carbon based fillers like carbon fibers or carbon nanofibers which require higher filler loading fractions to exhibit similar levels of electrical conductivity. Graphene tipped to be a strong competitor for the CNTs on the other hand is still in its infancy, and large production volumes of these materials is still challenging. CNT based polymer composites have been a significant area of research in the last couple of decades owing to CNTs high aspect ratios with nanometric dimensions, low mass density and intrinsically superior electrical [5–8], mechanical [8–11], and thermal properties [8, 12, 13].

Continuously lowering costs of CNTs, especially multiwalled carbon nanotubes (MWCNTs) with increase in demand and production capabilities augments favorably for a huge polymer-CNT nanocomposite market. Interesting observations on MWCNT based composites have been plenty starting from different ways of MWCNT synthesis to its application in electromagnetic shielding (EMI) [14–17], sensors [18–20], electrostatic charge dissipation (ESD) [21], flame retardancy [22], wind turbine blades [23], photovoltaic packaging [24], electrically conducting cables [25] etc.

The potential of CNTs as fillers for multifaceted product development in polymer matrices certainly tilts the tide in its favour compared to its competitors. Due to the renown challenge involved with dispersing the CNTs in thermoplastic matrices for good macroscopic properties of the composites, thermoplastic composites with CNTs (primarily MWCNTs) is the main focus of attention in this chapter.

2 Carbon Nanotubes: An Overview

The interest in the special allotrope of carbon called CNTs was greatly stimulated by the observations of tubular forms of carbon with nanometric dimensions by Iljima [26], and succeeding reports on the capabilities to synthesize these nanoparticles in large quantities [27, 28]. CNTs belonging to the fullerene family are geometrically idealized to be a cylinder formed by concentrically rolled graphene sheets. They are typically of a few nanometers in diameter and microns in length.

2.1 Structure of CNT

Similar to the structure of graphite, the sp^2 hybridized CNT has each of its atoms bonded to three neighboring atoms in an hexagonal array. As the cylindrical structure of the CNT is brought about by the rolling up of the graphene sheets, the type of CNT would depend upon the number of concentric cylinders. CNTs

Fig. 1 Different types of CNTs—based on the number of tube walls. Reprinted from [141], Copyright (2003) with permission from Elsevier

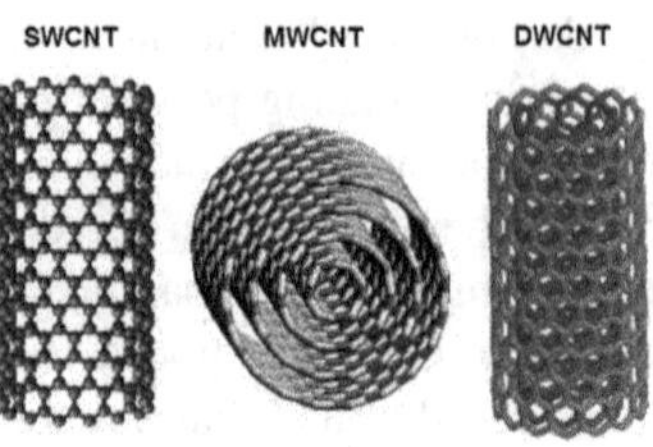

principally exist as SWCNTs, double-walled carbon nanotubes (DWCNTs) or as MWCNTs (Fig. 1).

The intrinsic properties of the CNTs would stem from their atomic arrangement (or in other terms the way their graphene sheets are rolled), the type of CNT and their individual geometric dimensions. The rolling up of the graphene sheets and hence the atomic structure of the CNTs is well defined by tube chirality. The chirality is given by the chiral vector $C_h = n\vec{a}_1 + n\vec{a}_2$ where $(\vec{a}_1, \vec{a}_2)$ are the unit vectors and (n,m) are the number of steps along the unit vector of the hexagonal lattice [29]. This gives rise to three possible orientation forms for the carbon atoms of the CNTs. If n = m, it leads to the "armchair" structure while the "zigzag" structure results when m = 0. Any other orientation is referred to as the "chiral" structure. The electronic conduction of the CNTs is dependent on its chirality. CNTs possessing an "armchair" chirality are metallic, semi-metallic when (n–m)/ 3 = i and n ≠ m, where i is an integer and semi-conducting otherwise [30]. The chirality of a MWCNT is extremely complex as each of the concentric nanotube walls could carry different chiralities. Figure 2 shows the nomenclature system of CNTs and the different chiral forms of SWCNTs [(a) arm chair, (b) zigzag, (c) chiral].

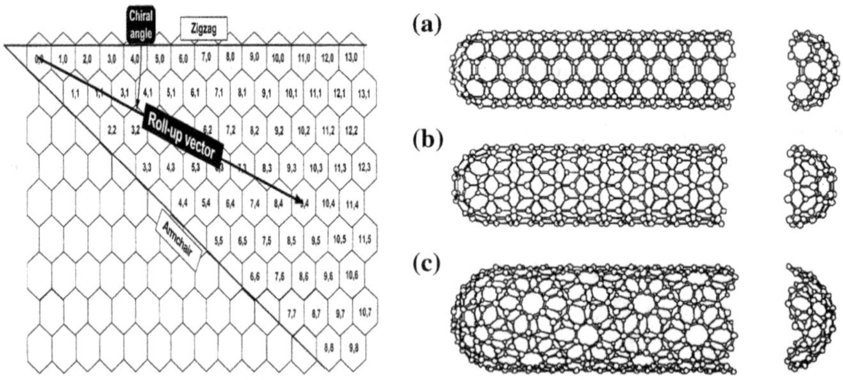

Fig. 2 System of CNT nomenclature [142] (*Left*); Chiral forms of SWCNTs (*Right*). Reprinted from [143], Copyright (1995) with permission from Elsevier

2.2 Synthesis of CNTs

Arc discharge, laser ablation and chemical vapor deposition (CVD) are three principal techniques by which CNTs are synthesized. They are briefly described below. The schematic of the different techniques are shown in Fig. 3.

The **arc discharge** approach is probably the easiest to synthesize MWCNTs, however the quality of the resulting CNTs and the pattern of growth is not the best while comparing to the production by other techniques [31]. In the arc-discharge approach, two graphite electrodes are placed 1 mm apart in an inert atmosphere and a direct current of about 50–100 A at a voltage of 20 V is applied across them. The vaporization of carbon at these conditions result in a high temperature discharge (also termed as plasma) resulting in the consumption of the anode and the formation of deposits on the surface of the cathode from where the CNT grows. The typical yield of this process ranges from 30 to 90 % and the principal advantage of this process is the economic production of MWCNTs without a catalyst. The doping of the anode with a metal catalyst would result in the formation of SWCNTs [30]. The principal drawback of the arc-discharge approach is the production of shorter and impure CNTs which mandates a purification step to generate high quality CNTs.

The **laser ablation** technique uses a high power laser to form the discharge product unlike the direct current (DC) generated electricity in the arc-discharge approach. Laser is shot on a graphite source that is placed in an oven under inert atmosphere at temperatures of around 1,100–1,200 °C. The carbon vapors formed under these conditions expands and cools resulting in the formation of CNTs and other members of the fullerene family depending upon the set process conditions. MWCNTs are the principal product and are shorter in length compared to arc-

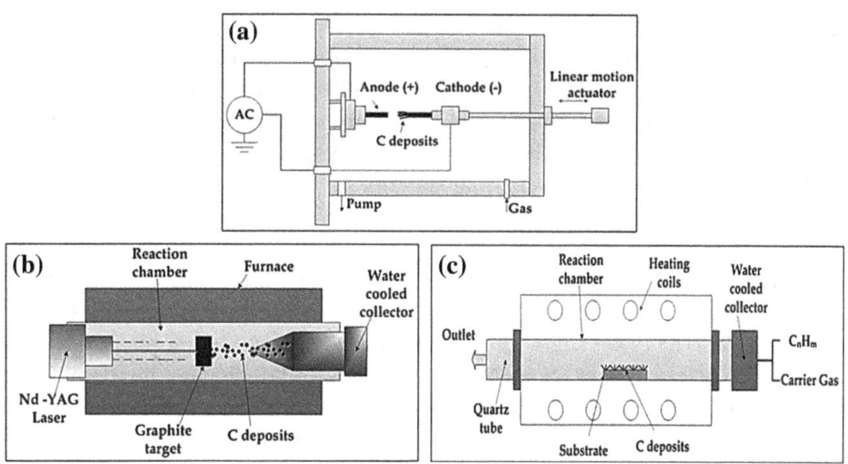

Fig. 3 Techniques for CNT synthesis [144]. **a** Arc discharge, **b** Laser ablation, **c** CVD

discharge process [30, 32], but the addition of a metal catalyst to graphite would result in SWCNTs. The typical yield of this process is around 70–80 % and the purity of the produced nanotubes is significantly higher. Good product control is possible; however this comes with the trade-off of high production cost due to higher power requirements for operating the laser.

CVD is the widely adopted industrial approach for producing CNTs owing to its scalability. When a carbon containing gaseous source like acetylene, methane etc. is passed over a metal catalyst at temperatures ranging from 550 to 1,200 °C it results in the freeing up of carbon atoms on the decomposition of the gas, which recombines in the form of CNTs on the metallic substrate. This method is widely adopted for flexibility in controlling the quality and exercising good control over the dimensions of the formed product. The yield varies from 20 to 100 % with minimal need for purification of the synthesized CNTs due to the negligible traces of amorphous carbon. Very long CNTs can be produced by this approach. The production of extremely pure SWCNTs is possible with controllable diameters, but the MWCNTs that are formed with the CVD process has more structural defects [33], and are highly entangled compared to the arc discharge and laser ablation approach.

2.3 Properties of CNT

CNTs are known for their intrinsically high mechanical, thermal and electrical properties. The sigma bonds which make up the C–C bonding in CNTs attributes to their excellent mechanical characteristics. Elastic modulus of 1.2 TPa and tensile strength of about 50–200 GPa have been reported for CNTs [34]. However, there does not seem to be a consensus on the reported mechanical behaviour among published literatures. Their high axial and low lateral thermal conductivity also add to their significance. Although theory predicts a room temperature thermal conductivity of 6,000 W/mK [35], Kim et al. estimate the thermal conductivity of MWCNTs to be 3,000 W/mK [36], and Pop et al. report 3,500 W/mK for SWCNTs [37]. This is significantly higher than the thermal conductivity of diamond (1,000–2,200 W/mK) which is reported to be one of the best known thermal conductors [38].The oxidative thermal stability for both MWCNTs and SWCNTs are greater than 600 °C which compares strongly with that of 450–650 °C for graphite. Their co-efficient of thermal expansion is very minimal which supports the excellent thermal conductivity of these materials. SWCNTs show electrical conductivities of the order of 10^2–10^6 S/cm, while it ranges between 10^3 and 10^5 for MWCNTs. The lower scales are very similar to the reported in-plane electrical conductivity of graphite of about 200–2,500 S/cm [39], indicating the dominative electrical properties of CNTs compared to one of the best reported electrical conductors in graphite. They also boast of extremely high intrinsic electron mobility of greater than 10^5 cm^2/Vs [40], which is very similar or even higher than graphite.

The extraordinary multi-faceted properties of CNT confer to these materials a significant potential to be used in a wide-variety of applications. The intrinsic tendencies of the CNTs to exist as agglomerates due to their van der Waals (vdW) forces of attraction (0.5 eV/nm) [41], by large limits the realization of the complete potential of these materials as nanoscale reinforcements in different host matrices.

2.4 Commercially Available CNTs

Commercially available CNTs that primarily results out of the CVD process are highly disorganized and have a highly agglomerated morphology with a significant agglomerate strength. This inability to produce individualized and aligned morphologies severely inhibits the theoretical potential of CNTs as to what could be expected out of CNT yarns or forests. Although it is highly complicated to achieve aligned CNT morphology in a commercially viable process, the cohesive or the vdW forces of attraction that results in the agglomerated morphology of CNTs could at least be minimized in order to gain a competitive advantage with CNTs as polymeric fillers. The initial or the primary CNT agglomerates have to be broken down theoretically into individual CNTs or in other terms dispersed well into the polymer host to translate the intrinsic characteristics of the CNTs to their composites. This has remained a significant challenge over the years and any effort to maximize the quality of dispersion on industrial scale could result in taking these nanocomposites close to application.

Due to its availability in bulk and lower costs, MWCNTs would be the ideal choice for high volume industrial applications; hence attention would be focused on MWCNT dispersion in thermoplastics in this Chapter. Scanning electron micrographs (SEM) images of two of the most widely reported type of MWCNTs in literature namely *Baytubes*® *C150P* (*Bayer Material Science, Germany*) and *NanocylTM NC7000* (*Nanocyl S.A., Belgium*) are shown in Fig. 4.

These pictures indicate a highly entangled network of MWCNTs with sizes ranging up to a few hundred microns, unusual for a material described as nanofiller. Alig et al. describe the structure of loosely packed larger agglomerated NC7000 as a "combed yarn" structure while that of C150P as a "birds nest" with smaller tightly held primary agglomerates [42]. Although the geometries of these MWCNTs do not differ significantly, the average deformation stress (at 25 % strain) of C150P with a bulk density of 120–170 kg/m^3 [43], is 0.64 MPa [44], while that of NC7000 with a bulk density of 60 kg/m^3 [45], is 0.39 MPa [44]. This difference in the deformation stresses of the MWCNTs will be a critical factor in determining the magnitude of processing parameters and consequently the extent of processing shear required to create equivalent dispersion qualities with these two different CNT types.

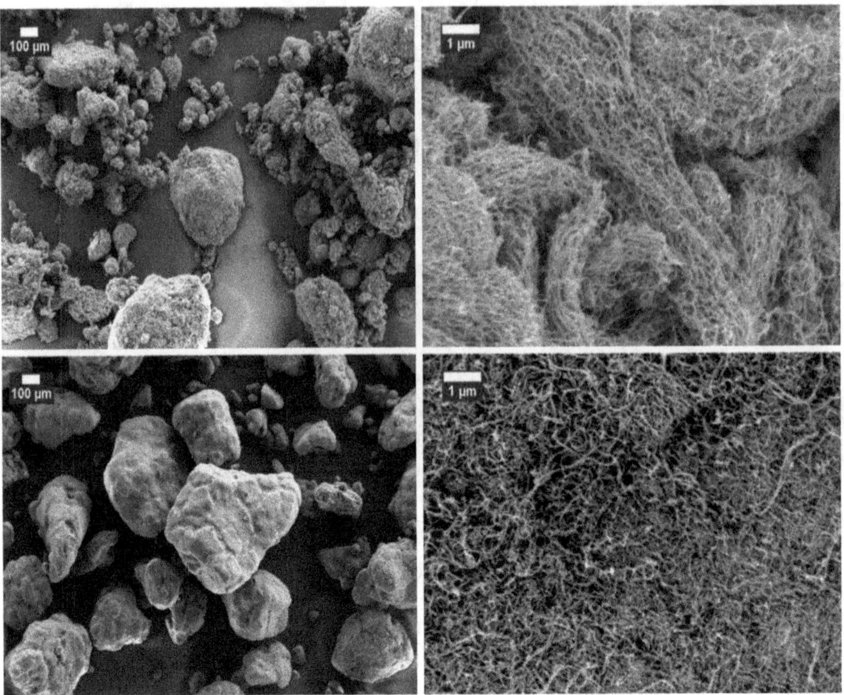

Fig. 4 Low and high magnification images of NC7000 (*Top*), C150P (*Bottom*)

2.5 Modification of CNT Surface

Modification of the CNT surface is an approach widely adopted to render the surface of the inert CNTs compatible to interact efficiently with the polymer and also weaken their intrinsic vdW forces of attraction. This is also termed as CNT functionalization. Functionalization both by chemical and physical means has been largely reported.

Chemical functionalization is done either on the side walls of the nanotubes or by creating defects on CNT surfaces by acid treatments. This is expected to result in a very strong matrix-CNT interaction with a trade off on the structural characteristics of CNTs. The sp^2 hybridization of carbon is transformed to sp^3 to create active sites for effective interaction with polymer accompanied with the loss of π conjugation bonds on the outer walls. Covalent functionalization of CNTs could modify CNT stacking morphologies by altering hydrogen bonds resulting in improved CNT solubility in solvents [46]. But, covalent functionalization tends to negatively affect the physical and chemical integrity of CNTs. Strong acid treatments affects the geometry of the CNTs resulting in increased filler percolation thresholds [47]. Covalent treatments could also negatively affect the intrinsic electrical properties of the CNTs and have a profounding effect on the metallic characteristics of metallic CNTs [48]. Enhanced reactivity between CNT and the polymer would result in wrapping or encapsulation of CNT surfaces by the polymer due to coupling or grafting lowering the bulk conductivity of the composite [49]. The phonon scattering length is also expected to be lowered leading to lower thermal conductivities [50]. Covalent treatments are mainly aimed at improving polymer-CNT compatibility and CNT dispersion. Effective achievement of this could lead to improved structural characteristics accompanied with losses in electrical and thermal conductivity of the composite.

Physical functionalization is done by polymer wrapping (vdW forces and π-π stacking), surfactant adsorption on CNT surface (physical adsorption) and endohedral techniques (capillary principle). Although this technique would not damage CNTs a weak CNT-polymer interaction could be expected out of these approaches. Non-covalent or physical treatment of CNTs result in the outer tube walls subjected to more treatment than the inner tubes [51]. This could result in the CNTs being bundled even after treatment. Hence mechanical treatments like milling, ultrasonication etc. should precede this treatment. Since physical methods don't result in significant CNT structural damage, their composites could be expected to show enhanced electrical conductivity as the π-conjugation of the CNTs and their electron transfer paths could maintain their identities. Polymer wrapping of CNT surfaces cannot be completely ruled out resulting in lower electrical conduction. Mechanical property enhancements cannot be expected to be on a larger scale because the chance for the interfacial strength to improve is minimal because of lack of chemistry.

The techniques of chemical and physical modifications are however, relatively expensive which limits modification of the CNT for large scale applications. A review by Bose et al. [52], discusses the positives and the drawbacks of both physical and chemical functionalization in detail on the properties of CNT based polymer composites.

2.6 Applications of CNTs

The intriguing properties of CNTs have led to many research efforts leading to uncovering the potential prospects of employing CNTs in practical applications. Although it has been very complex to translate the theoretical potential of CNTs till date into potential applications, efforts are still ongoing towards understanding the intrinsic properties of CNT and ways to implement them in real life. The benefits of CNTs are expected to be significantly realized from an economical perspective when they are used as functional fillers for materials like plastics, ceramics etc., however their potential to be used as such (especially SWCNTs in flexible electronics applications) is also being widely worked on. This is supported from the finding that 69 % of the global CNT market share of \$472.9 million in 2010 was the contribution of the plastics and composites industry [53].

Although a wide variety of applications could result owing to superior CNT properties, they can principally be classified into three main categories of mechanical (structural composites for aerospace and automotive industries, textiles, sport goods, etc.), thermal (flame retardant additives, coatings, electronic circuitry etc.) and electrical & electronics (semi-conducting materials, electronic circuitry, fuel supply systems of automobiles, navigation systems etc.). The multifunctional derivates that stems out by a combination of these properties on the final part application adds significant value to the applications of CNTs.

The European Chemical Industrial Council (CEFIC) classifies the potential applications of CNT on the basis of time into short term—those products based on CNT already available or will shortly be available (e.g. conductive composites, sensors, electromagnetic shielding, sport goods), mid-term (e.g. coatings, lithium-ion battery, fuel cells, semiconducting materials, petrochemical catalysts etc.) and long term—those currently under the scope of research & development (e.g. drug delivery, microwave antennas, medical implants etc.) [54].

Some of the existing commercial applications of the CNTs include the long range vessel LRV-17 from *Zyvex Marine* made of a carbon fiber-nanocomposite system consisting of carbon fiber reinforced plastic (CFRP) and CNTs which enables reduction in structural weight, efficient fuel usage and increased range [55], and a 25 times stronger tennis racket by *Völkl* made of CNT compared to carbon fibers (at same weight) which would eventually result in more kinetic energy returned to the ball [56]. *Aldila* have also reported to have used CNT based epoxy composites for golf shafts for vibration damping [57].

2.7 CNTs: A Market Outlook

The production capacity of CNTs is expected to grow globally from 3,141 to 12,806 metric tons in 2016 at a Compound Annual Growth Rate (CAGR) of 10.5 % according to Centre for Knowledge Management of Nanoscience and Technology (CKMNT), India [53]. As per their 2011 forecast the total CNT market is to yield around $1.1 billion by 2016 at a CAGR of 10.5 %.

A Frost and Sullivan report published in 2011 forecasts a $35.52 million market for CNT in the automotive composites by 2015, indicative of a market share of 3.6 % at 1 % CNT loading [58]. The report also predicts a 10 % penetration of CNT in the construction industry, 15 % in the fibers and textile domain, and about 25 % in the intumescent coating market. From an electronics industries' standpoint, the largest share for CNT is expected for the displays (1 - >5 %) and flexible displays (1–10 %), whereas only 1 % penetration in the sensor market is foreseen. Significantly, CNTs are also expected to gain a 1–5 % share in the total authentication market. With recent developments in research, it is expected that these numbers will continue to rise in spite of a foreseen potential threat by graphene which is expected to gain a significant market share especially in the electronics industry.

3 Processing of CNT Incorporated Thermoplastic Nanocomposites

The level of CNT dispersion visible on the composite morphology is a direct function of the type of polymer and CNT employed, the processing approach and the process factors. The morphology and therein the macroscopic properties of the composite are dictated by the thermo-mechanical history during processing. Several processing methodologies such as solution casting [59–61], melt mixing [62–67], solution mixing [59, 60, 68], different methods of in situ polymerization of the monomer in the presence of CNTs [69, 70], coagulation spinning [71], mechano-chemical pulverization [72], solid-state shear pulverization [73], electro spinning [74] etc. have been adopted for the synthesis of polymer-MWCNT composites. In this section only the predominantly preferred approaches namely solution processing, in situ polymerization and melt mixing would be discussed in brief.

Solution processing is one of the widely utilized approaches owing to its simplicity. It involves dispersion of the CNT in a suitable solvent by sonication and/or stirring followed by mixing with a solution of the polymer host, evaporation of the solvent and drying with or without vacuum. Ultrasonication is widely employed to break the CNT agglomerates as the shear in the solution mixing process is significantly low. The type of ultrasonication (ultrasonic bath or horn), bath temperature, rate of sonication (frequency and time) and the nature of the solvent could have a strong influence on the properties of the product. This method can be successful with the right choice of solvent and complete removal of the solvent during the drying stage, however it is not easily scalable. It is widely adopted for preparing composites based on thermosetting matrices where a significant magnitude of shear is not required as the low viscosity of the host would result in good infiltration of the CNT agglomerates facilitating dispersion.

In situ polymerization of monomers in the presence of CNTs is another widely used technique for composite preparation. This technique is proposed to form a very good chemical affinity between the polymer chains and the CNTs in most cases, but depending on the nature of the reactants a non covalent interaction is a definite probability. The polymerization of the monomer in the presence of an initiator is carried out with CNTs in the vicinity enabling the production of composites with high CNT loadings. A mix of polymer grafted CNTs and free polymer chains are obtained which creates a favorable environment for the development of a highly compatible polymer-CNT interface. The low viscosity of the starting monomer facilitates better infiltration into the CNTs and consequently their dispersion. CNT based composites of polystyrene (PS), poly (methyl methacrylate) (PMMA), and vinyl monomers are commonly produced using this approach. Yuan et al. show significant enhancement in the mechanical properties of PS-MWCNT composites prepared by a combination of in situ polymerization followed by melt mixing owing to strong interfacial adhesion between the CNT grafted PS and the PS matrix [75].

Owing to its simplicity and adaptability for a commercial scale up **melt mixing/ compounding** seems to be the most commonly employed approach for thermoplastic polymer/MWCNT nanocomposites. This method is most suitable for polymers that cannot be processed with the solution processing approach owing to its inability to dissolve in commonly employed solvents. The higher magnitude of shear during the melt mixing process facilitates the breakup of the CNT agglomerates followed by simultaneous dispersion and distribution in the polymer melt. For a melt mixing process to give optimum dispersion quality optimization of all the process parameters like the screw configuration, screw speed, throughput/residence time, barrel temperatures, filler feeding position etc. is imperative.

3.1 Mechanism of CNT Dispersion in Thermoplastics

Dispersion of CNT in thermoplastics is a simultaneous sequential process starting with the wetting of the CNT agglomerates by the polymer melt, infiltration of the polymer melt into the CNT agglomerates, disintegration of agglomerate fractions weakened by the infiltration process and shear forces into small fractals by mechanisms of erosion and/or rupture followed by their distribution in the polymer host (Fig. 5). This mechanism of filler dispersion is strongly influenced by the processing approach, associated process parameters and the nature of the polymer and the CNTs.

For effective wetting of the primary CNT agglomerates by the polymer melt, the interfacial energy difference between the polymer and the CNT should be at its minimum. As the interfacial energy difference between polar thermoplastics like polycarbonate (PC), polyvinyl chloride (PVC) etc. with the hydrophilic CNTs is significantly low, excellent CNT dispersion is achievable with these matrices resulting in extremely low filler percolation thresholds for the final macroscopic properties [67, 76–78]. On the other hand polyolefins exhibit a very high interfacial energy difference with the CNT limiting the level of CNT dispersion in the composite [62, 79]. Incorporation of surface functionalities to the CNT is a solution to render compatibility with the polymer; however this is an expensive approach as detailed earlier.

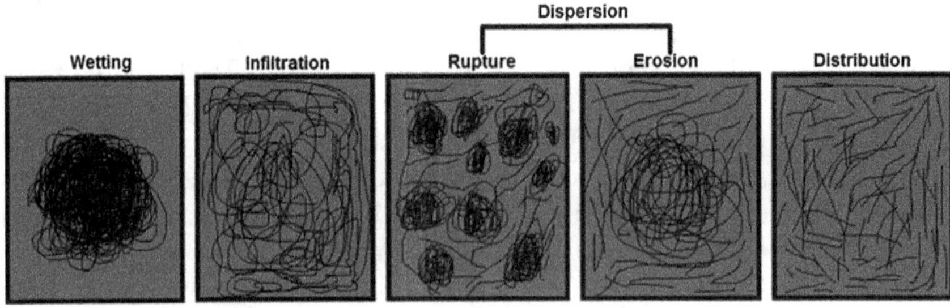

Fig. 5 Mechanism of CNT dispersion

Infiltration of the polymer melt is principally governed by the mobility of the polymer chains (dependent on viscosity), the available pore radius or in other terms the agglomerate density of the CNTs, the strength and size of the CNT agglomerates and surface tension variations between the polymer and the CNT. Lower polymer viscosity or high melt temperatures and low agglomerate density would aid faster infiltration, while minimal interfacial energy difference between the polymer and the CNT would enhance the efficiency of infiltration. The infiltration process can be manipulated by the choice of the raw-materials and the process parameters which will be discussed in detail later in Sects. 3.2 and 3.3.

The magnitude of applied shear stresses arising due to the viscous flow on the CNT agglomerates during processing must be above the inherent strength of the agglomerates (σ_a) for dispersion to occur. The dispersion of the weakened primary agglomerates by preceeding wetting and infiltration would take place by mechanisms of rupture/and or erosion governed by a dimensionless fragmentation number (F) which is directly proportional to the product of the melt viscosity (η) and the shear rate (γ_r) and inversely proportional to the inherent maximum agglomerate strength (σ_{am}). The magnitude of contribution of the infiltration step towards the reduction in σ_a could play a role in regulating the maximum shear stress that has to be exerted on the agglomerates. It is to be remembered that enhanced shear stress could result in tube breakage and also could lead to possible polymer degradation. Though rupture and erosion would co-occur, a significantly larger magnitude of F (F $\gg$ 1) would result in a situation where rupture would dominate while lower F (F $\ll$ 1) would result in dispersion dominated by erosion [80]. The mechanism of rupture is comparatively faster than erosion owing to the instantaneous breakage of primary agglomerates into smaller fractals in the former, while a slowly peeling of individualized CNTs occurs from the outer surface of the primary agglomerate fraction in the latter. Distribution of the dispersed agglomerate fractals takes place simultaneously with the disintegration or the rupture of remaining agglomerates over the course of the process. Idealizing this in the context of twin-screw extrusion, the residence time in the extruder would determine the extent of filler dispersion for a given set of process parameters and material combination.

3.2 Influence of Raw Materials

The selection of a suitable polymer host and the intrinsic characteristics of that polymer host is one of the critical factors affecting the quality of filler dispersion in the final composite. An increased affinity between the polymer and the CNT would result in better wetting of the CNT agglomerates which is the preliminary step in the mechanism of dispersion. Polymers like PC and PA which has lower interfacial energy difference could result in better wetting of the CNT agglomerates and hence better dispersion while the quality of dispersion in polyolefins is poor owing to its hydrophobic nature and higher interfacial energy difference with CNT.

Exceptionally polystyrene (PS) presenting a polar characteristic shows poor CNT dispersion morphology owing to the high chain stiffness due to its phenyl groups.

The viscosity of the polymer melt governs the extent of melt infiltration into the agglomerates and the dominating mechanism of filler dispersion via rupture or erosion. Lower polymer viscosity would result in better melt infiltration, but could result in lower magnitude of viscous flow shear stresses during processing. Higher polymer viscosity would hinder easier and faster melt infiltration but can result in increased shear stresses for agglomerate breakage. Magnitude of the shear stresses required for dispersion of CNT is also dictated by the intrinsic characteristics of the CNT. Kasaliwal et al. discuss the influence of PC melt viscosity and molecular weight on the level of MWCNT dispersion and conclude that lower matrix viscosity resulted in better melt infiltration leading to better MWCNT dispersion and hence lower electrical resistivities in the composite [81].

The type of CNT especially its number of walls, length, aspect ratio, agglomerate strength and morphology, and the nature of surface functionalization are factors to be considered while designing the process and the associate parameters for its dispersion in a polymer matrix. A CNT with higher agglomerate strength would present increased difficulties to melt infiltration compared to a CNT type with more loosely packed agglomerates even if larger agglomerate size fractions are present in the latter. High aspect ratios of CNT are beneficial for efficient load transfer and hence improved mechanical properties at low percolation thresholds. However, it could complicate dispersion owing to the enhanced available surface area for intra-CNT interactions compared to employing CNTs with low aspect ratios. A balance between the level of dispersion and the aspect ratios of the CNT are quite critical for the end properties as elucidated by Castillo et al. on their work on PC incorporated with different types of CNTs [78]. The enhancement in the compatibility of the CNT with the polymer by functionalization techniques also plays a role in dictating the level of MWCNT dispersion in the polymer host. Menzer et al. found that the altering of the CNT lengths by ball milling resulted in compact agglomerate morphologies and reduction in CNT lengths contributing to higher electrical and rheological percolation thresholds in the composite [62]. Idealizing this to the reduction in the length of CNTs as a function of processing, it would be wise to prefer CNTs with longer average lengths.

3.3 Influence of Process Parameters

The typical melt mixing or compounding parameters namely screw speed, barrel temperature, material throughput, screw configuration and residence time would have to be tailored to specific thermoplastic polymer-CNT system in order to achieve the desired CNT dispersion quality for maximum composite performance [65, 67, 82]. As dispersion is a continuous mechanism influenced by the shear stresses generated by the viscous flow governed by these parameters, addressing the individual contribution of these factors is complicated. This section briefly outlines the individual effects and complementary influences these factors would have on the level of CNT dispersion from a processing perspective.

A higher screw speed would result in an increased specific mechanical energy input (SME) of the process resulting in enhanced shear stresses and thereby dominant dispersion mechanism by rupture. A higher screw speed also would result in a decreased level of screw fill and a lower residence time for the melt in the extruder at a condition where the other parameters are kept constant. Higher screw speeds could also result lower polymer viscosity aiding faster melt infiltration, but at the same time can result in polymer degradation and also negatively affect the aspect ratio of the CNTs [82]. Lower speeds on the other hand would act opposite to what has been detailed with significant limitation on the level of generated shear stresses. A higher screw speed was found to be helpful in achieving good filler dispersion when employing matrices that are not known to be compatible with the CNTs like PS, PP and when using CNTs with a high agglomerate strength [65, 83].

Melt or barrel temperatures influence the melt viscosity of the polymer and thereby the level of infiltration and the magnitude of generated shear stresses for agglomerate dispersion. Lower melt viscosities are highly recommended for faster and better infiltration of the CNT agglomerates while higher melt viscosities are important for enhanced shear stresses. The desired level of dispersion would determine the domination of the melt infiltration step or the creation of enhanced shear stresses for higher fragmentation numbers.

Low material throughputs would result in higher SMEs, lower degree of screw fill, and higher local melt residence time leading to improved CNT dispersion quality with trade-off on production volume. Throughput levels are also regulated by the feeding capacity of the nanomaterial feeder owing to very low bulk density of the CNT along with a very low volume fraction of filler feed.

An optimum screw design must be able to provide the desired level of SME for the pre-set extrusion parameters and is generally a combination of mixing, kneading and transportation elements. The magnitude of SME, the nature of the CNT agglomerates and the desired residence time of the melt in the extruder would dictate the screw configuration. The screw configuration should be designed keeping in mind of the other process factors. When a high speed is required for processing composites containing high agglomerate strength CNT, the screw design must be able to accommodate this process with a sufficient residence time to allow for other dispersion steps like wetting and infiltration. An enhanced residence time is bound to result in polymer degradation while shorter residence times may not provide the desired level of dispersion.

The position of filler feeding becomes important when dealing with CNTs. CNTs with enhanced intrinsic agglomerate strength like the Baytubes® C150P are recommended to be fed along with the polymer in the principle feeding port in order to allow for increased residence time for agglomerate dispersion. Side feeding of the filler Nanocyl™ NC7000 which has a lower agglomerate strength is recommended owing to its loosely packed agglomerate structure with better dispersion envisioned by a dominant melt infiltration [84].

We have carried out comprehensive investigations on the influence of the twin-screw process parameters on MWCNT dispersion in PP, PS, PPE/PS and PC and have understood that the ideal processing parameters is strongly dependant on the nature of the matrix and its compatibility with the CNTs [65–67, 83].

3.4 Strategies for Improving CNT Dispersion During Processing

Functionalization of CNTs by physical or chemical means as discussed earlier has been reported to result in improved CNT dispersion in polymer matrices. This approach is however not commercially viable owing to the high costs, scale up limitations and environmental issues. Hence, it becomes important to focus on alternative strategies via which CNT dispersion in a polymer matrix can be improved.

"Nanodirekt" is an approach involving the dispersion of nanomaterials in water or solvents followed by a direct addition of this homogeneous dispersion into the polymer melt [85]. The dispersing medium will evaporate immediately and the nanoparticles will stay finely dispersed in the polymer matrix. This method ensures that the dry nanofiller is not fed directly into the extruder eliminating potential health hazards associated with handling of nanofillers.

Subjecting the polymer-CNT melt to ultrasonic waves in the extruder has resulted in improved CNT dispersion in PEEK (polyether ether ketone) in the work of Lewis and Isayev [86]. Isayev et al. have demonstrated this approach also on polyetherimide (PEI)/MWCNT composites [87]. Weiss et al. on their work on PP-CNT composites with two different types of CNT however state that the efficiency of employing an ultrasound during the extrusion process is strongly dependant on the type of the CNT employed [88].

Enhanced nanofiller dispersion in a polymer matrix is aided by supercritical carbon dioxide (sc-CO_2) assisted processing. Different variants of this method have been adopted by research groups. Nguyen and Baird propose an approach in which a pressurized CO_2 chamber is employed to assist in the exfoliation and delivery of the clay into a stream of polymer melt in the extruder [89]. A modified hopper in the feed section of the extruder to allow polymer and clay to interact with sc-CO_2 before processing was proposed in the work of Zerda et al. [90]. Clay particles were pre-treated with sc-CO_2 in a pressurized vessel and then rapidly depressurized into another vessel at atmospheric pressure to force the clay platelets apart in the work of Manke et al. [91]. Though this approach has been widely reported for nanoclays, there exists a possibility for extending it to the CNTs. Chen et al. demonstrated an identical process employed by Nguyen and Baird for the processing of CNT/poly(phenylsulfone) (PPSF) composites [92].

Incorporation of processing additives is a simple and economical strategy for enhancing CNT dispersion in thermoplastics. The addition of peroxide during twin-screw compounding led to a substantial improvement in the quality of MWCNT dispersion in PP owing to the reduction in PP melt viscosity (facilitating better melt infiltration into the agglomerates), reduction in ther interfacial tension between PP and CNTs (leading to better melt wetting of the agglomerates), and surface functionalization of carbon nanotubes (resulting in weakening the intra-filler attractive forces) [64]. Better wetting and infiltration of the MWCNTs by polyethylene glycol resulted in a significant improvement in MWCNT dispersion in PE [93].

3.5 Influence of Secondary Processing on Composite Properties

Though good primary filler dispersion is a pre-requisite for good macroscopic characteristics of the composite, widely employed secondary processing or finishing operations for nanocomposites like injection or compression molding also have a major role in regulating final composite properties. The thermo-mechanical history generated on the composites due to these operations has an effect on the level of primary CNT dispersion (created with the extrusion process). Compression molding results in the re-agglomeration of the previously dispersed CNTs facilitated by the reduction in the viscosity of the matrix at a specific process temperature, pressure and holding time. A network of re-agglomerated CNTs forms multiple conductive pathways in the polymer leading to good composite electrical properties, though theoretically one such pathway is sufficient for achieving conductive properties. Higher melt temperatures and longer holding times have been reported to result in lower electrical percolation threshold and better composite conductivity [76]. It must however be remembered that the re-agglomeration of the dispersed CNTs could negatively affect the mechanical characteristics. Injection molding on the other hand is associated with a significant magnitude of secondary shear compared to the compression molding process resulting in filler orientation. The temperature gradient existing between the low temperature mold wall and the high temperature melt results in the freezing of the outer core, lowering the bulk conductivity of the composite. Injection molding is widely preferred from an industrial perspective being a high volume manufacturing process and hence the tailoring of the process parameters to achieve good macroscopic characteristics of the composite becomes important. Villmow et al. recommend lower injection velocities to limit orientation effects and higher melt temperatures for good electrical conductivity [94]. Mold temperature and holding pressure had a minimal influence in their work. Elevated mold temperatures could minimize the effect of temperature gradient. Some of the important factors to be considered during the production of a CNT incorporated thermoplastic composite are summarized in Fig. 6.

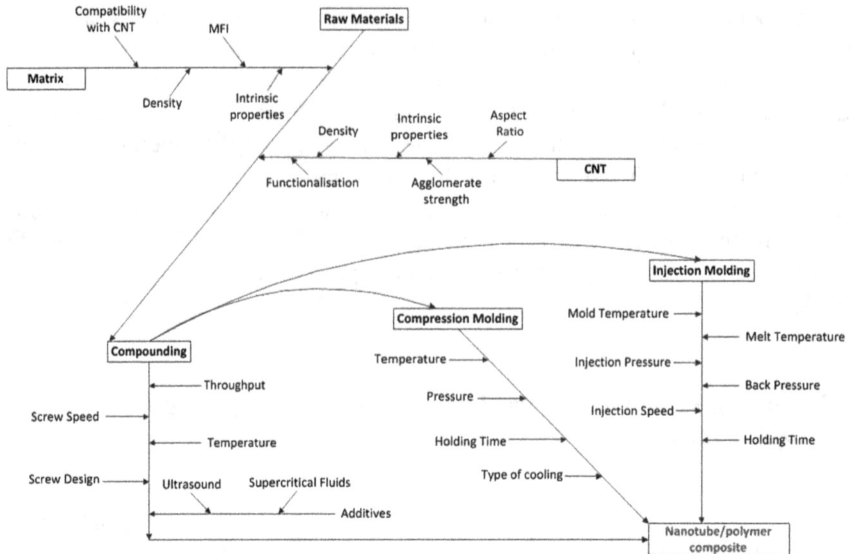

Fig. 6 Tailoring a CNT incorporated thermoplastic composite

4 Effect of CNT Dispersion and Functionalization on Composite Properties

The properties of CNT incorporated thermoplastic composites are mainly dependent on the level of CNT dispersion in the host matrix and the nature of polymer-CNT interaction. These two factors in turn is a complicated function of the type of host matrix and its inherent characteristics, the type and surface functionality of the CNTs, processing methodology and associated parameters as detailed in the previous section. Hence, the important macroscopic properties of the CNT filled nanocomposites namely mechanical, electrical, and thermal properties would be elucidated upon in this section.

4.1 Mechanical Properties

Carbon nanotubes are envisaged to be ideal reinforcements for polymeric matrices owing to their exceptional mechanical properties. However, this does not alone ensure composites with excellent structural characteristics. The extent of dispersion and distribution of CNTs in the polymer host, alignment/orientation of CNTs in the matrix, the aspect ratio of the CNTs in the final composite, and the nature of the interface between the polymer and CNTs are considered to be crucial factors having an effect on the mechanical properties of the polymer-CNT composite.

In the case of reinforcements with nanofillers, the **quality of filler dispersion** is extremely important as the presence of large scale agglomerates would substantially lower the efficiency of load transfer. The level of CNT dispersion is primarily determined by the compatibility between the polymer and the CNTs, in addition to the methodology of composite processing. Achieving a good dispersion of CNTs in a thermoplastic matrix has been a perennial problem, but matrices like PC, PVA, PA etc. [67, 95, 96], show very good CNT dispersion morphologies. This is probably due to their polar nature and lower interfacial energy difference with the CNTs. On the other hand, the polyolefins have been so far considered the hardest of the thermoplastic matrices to achieve a good CNT dispersion. But recent results on PP and PE [64, 93], show that this challenge could be overcome by the employment of functional additives. Controlling or characterizing **distribution** of nanofillers in the matrix has so far been almost impossible, but is very critical to tune composite functionality. Prashantha et al. report an increase of the modulus and tensile strength of PP containing 1 wt% MWCNT (due to significantly enhanced MWCNT dispersion in PP processed by masterbatch containing 20 wt% MWCNT) by 1.27 and 1.18 times respectively [97], further additions of MWCNT did not improve the structural characteristics significantly. Castillo et al. [78], and Sathyanarayana et al. [98], report minimal/no increase in the tensile properties of PC-CNT composites, inspite of achieving excellent CNT dispersion in PC.

The interfacial shear strength (IFSS) which governs the maximum stress transfer from the polymer to the CNTs is strongly dependant on the nature of the **polymer-CNT interface**. If the interface is strong (due to good polymer-CNT adhesion), the external load applied on the composite is efficiently transferred across the interface to be borne by the CNTs. Coleman et al. report an increase in the Young's modulus, tensile strength and toughness of PP by 3.1, 3.9 and 4.4 times respectively on the addition of 1 wt% thin-walled MWCNTs. The CNTs were chemically grafted with chlorinated PP chains [95], which could have possibly contributed to enhanced load transfer via a strong interface. In another example of a development of a strong interface, the CNTs modified by grafting with PS via in situ polymerization and melt mixed with PS resulted in an increase of impact strength by 250 % at 0.32 wt% MWCNT loading compared to pristine PS while the un-modified CNTs resulted in 150 % increase of impact properties at an identical CNT loading [75]. However, there are many instances in which functionalization have resulted in the deterioration of composite properties [52].

As the load transfer efficiency is also determined by the length of the CNTs, it is important to maintain the **aspect ratio of CNTs** as high as possible after composite processing. Krause et al. [99], and Castillo et al. [78], have demonstrated the reduction in the length (and subsequently the aspect ratio of the CNTs) on melt processing. Figure 7 shows the comparison of length distribution of as-received MWCNTs and that of the PC-MWCNT composite processed by melt mixing containing 2 wt% of identical MWCNTs. Reduction in aspect ratio would lead to a case where long fibers would have to be treated as short fiber reinforcements and hence substantial increase in the composite modulus cannot be expected [100]. Incorporation of CNTs with high aspect ratios in PLA resulted in improved

                                                        S. Sathyanarayana and C. Hübner

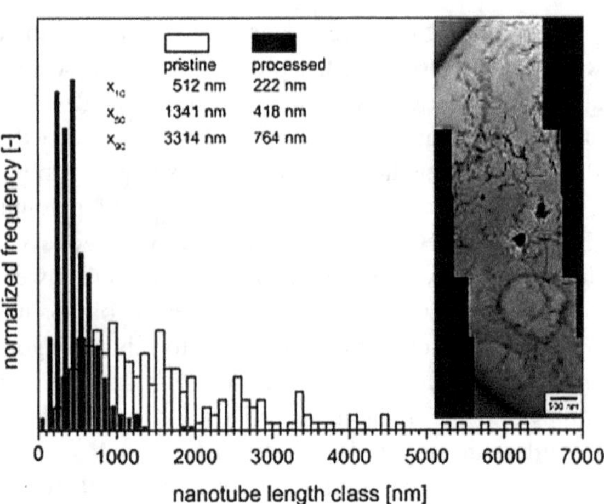

Fig. 7 Length distribution comparsion of Nanocyl™ NC7000: before (pristine MWCNTs) and after processing (as recovered from a melt processed PC composite with 2 wt% MWCNTs. x_{10}, x_{50} and x_{90} were calculated indicating that 10, 50, and 90 % of the nanotubes lengths are smaller than the given value. Reprinted from [99], Copyright (2011) with permission from Elsevier

structural characteristics of the melt processed PLA-CNT composites than those observed on the composites containing low aspect ratio CNTs [101]. This holds true over the whole range of investigated CNT concentrations from 0.5 to 3 wt%. The authors claim that the higher hydrodynamic radii of the high aspect ratio CNTs present as randomly bent fibers (to an extent) and as self-entangled flocs is responsible for higher effective filler volume fraction and hence improved mechanical properties.

Idealizing reinforcements with endless glass/carbon fibers, it is known that long aligned fibers are important for high structural characteristics. Hence, it becomes important to create and/preserve the **alignment** of the CNTs in the composite. It is important to mention here that aligned CNTs are seldom available commercially due to the complexities and costs involved with their production. Orientation of CNTs during composite processing on the other hand has also been reported to negatively affect the electrical properties of the composites [94]. The importance of straightening and aligning the CNTs on the mechanical properties of the PA66-CNT composites produced by rotational winding is illustrated in the work of Wang et al. [102] (Fig. 8). The reduction in CNT waviness and their presence as individual filaments on stretching the composite by 7 % resulted in a substantial improvement in the elastic modulus of the un-stretched composite (E = 14 GPa) by 290 %. The corresponding tensile strength increased from 225 to 630 MPa, a 190 % improvement on stretching. CNT alignment, long CNTs and decreased waviness all simultaneously result in significant improvements in mechanical behavior. These results significantly outnumber the properties achieved on other CNT incorporated PA composites processed by conventional composite processing techniques [103, 104]. However, the production of composites by this approach is not conventional.

Reports on substantial improvements on mechanical properties of the composites do exist, but are very rare. In most cases it is not close to anywhere as to what theories predict. There is no doubt that the researchers working in this area

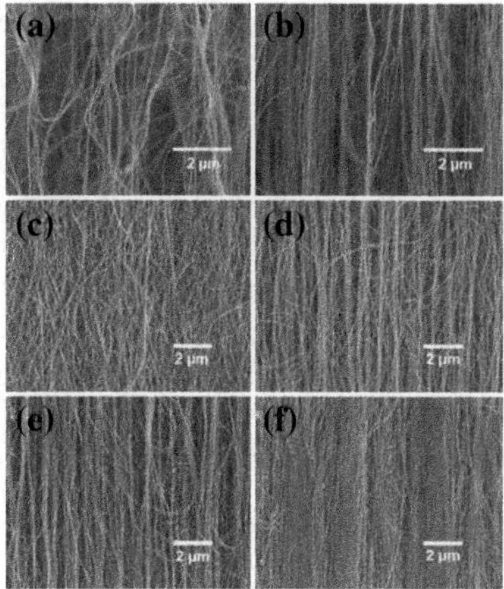

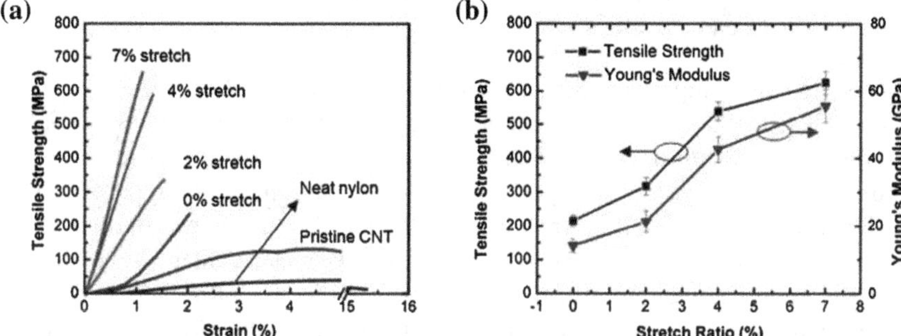

Fig. 8 Top: SEM images of **a** as-drawn CNT dry ribbon showing wavy nanotubes, **b** stretched CNT dry ribbon representing reduced CNT waviness, **c** non-stretched CNT/nylon 6,6 composites showing wavy nanotubes, and CNT/nylon 6,6 composite after stretching with different ratios: **d** stretched for 2 %, **e** stretched for 4 % and **f** stretched for 7 %. Bottom: Enhancement in the mechanical properties of the composites as a function of stretching. Reprinted from [102], Copyright (2011) with permission from Elsevier

would unanimously agree that most literatures on thermoplastic CNT nanocomposites do not tend to report on the mechanical properties of the composites. Results on melt processing of composites especially with those considering MWCNTs however are not very encouraging. Analyzing the reviews on mechanical properties and other related literature on thermoplastic-CNT composites reports it is very evident that the theoretical potential of CNTs has so far not been transferred to substantially enhance structural characteristics of its composite. This could be attributed to the following reasons:

- Irrespective of how well the composites have been processed, achieving individually dispersed CNTs is not realistic.
- Aggregated morphologies of the CNTs in the polymer host are ubiquitous. Due to the cohesive strength of the agglomerates, there exists a small agglomerate

fraction irrespective of the extent of shear stresses exerted on the composite melt during processing.

- Increasing magnitude of shear stresses applied to break up the CNT agglomerates could result in a trade-off of the aspect ratio of the CNTs. Reduction in the CNT length would have a significant effect on the surface area and the effective load bearing length (or the critical length) of the CNTs.
- Functionalization of CNTs carried out to weaken their intrinsic vdW forces could lead to increasing defect density on the CNTs (or disturbing the perfect C–C bond of graphene responsible for high mechanical characteristics), thereby reducing the efficiency of reinforcement.
- The polymer-CNT interface is weak/not significantly developed to efficiently transfer the external load from the polymer to the reinforcing filler.
- Small, wavy and randomly oriented CNTs do not carry the same inherent mechanical characteristics as long, straight and unidirectional CNTs, thereby idealizing CNT reinforcement in a thermoplastic as reinforcement with short fibers instead endless fibers.

It is also very important to point out that the mechanical properties reported here are just to illustrate the potential of CNTs as reinforcements. While interpreting the results from mechanical testing, it is important to consider all the accompanying factors like the type of processing, raw materials, type of testing etc. Bryne and Gun'ko in their review have summarized the huge variations in the mechanical property observations on a similar matrix material that can be brought about by employing different processing routes [105]. Castillo et al. report the influence of five different types of commercially available MWCNTs on the mechanical properties of PC-MWCNT composites at different filler loadings [78]. For a comprehensive reading on the mechanical properties of polymer-CNT composites, and positive and negative influences of functionalization on composite properties we would like to divert the attention of the readers to some interesting reviews [52, 105–108].

4.2 Electrical Properties

The extremely attractive electrical characteristics of CNTs have lived up to their potential as excellent conductive fillers for the insulating polymer matrices. Substantially high electrical properties have been widely reported on polymer-CNT composites. CNTs claim their advantage over conventional carbon black as conductive fillers for polymer matrices due to their high aspect ratio and consequently electrical properties at much lower filler percolation thresholds without compromising on the density, mechanical properties, thermal properties etc. of their host matrices.

The electrical percolation threshold is the critical filler concentration at which the insulator to conductor transition occurs. At this juncture, the electrical conductivity of the composite jumps significantly due to the formation of a conductive pathway. The electrical percolation can be at its theoretical lowest with only one such conductive pathway when well dispersed CNTs are present with a shorter tunneling distance of less than 1.8 nm [109], in a specified matrix free volume. Figure 9 shows the evolution of the conductive characteristics in CNT filled polymer composites. While the electrical properties of the composites is a result of dominative behaviour of the insulating polymer below the percolation threshold, the formation of multiple conductive pathways by individual CNTs and conglomeration of agglomerates beyond percolation saturates the conductive characteristic of the composites.

The percolation threshold is estimated using the scaling law which describes the statistical percolation behavior in the vicinity of percolation. It is given by

$$\sigma \propto (w - w_c)^t \tag{1}$$

where: σ—experimental volume conductivity for $w \geq w_c$, w—MWCNT concentration (wt%), w_c—critical/percolation MWCNT concentration (wt%), t—critical exponent governing the dimensionality of the system.

Munson-McGee predict that the critical volume fraction for the formation of the percolation network will vary from less than 1 % to more than 20 % depending on filler orientation and aspect ratio from a formulation based on statistical arguments [110]. Celzard et al. also emphasize the fact that the theoretical percolation equation must be modified for fillers having higher aspect ratios based on their interpretations of the excluded volume concept [111]. Although the usage of the equation for statistical percolation is common, possibilities of kinetic percolation (filler mobility due to mechanisms like diffusion, convection etc.) should not be

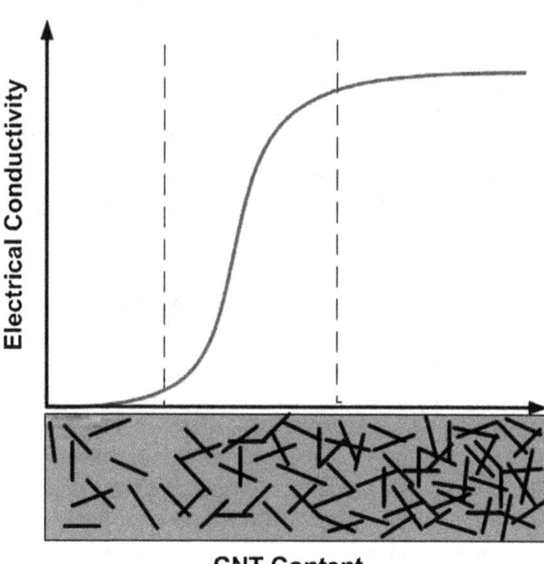

Fig. 9 Evolution of electrical conductivity as a function of CNT content in a composite

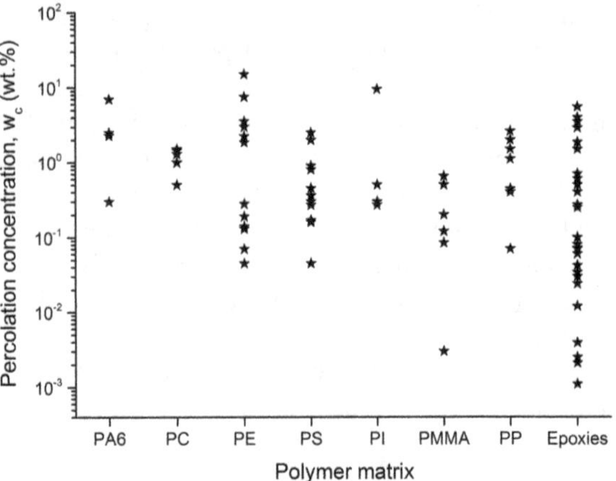

Fig. 10 Electrical percolation threshold of MWCNTs in commonly employed thermoplastic polymer matrices independent of the composite processing approach and MWCNT surface functionality. Data on epoxies (a thermoset) is also reported for comparison. Data compiled from [117, 145]

excluded which was the main reason for contradicting results reported in the work on Bai and Allaoui [112], and Martin et al. [113]. Electrical percolation of CNTs in polymer matrices vary from as little as 0.0025 wt% to as high as 15 wt% as could be seen from Fig. 10. This huge variation is due to the fact that electrical percolation is dependent on the nature of matrix [63–67] and the type of CNT [78, 84], aspect ratio of the CNTs [47, 114], extent of CNT dispersion [64, 65], CNT functionalization [49, 115], processing approach and parameters [65, 84], and alignment and orientation of the fillers [94, 102].

Higher filler percolation thresholds in the range of 1–3 wt% CNT content are common in melt mixed thermoplastic-CNT composites processed with commercially available CNTs [65, 93, 97, 98]. It is generally lower for matrices presenting a favorable environment for a good CNT dispersion. The variation in the electrical resistivity in different thermoplastic matrices (with varying viscosities) containing an identical CNT type is illustrated in Fig. 11. The significant reduction in the aspect ratio of the composites due to their exposure to higher magnitudes of shear stress whilst melt processing as compared to other production approaches is also attributed to be another major reason for higher filler percolation thresholds. This shear stress is however important for achieving good filler dispersion, also an important criterion for good electrical properties of the composites at lower filler loadings.

Although, the dispersion morphology of the CNTs is primarily affected by extrusion, the widely employed secondary processing or the final shaping step for thermoplastic matrices namely compression molding or injection molding have been reported to have a substantial influence on the electrical properties of the composites. The reasons for difference in the electrical properties of the composites resulting out of these production techniques had been discussed earlier. The influence of process parameters of the compression molding [64, 76] and the injection molding [94, 116] have been studied in detail, and has been reported to have an overwhelming influence on the electrical properties of the composites.

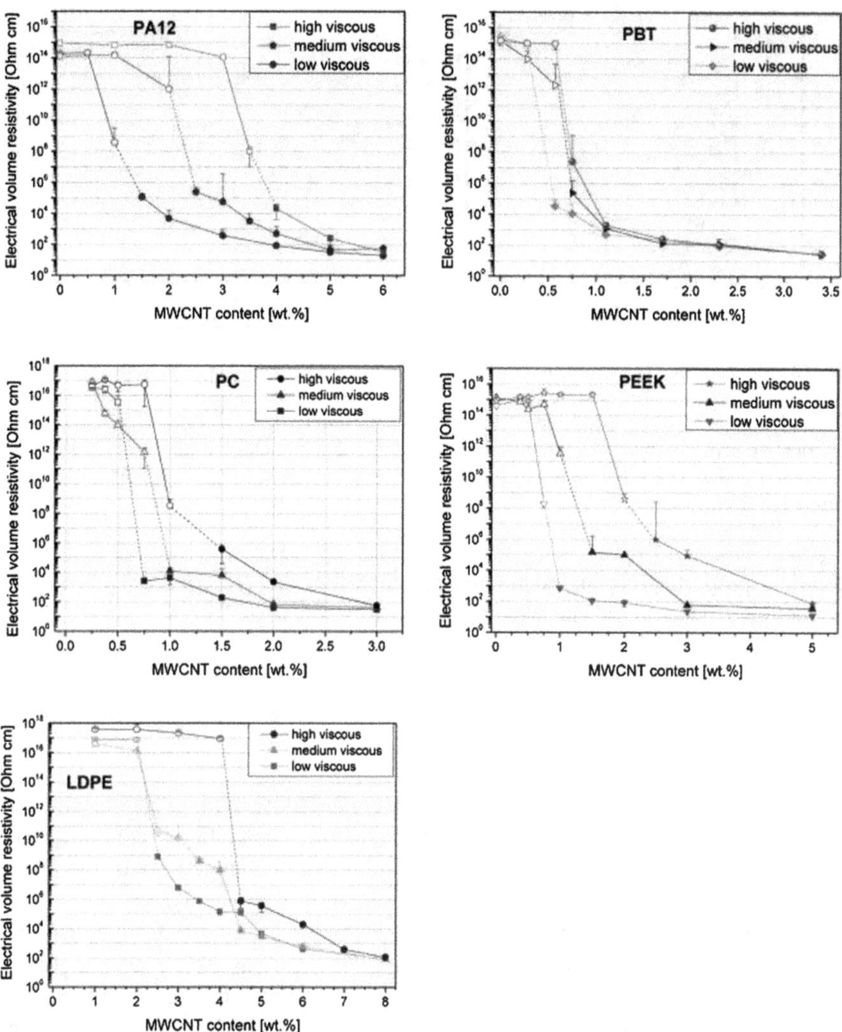

Fig. 11 Electrical volume resistivity of different thermoplastic matrices containing Baytubes® C150P. Reprinted from [63], Copyright (2012) with permission from Elsevier

Figure 12 shows the influence of CNT dispersion (as a function of extruder process conditions) and the nature of the secondary shaping operation on the electrical properties of the composites

Although a wide variety of factors have been listed for its influence on the electrical properties of the composites, consensus cannot be arrived on which has the pivotal role. Lot of contradicting reports does exist in literature. Bauhofer and Kovacs in their review find that the nature and type of the polymer host along with the technique of composite production are much more significant for good

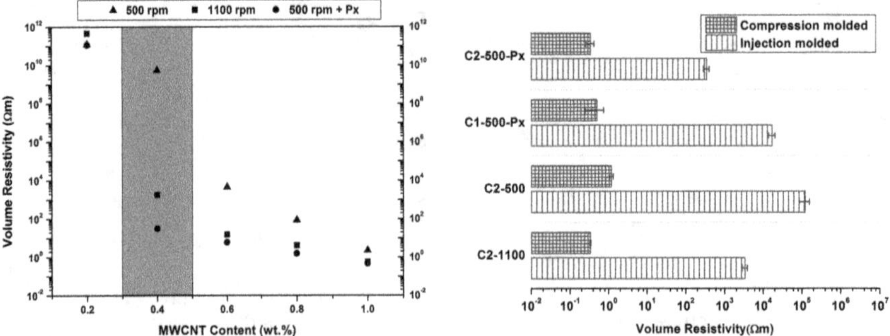

Fig. 12 Influence of CNT dispersion in PP from the compounding process on volume resistivity (=1/conductivity) of the compression molded composites—composites were processed at 500 rpm with and without peroxides "Px" as additives and 1,100 rpm (*Left*). Difference in the electrical properties between injection molded and compression molded composites—C1 and C2 represent 1 and 2 wt% CNT content in PP (*Right*). Adapted from [64], Copyright (2013) with permission from Elsevier

electrical properties compared to the influence that the type of CNTs and their production process could have [117].

4.3 Thermal Properties

The extraordinarily high intrinsic thermal conductivity and excellent thermal stability of the CNTs are of considerable interest for applications like nano-electronics and conductive polymer films. The high thermal conductivity of the CNTs is owing to the atomic vibrations or phonons while the increase in the thermal stability of the composites due to the incorporation of CNTs is a result of its high electron affinity (of 2.65 eV [118]) enabling them to act as radical scavengers.

The expected improvements in the thermal properties of the polymers with CNT addition, especially in the context of good thermal conductivity has however not been achieved. The transportation of phonons is more likely to occur through the insulating matrix rather than that of the ideally visualized case through CNTs as the difference in the thermal conductivities between the CNT and polymer is very small, i.e. of the order 10^4 W/(m.K) [119]. Compare this to the 15–19 orders of magnitude difference in the electrical conductivity! The interfacial or the boundary resistance between the CNTs and the matrix acts as a barrier to heat flow decreasing the overall conductivity of the composite. CNT surface functionalization [120], and aspect ratio [121], along with CNT dispersion in the polymer host are considered to be important factors regulating the thermal conductivity of the composites. Reports on thermal conductivities of epoxy-SWCNT composites are handful [122–124], while on thermoplastic matrices with MWCNTs are seldom to illustrate any conductivity increase/decrease.

The strong radical accepting capacities of the CNTs interrupt the released radicals from oxidation of the polymer, delay the rate of radical propagation and hence decrease the rate of degradation. This results in substantially enhanced thermal stability of the composites containing CNTs. The onset of degradation and temperature of maximum degradation of PS at 300 and 365 °C respectively were enhanced to 313 and 398 °C and 337 and 407 °C respectively on the addition of 1 and 2 wt% MWCNT [65]. Increasing additions of CNT did not further enhance the thermal stability of the composites as enhanced intra-filler interactions due to CNT agglomeration at higher filler loading reduces the effective CNT surface area available for accepting free radicals. Similarly, the addition of 2 wt% MWCNT to PP enhances the thermal stability of the composite with an increase in the onset and maximum temperature of degradation by 40 and 85 °C, the increase also dictated by the enhanced quality of MWCNT dispersion in PP [64]. These results also highlight the importance of having a good CNT dispersion in the matrix.

CNTs have been reported to act as nucleating agents in semi-crystalline matrices [47, 82, 95]. The incorporation of CNTs to the polymers also affects the glass transition temperature, as the part of the polymer in close proximity to the CNTs will have properties much different than the bulk matrix owing to the restrictions imposed on its mobility due to the alterations of the conformational entropy and chain kinetics. An increase of glass transition temperature at lower filler loadings and a decrease at higher loading fractions (due to predominant intra-filler interactions contributing to the development of an interphase) has also been reported [65].

Wu et al. demonstrate a significant improvement in the flame retardancy of melt mixed PET (polyethylene terephthalate)-MWCNT composites with liquid bi-sphenol A bis(diphenyl phosphate) (BDP) as a dispersing additive. The uniformly dispersed CNTs in PET acted as a support for the char released by the oxidation of PET and BDP. The network layer consisting of the CNTs and char acted as a heat shield effectively reducing the exposure of PET to external atmosphere and the heat feedback of the heat flux [125] (Fig. 13). Kashiwagi et al. have also demonstrated similar effects with MWCNT incorporated PP composites [126]. A good CNT dispersion and a better polymer-CNT interface along with increasing filler content could augur well for excellent flame retardant characteristics [127].

In addition to the aforementioned commonly listed characteristics of CNTs in composites, their functionalities also extend to tuning the non-linear property [128] and improve the photoluminescence of polymer-CNT composites [129]. These areas are subjected to very less research inspite of their tremendous potential for engineering applications.

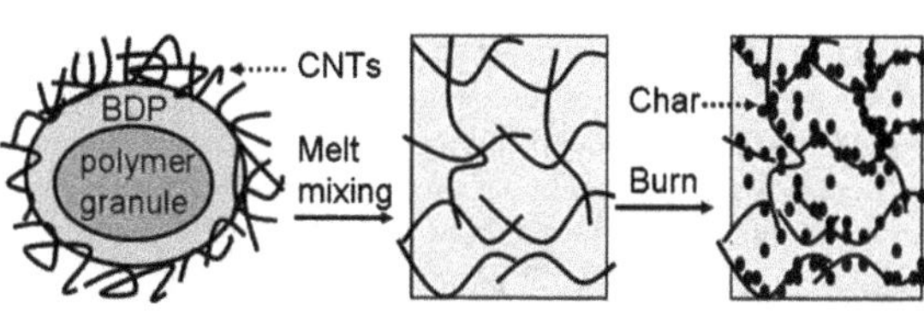

Fig. 13 Schematic of the network of CNT-char network acting as a heat shield for PET. Reprinted from [125], Copyright (2013) with permission from Elsevier

From an outlook on the analysis of the mechanical, electrical and thermal properties of CNT incorporated polymer matrices in this chapter and also by analyzing reviews (recommended for further reading in this manuscript) it seems that the potential of CNTs as fillers for polymer matrices is not fully realized. CNT dispersion in thermoplastics and consequently improved polymer-filler interactions seem to be key criterion influencing the macroscopic properties of the composites. Employing as-received CNTs mandates high shear stresses on processing leading to reduction in CNT aspect ratio. In order to compromise this, CNT functionalization or the use of dispersing additives are widely recommended. The statistic presented in Fig. 14 is from a review on the effect of functionalization on the electrical and mechanical properties of polymer-CNT composites. It is seen that in most cases the increase in electrical property comes with a trade-off of a potential reduction in structural characteristics and vice versa. This complicates scenario for advising a particular approach to tailor the desired composite functionality with CNTs as fillers.

At this juncture, the reader ponders about the potential of CNTs as fillers for thermoplastic matrices. Although excellent electrical properties and (to an extent) thermal properties have been achieved on thermoplastic polymer-CNT composites, substantial improvements in the mechanical properties of these composites are far from being realized due to the various reasons detailed earlier. Long, aligned, perfectly grown infinite CNTs would be ideal; current research has been led in this

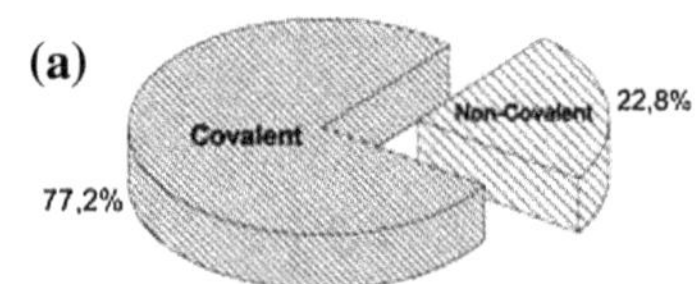

Journal papers reviewed in this article

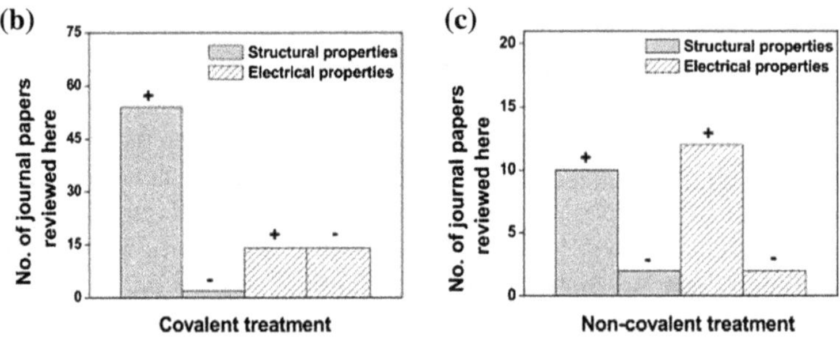

Fig. 14 a Overall statistics of the journal papers reviewed in this article which addresses the influence of various pre-treatment in polymer/CNT composites and compares with respect to pristine CNT; **b** Statistics showing the strengths (+) and weaknesses (−) of covalent and non-covalent types of pre-treatment on the composite properties (statistics also includes paper which report simultaneous improvement in both structural and electrical properties). Reprinted from [52], Copyright (2010) with permission from Elsevier

direction however the economics of scale up and production on a sustainable scale have not been successfully realized thus far. Hence, there is now a desperate need to look for alternate solutions to tap the multifaceted properties of CNTs into commercial applications. The fact that enhancing the structural properties of the composites is the challenge focuses our attention towards multi-scale reinforcements. The established success of conventional glass or carbon fibers as fillers for structural reinforcements positions them as strong contenders for incorporation as secondary fillers in addition to CNTs for the development of a composite with multi-functional properties. It is however important to ensure that the addition of glass/carbon fibers do not significantly alter the electrical and thermal characteristics of the composites brought in by the presence of CNT networks.

5 Multifunctional Composites with Multi-Scale Reinforcements

Fiber reinforced composites (FRC) have revolutionized engineering structures over the last three decades owing to high-strength to weight ratio, flexibility and ease of processing with conventional techniques, and prospects of tuning their functionality. Although the in-plane properties of the FRC's are extremely high, the contribution to its load bearing capacity along the z-direction primarily arises from that of the matrix (which has much lower mechanical properties). Nano-metric dimensions of the nanofillers could facilitate their easy infiltration between the micron-sized fillers leading to better interlaminar properties. In addition to this, the attractive nature of CNTs to impart good electrical and thermal properties offers enormous scope for development of multi-functional light weight composites structures to compete for engineering dominance with metallic and ceramic

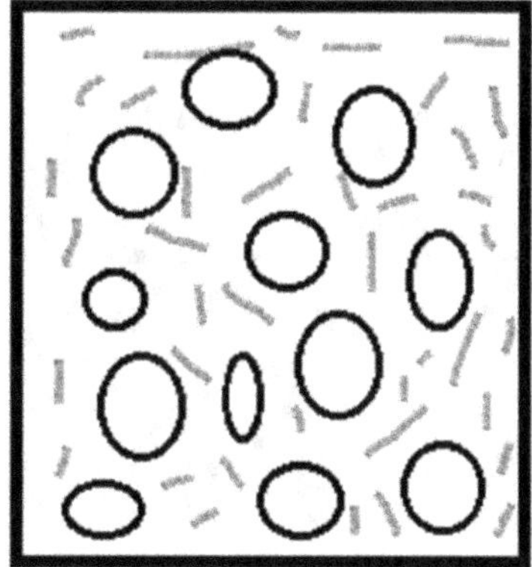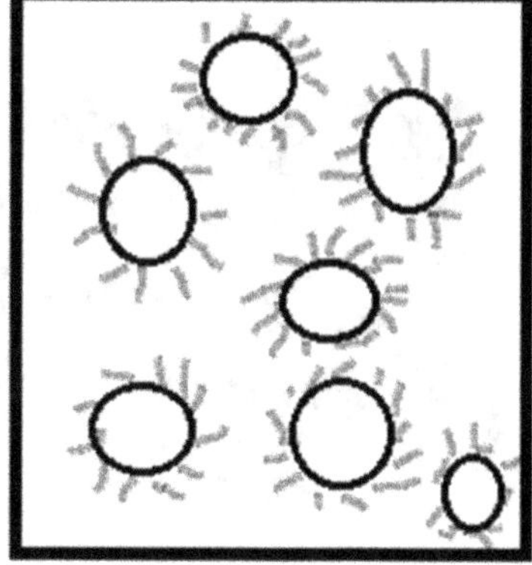

Fig. 15 Multi-scale reinforcement concept—Individual addition of fillers (*Left*), Hybrid fillers (*Right*)

reinforcements. Two possible forms of multi-scale reinforcements could be thought of (Fig. 15):

- Individual addition of the fillers into the polymer host [98, 130].
- Employment of hybrid fillers wherein the nanostructures are grown insitu on the surface of the carbon/glass fibers (parent structures) via the CVD process [131–133], coated on woven parent surfaces [134], and deposited on fiber surfaces using electrophoresis [135, 136].

Figures (16, 17, 18) illustrates micrographs of hybrid fillers processed via different techniques.

The employment of hybrid filler structures in literature is widely discussed in the context of composites based on thermoset matrices. Considering the scope of production of multi-scale composites, the use of these hybrid fillers for thermoset composites is justifiable as the integrity of the nanostructure on the parent could be easily preserved during composite processing. On the contrary, with the processing approach for thermoplastic composites (especially polyolefins) and the harsh

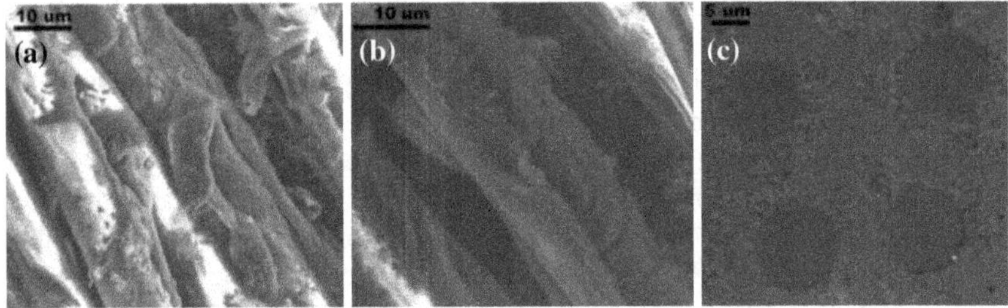

Fig. 16 MWCNTs grown on glass fiber surface through CVD—morphology of the hybrid fibers is observed in different forms: through scanning electron microscopy (SEM) of the fibers (**a** and **b**) and optical microscopy of the cured, polished composite (**c**). Reprinted from [132], Copyright (2012) with permission from Elsevier

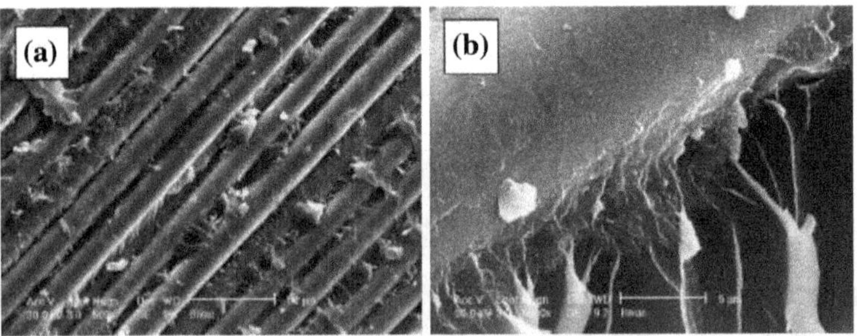

Fig. 17 Hybrid fillers prepared via coating of nanofillers on parent surfaces—fracture surface of fiberglass/vinyl ester composites with silane-SWNTs: **a** 50 μm scale bar magnification and **b** 5 μm scale bar magnification. Reprinted from [134], Copyright (2007) with permission from Elsevier

Fig. 18 SEM images of carbon fibers with (**a**) SWNTs and (**b, c**) MWNTs deposited by electrophoresis. Reprinted from [135], Copyright (2007) with permission from American Chemical Society

conditions involved there is a strong possibility for the nanostructures to detach from their parent surfaces consequently leading to reduction in the reinforcement efficiency. The addition of glass fibers to the polymer-CNT masterbatch stands out as the most feasible approach for the production of thermoplastic composites taking advantage of the properties of both glass fibers and CNTs.

Sathyanarayana et al. demonstrated individual addition of the bi-filler system of MWCNTs and short glass fibers (two variants: one with a compatible sizing to PC and the other with an incompatible sizing) to PC via melt compounding and injection molding [98]. Whilst 2 wt% MWCNT addition to PC resulted in a elastic modulus and tensile strength increase of a meager 10 % and 5 % respectively, addition of 20 wt% short glass fibers to this composite resulted in a substantial 187 % and 86 % increase of the corresponding properties. The increase in the modulus was independent of the type of the glass used, but the tensile strength improvement was significant only when the glass fiber with a compatible sizing with PC was employed. Glass fiber addition however lowered the impact properties of PC, but the presence CNTs compensated for the decrease. The contribution to enhancement in the thermal and electrical properties of the bi-filler composite arose due to the presence of CNTs, with increasing glass fiber additions slightly lowering the efficiency of reinforcement. The SEM micrographs of these composites are presented in Fig. 19.

A short reference to some interesting results with thermoset based multi-scale composites is being made here to illustrate the potential of these materials. Lv et al. grafted MWCNTs of varying lengths on carbon fibers using the injection CVD approach [137]. The MWCNT lengths and the orientation were controlled by the nanostructure growth time and the adopted surface treatment on the carbon fibers. Single fiber fragmentation tests showed an 175 % improvement in the interfacial shear strength of the hybrid fiber (containing 47.2 μm of aligned nanostructure on the parent) compare to the pristine carbon fiber. The SEM morphologies from their work are presented in Fig. 20. Zhang et al. achieved 150 % improvement in the IFSS of CNT-CF reinforced epoxy composites [138]. Zhu et al. report an increase of 45 % in the shear strength in their effort to increase the properties along the z-axis for carrying transverse loads on a glass fiber reinforced vinyl ester composite with nanotube integration. The CNT loading was

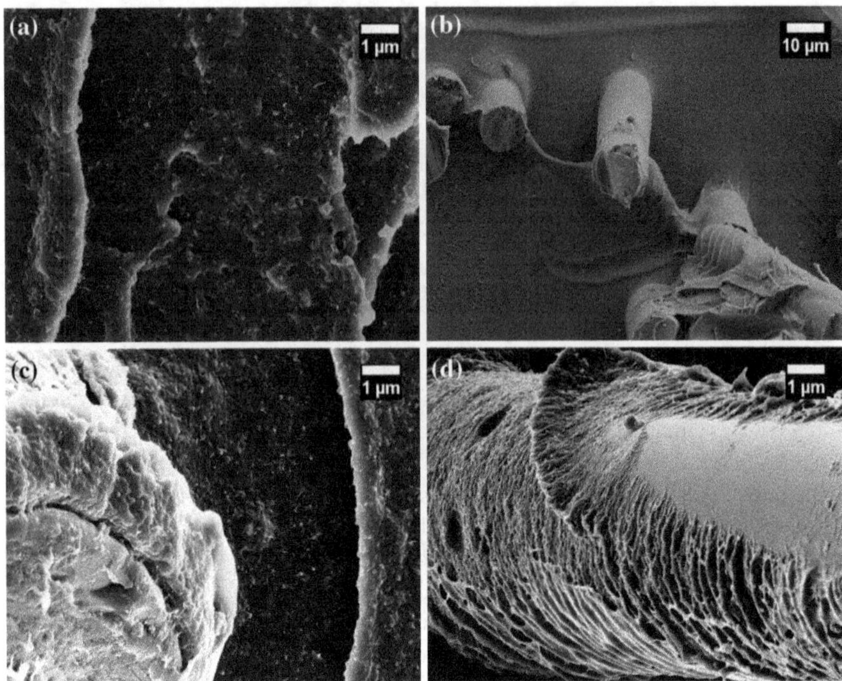

Fig. 19 a Good CNT dispersion in PC, **b** PC with 20 wt% glass fibers with PC compatible sizing, **c** Individual morphologies of the filler kept in the bi-filler composite, **d** Magnified view of a single glass fiber with a polymer sheathing (some CNTs were also identified on the sheathing) from the bi-filler composite [98]

0.015 wt% in the midplane ply [134]. Gojny et al. carried out resin transfer molding for the production of glass-fibre-reinforced polymers (GFRP) with nanotube/epoxy matrix. They report enhancements in the modulus, tensile strength, fracture toughness and the anisotropic electrical properties of the composites containing 0.3 wt% MWCNTs (modified by amine functionalization) [139].

Irrespective of how good the hybrid-filler is, the macroscopic property enhancements of the composites would strongly depend on the extent to which the nanostructures are dispersed in the matrix. The strength of the nano-filler-parent interface and the linkage between the nanostructure and the host matrix are considered to be crucial. The biggest challenge facing the multi-scale composites especially on thermoset matrices is the fact that high volume fraction of nanofillers would lead to processing difficulties due to its ability to enhance the viscosity of the resin significantly. The need for environmentally benign applications for example light weight structures with multifunctional characteristics will no doubt drive the market towards the development of composites with multi-faceted properties.

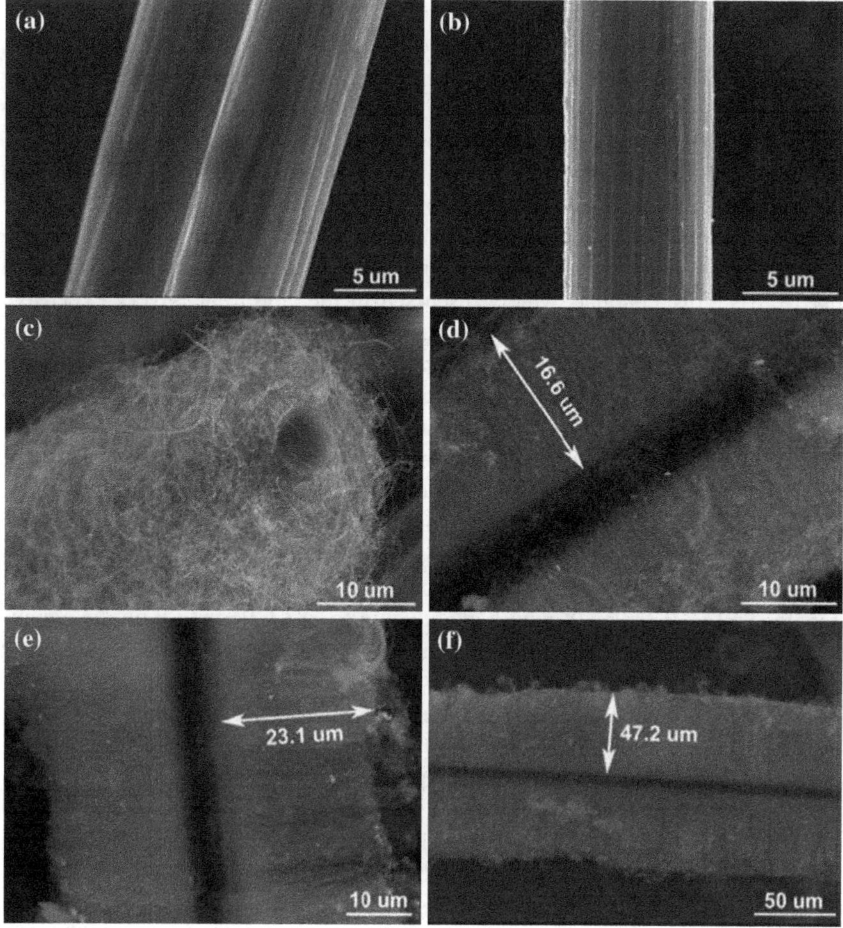

Fig. 20 SEM images of unsized CFs (**a**) before and (**b**) after the surface treatment, and CFs grafted by (**c**) entangled MWCNTs and (**d–f**) aligned CNTs. Reprinted from [137], Copyright (2011) with permission from Elsevier

6 Synopsis

The principal objective of this chapter was to give a comprehensive overview on the potential of CNTs as fillers for thermoplastic polymer matrices processed by techniques suitable for large scale production. Thermoplastic matrices present significant challenges for achieving a good CNT dispersion primarily due to its high melt viscosity and high interfacial tension between the polymer and the CNTs as compared to thermosets. As a consequence the efficiency of CNTs as fillers for thermoplastic matrices is far off from what theory would predict.

We have elucidated in short the world of CNTs starting from its production, properties, surface modification, applications and a brief market outlook. Although understanding of the complicated mechanism of CNT dispersion in thermoplastic is crucial, equally important is to understand the influence of raw-materials and the process parameters. The complex picture of achieving a good CNT dispersion that was presented in terms of processing with the twin-screw extruder, and the influence that the secondary shaping operation could have on the composite properties present a strong case that mandates a careful selection of the raw-materials and process conditions. Functionalization of CNTs as a technique to improve CNT dispersion is widely discussed in literature, but is very expensive and present problems of scale up for large scale production. Hence, we focused on special techniques like the use of ultrasound, supercritical fluids, processing additives etc. whilst processing in the extruder as non-conventional approaches to enhance CNT dispersion in a thermoplastic matrix.

Although the potential of CNTs as fillers for mechanical reinforcements do not look promising after analyzing many literature reports, Loos and Schulte claim that CNTs are the most viable strengthening option for composites with strengths up to 11.61 GPa on analysis of their cost versus property relation compared to carbon fibers [140]. The latter could be beneficial for producing composites with strengths up to 4.18 GPa, Young's modulus up to 383 GPa with a possible compromise on toughness. However, according to them and most literatures, in order for CNTs to be competitive enough they would have to be produced with high purity as long aligned structures with high aspect ratio, in large volumes and at low cost. Also, issues such as CNT dispersion, distribution and orientation in the matrix and enhanced polymer-filler compatibility must be addressed quickly for competitiveness and commercial success.

The addition of CNTs has contributed immensely for enhancing the electrical conductivity of the composites at substantially low filler percolation thresholds. The review of literature indicates that this is influenced by many factors. Thermal properties of the composites have also been reported to be enhanced by the addition of CNTs, but not many literature reports are available to claim this strongly. It could thus be generalized that the presence of CNTs as fillers for polymer matrices promises to have a substantial influence in enhancing the electrical, and to an extent thermal properties of the composites, however the improvement in mechanical properties is far from existing imagination.

Multi-scale reinforcements with CNTs, principally for electrical and thermal properties, and traditional fillers like glass or carbon fibers, for structural properties, look to be a promising solution towards the development of multifunctional composites. The individual addition of these fillers or the development of a hybrid filler, wherein the carbon nanostructures are grown on the parent glass/carbon fiber surface, to be incorporated into the polymer host are two possible approaches for achieving the multi-scale reinforcements. These materials show some promise, however they are predominantly reported on thermosetting matrices. Going forward, there is a strong belief that the potential of these fillers would also translate well into excellent macroscopic properties on a thermoplastic composite.

Acknowledgments Shyam Sathyanarayana thanks the European Community's Seventh Framework Programme (FP7-PEOPLE-ITN-2008) for his funding through the Marie Curie Fellowship under grant agreement number 238363 during his stay at the Fraunhofer Institute für Chemische Technologie ICT.

References

1. Usuki, A., Kojima, Y., Kawasumi, M., Okada, A., Fukushima, Y., Kurauchi, T., Kamigaito, O.: Synthesis of nylon 6-clay hybrid. J. Mater. Res. **8**, 1179–1184 (1993)
2. Lan, T., Pinnavaia, T.: Clay-reinforced epoxy nanocomposites. Chem. Mater. **6**, 2216–2219 (1994)
3. Ajayan, P.M., Schadler, L.S., Braun, P.V.: (eds.) Nanocomposite Science and Technology. Wiley VCH, Weinheim, (2003)
4. ISO/TS27687 (2008). Nanotechnologies—Terminology and Definitions for Nano-Objects—Nanoparticle, Nanofibre and Nanoplate
5. McEuen, P.L., Fuhrer, M.S., Park, H.: Single-walled carbon nanotube electronics. IEEE Trans. Nanotechnol. **1**, 78–85 (2002)
6. Gao, B., Chen, Y.F., Fuhrer, M.S., Glattli, D.C., Bachtold, A.: Four-point resistance of individual single-wall carbon nanotubes. Phys. Rev. Lett. **95**, 196802 (2005). 4pp
7. Ebbesen, T.W., Lezec, H.J., Hiura, H., Bennett, J.W., Ghaemi, H.F., Thio, T.: Electrical conductivity of individual carbon nanotubes. Nature **382**, 54–56 (1996)
8. Dresselhaus, M.S., Dresselhaus, G., Charlier, J.C., Hernández, E.: Electronic, thermal and mechanical properties of carbon nanotubes. Philos. Trans. R. Soc. Lond. A. **362**, 2065–2098 (2004)
9. Treacy, M.M.J., Ebbesen, T.W., Gibson, J.M.: Exceptionally high Young's modulus observed for individual carbon nanotubes. Nature **381**, 678–680 (1996)
10. Yao, N., Lordi, V.: Young's modulus of single-walled carbon nanotubes. J. Appl. Phys. **84**, 1939 (1998). 5pp
11. Demczyk, B.G., Wang, Y.M., Cumings, J., Hetman, M., Han, W., Zettl, A., Ritchie, R.O.: Direct mechanical measurement of the tensile strength and elastic modulus of multiwalled carbon nanotubes. Mater. Sci. Eng., A **334**, 173–178 (2002)
12. Hone, J., Llaguno, M.C., Biercuk, M.J., Johnson, A.T., Batlogg, B., Benes, Z., Fischer, J.E.: Thermal properties of carbon nanotubes and nanotube-based materials. Appl. Phys. A **74**, 339–343 (2002)
13. Berber, S., Kwon, Y.K., Tománek, D.: Unusually high thermal conductivity of carbon nanotubes. Phys. Rev. Lett. **84**, 4613–4616 (2000)
14. Jou, W.S., Cheng, H.Z., Hsu, C.F.: A carbon nanotube polymer-based composite with high electromagnetic shielding. J. Electron. Mater. **35**, 462–470 (2006)
15. Li, N., Huang, Y., Du, F., He, X., Lin, X., Gao, H., Ma, Y., Li, F., Chen, Y., Eklund, P.C.: Electromagnetic interference (EMI) shielding of single-walled carbon nanotube epoxy composites. Nano Lett. **6**, 1141–1145 (2006)
16. Yang, Y., Gupta, M.C., Dudley, K.L., Lawrence, R.W.: Novel carbon nanotube-polystyrene foam composites for electromagnetic interference shielding. Nano Lett. **5**, 2131–2134 (2005)
17. Hoang, A.S.: Electrical conductivity and electromagnetic interference shielding characteristics of multiwalled carbon nanotube filled polyurethane composite films. Adv. Nat. Sci. Nanosci. Nanotechnol. **2**, 025007 (2011). 5pp
18. Philip, B., Abraham, J.K., Chandrasekhar, A., Varadan, V.K.: Carbon nanotubes/PMMA composite thin films for gas-sensing applications. Smart Mater. Struct. **12**, 935–939 (2003)
19. Lee, S., Müller, A.M., Al-Kaysi, R., Bardeen, C.J.: Using perylene-doped polymer nanotubes as fluorescence sensors. Nano Lett. **6**, 1420–1424 (2006)

20. Li, J., Lu, Y., Meyyappan, M.: Nano chemical sensors with polymer-coated carbon nanotubes. IEEE Sens. J. **6**, 1047–1051 (2006)
21. Dervishi, E., Li, Z., Saini, V., Sharma, R., Xu, Y., Mazumder, M.K., Biris, A.S., Trigwell, S., Biris, A.R., Saini, D., Lupu, D.: Multifunctional coatings with carbon nanotubes for electrostatic charge mitigation. IEEE Trans. Ind. Appl. **45**, 1547–1552 (2009)
22. Fu, X., Zhang, C., Liu, T., Liang, R., Wang, B.: Carbon nanotube buckypaper to improve fire retardancy of high-temperature/high performance polymer composites. Nanotechnology **21**, 235701 (2010). 8pp
23. World's first carbon nanotube reinforced polyurethane wind blades, Accessed on 15th October 2011 http://polymers.case.edu/stories/World's%20First%20PU%20CNT%20 Blades.html
24. Ravichandran, J., Manoj, A.G., Liu, J., Manna, I., Carroll, D.L.: A novel polymer nanotube composite for photovoltaic packaging applications. Nanotechnology **19**, 085712 (2008). 5pp
25. Zhao, Y., Wei, J., Vajtai, R., Ajayan, P.M., Barrera, E.V.: Iodine doped carbon nanotube cables exceeding specific electrical conductivity of metals. Sci. Rep. **1**, 1 (2011)
26. Iijima, S.: Helical microtubules of graphitic carbon. Nature **354**, 56–58 (1991)
27. Ebbesen, T.W., Ajayan, P.M.: Large-scale synthesis of carbon nanotubes. Nature **358**, 220–222 (1992)
28. Ebbesen, T.W., Hiura, H., Fujita, J., Ochiai, Y., Matsui, S., Tanigaki, K.: Patterns in the bulk growth of carbon nanotubes. Chem. Phys. Lett. **209**, 83–90 (1993)
29. Thostenson, E.T., Ren, Z., Chou, T.W.: Advances in the science and technology of carbon nanotubes and their composites: A review. Compos. Sci. Technol. **31**, 1899–1912 (2001)
30. O'Connell, M.J.: (ed.) Carbon Nanotubes: Properties and Applications. Taylor & Francis, Boca Raton, (2006)
31. Loiseau, A., Launois-Bernede, P., Petit, P., Roche, S., Salvetat, J.P.: (eds.) Understanding Carbon Nanotubes from Basics to Application. Springer, Berlin, (2006)
32. Tanaka, K., Yamabe, T., Fukui, K.: The Science and Technology of Carbon Nanotubes. Elsevier, Amsterdam, (1999)
33. Meyyappan, M.: (ed.) Carbon Nanotubes: Science and Applications. Taylor & Francis, Boca Raton, (2005)
34. Qian, D., Wagner, G.J., Liu, W.K., Yu, M.F., Ruoff, R.S.: Mechanics of carbon nanotubes. Appl. Mech. Rev. **55**, 495–533 (2002)
35. Che, J.W., Cagin, T., Goddard, W.A.: Thermal conductivity of carbon nanotubes. Nanotechnology **11**, 65–69 (2000)
36. Kim, P., Shi, L., Majumdar, A., McEuen, P.L.: Thermal transport measurements of individual multiwalled nanotubes. Phys. Rev. Lett. **87**, 215502 (2001). 1–4
37. Pop, E., Mann, D., Wang, Q., Goodson, K., Dai, H.: Thermal conductance of an individual single-walled carbon nanotube above room temperature. Nano Lett. **6**, 96–100 (2006)
38. Sukhadolou, A.V., Ivakin, E.V., Ralchenko, V.G., Khomich, A.V., Vlasov, A.V., Popovich, A.F.: Thermal conductivity of CVD diamond at elevated temperatures. Diam. Relat. Mater. **14**, 589–593 (2005)
39. Pantea, D., Darmstadt, H., Kaliaguine, S., Summerchen, L., Christian, R.: Electrical conductivity of thermal carbon blacks: Influence of surface chemistry. Carbon **39**, 1147–1158 (2001)
40. Durkop, T., Getty, S.A., Cobas, E., Fuhrer, M.S.: Extraordinary mobility in semiconducting carbon nanotubes. Nano Lett. **4**, 35–39 (2004)
41. Walters, D.A., Casavant, M.J., Qin, X.C., Huffman, C.B., Boul, P.J., Ericson, L.M., Haroz, E.H., O'Connell, M.J., Smith, K., Colbert, D.T., Smalley, R.E.: In-plane-aligned membranes of carbon nanotubes. Chem. Phys. Lett. **338**, 14–20 (2001)
42. Alig, I., Pötschke, P., Lellinger, D., Skipa, T., Pegel, S., Kasaliwal, G.R., Villmow, T.: Establishment, morphology and properties of carbon nanotube networks in polymer melts. Polymer **53**, 4–28 (2012)
43. Datasheet Baytubes® C150P: Bayer Material Science AG, Leverkusen, Germany (2009)

44. Krause, B., Mende, M., Pötschke, P., Petzold, G.: Dispersability and particle size distribution of CNTs in an aqueous surfactant dispersion as a function of ultrasonic treatment time. Carbon **48**, 2746–2754 (2010)
45. Datasheet Nanocyl^{TM} NC7000: Nanocyl S.A., Sambreville, Belgium (2010)
46. Dyke, C.A., Tour, J.M.: Covalent functionalization of single-walled carbon nanotubes for material applications. J. Phys. Chem. A **108**, 11151–11159 (2004)
47. Bose, S., Bhattacharyya, A.R., Bondre, A.P., Kulkarni, A.R., Pötschke, P.: Rheology, electrical conductivity, and the phase behavior of cocontinuous PA6/ABS blends with MWNT: Correlating the aspect ratio of MWNT with the percolation threshold. J. Polym. Sci. B **46**, 1619–1631 (2008)
48. Kamaras, K., Itkis, M.E., Hu, H., Zhao, B., Haddon, R.C.: Covalent bond formation to a carbon nanotube metal. Science **301**, 1501 (2003)
49. Kang, X., Ma, W., Zhang, H., Xu, Z., Guo, Y., Xiong, Y.: Vinyl-carbon nanotubes for composite polymer materials. J. Appl. Polym. Sci. **110**, 1915–1920 (2008)
50. Padgett, C.W., Brenner, D.W.: Influence of chemisorption on the thermal conductivity of single-wall carbon nanotubes. Nano Lett. **4**, 1051–1053 (2004)
51. Cohen, R.S., Kalisman, L.Y., Roth, E.N., Rozen, R.Y.: Generic approach for dispersing single-walled carbon nanotubes: The strength of a weak interaction. Langmuir **20**, 6085–6088 (2004)
52. Bose, S., Khare, R.A., Moldenaers, P.: Assessing the strengths and weaknesses of various types of pre-treatments of carbon nanotubes on the properties of polymer/carbon nanotubes composites: A critical review. Polymer **51**, 975–993 (2010)
53. Global carbon nanotube market—industry beckons. Accessed on 1st Oct 2012. http://www.nanowerk.com/spotlight/spotid=23118.php
54. Applications and benefits of multiwalled carbon nanotubes (MWCNTs). Accessed 2nd Dec 2012. http://www.cefic.org/Documents/Other/Benefits%20of%20Carbon%20Nanotubes.pdf
55. Zyvex Marine launches LRV-17 long range vessel as the first nano-composite manned boat. Accessed 22nd Aug 2012 http://www.zyvexmarine.com/news/2012/7/18/zyvex-marine-launches-lrv-17-long-range-vessel-as-the-first.html
56. Völkl Racquets. Accessed 2nd Jan 2013. http://www.tennisexpress.com/category.cfm/tennis/volkl-tennis-racquets
57. Aldila. Accessed 14th Nov 2012. http://www.aldila.com/portals/12909/images/logos/aldila_2010_catalog.pdf
58. Frost and Sullivan's study on potential market for carbon nanomaterials' applications (2011). Accessed 26th June 2012. http://www.nist.gov/cnst/upload/Valenti-NIST.pdf
59. Safadi, B., Andrews, R., Grulke, E.A.: Multiwalled carbon nanotube polymer composites: Synthesis and characterization of thin films. J. Appl. Polym. Sci. **84**, 2660–2669 (2002)
60. Qian, D., Dickey, E.C., Andrews, R., Rantell, T.: Load transfer and deformation mechanisms in carbon nanotube-polystyrene composites. Appl. Phys. Lett. **76**, 2868 (2000). 3pp
61. Cadek, M., Coleman, J.N., Barron, V., Hedicke, K., Blau, W.J.: Morphological and mechanical properties of carbon-nanotube-reinforced semicrystalline and amorphous polymer composites. Appl. Phys. Lett. **81**, 5123 (2002). 3pp
62. Menzer, K., Krause, B., Boldt, R., Kretzschmar, B., Weidisch, R., Pötschke, P.: Percolation behaviour of multiwalled carbon nanotubes of altered length and primary agglomerate morphology in melt mixed isotactic polypropylene-based composites. Compos. Sci. Technol. **71**, 1936–1943 (2011)
63. Socher, R., Krause, B., Müller, M.T., Boldt, R., Pötschke, P.: The influence of matrix viscosity on MWCNT dispersion and electrical properties in different thermoplastic nanocomposites. Polymer **53**, 495–504 (2012)
64. Sathyanarayana, S., Hübner, C., Diemert, J., Pötschke, P., Henning, F.: Influence of peroxide addition on the morphology and properties of polypropylene—multiwalled carbon nanotube composites. Compos. Sci. Technol. **84**, 78–85 (2013)

65. Sathyanarayana, S., Olowojoba, G., Weiss, P., Caglar, B., Pataki, B., Mikonsaari, I., Hübner, C., Henning, F.: Compounding of MWCNT with PS in a twin-screw extruder with varying process parameters: Morphology, interfacial behaviour, thermal stability, rheology and volume resistivity. Macromol. Mat. Eng. **298**, 89–105 (2013)
66. Sathyanarayana, S., Wegrzyn, M., Benedito, A., Giménez, E., Hübner, C., Henning, F.: Multiwalled carbon nanotubes incorporated miscible blend of poly(phenylenether)/polystyrene—processing and characterization. Express. Polym. Lett. **7**, 621–635 (2013)
67. Mack, C., Sathyanarayana, S., Weiss, P., Mikonsaari, I., Hübner, C., Henning, F., Elsner, P.: Twin-screw extrusion of multi walled carbon nanotubes reinforced polycarbonate composites: Investigation of electrical and mechanical properties. IOP Conf. Ser. Mat. Sci. Eng. **40**, 012020 (2012)
68. Chen, L., Pang, X.J., Qu, M.Z., Zhang, Q.T., Wang, B., Zhang, B.L., Yu, Z.L.: Fabrication and characterization of polycarbonate/carbon nanotubes composites. Compos. A **37**, 1485–1489 (2006)
69. Lin, T.S., Cheng, L.Y., Hsiao, C.C., Yang, A.C.M.: Percolated network of entangled multi-walled carbon nanotubes dispersed in polystyrene thin films through surface grafting polymerization. Mat. Chem. Phys. **94**, 438–443 (2005)
70. Wu, T.M., Chen, E.C.: Preparation and characterization of conductive carbon nanotube-polystyrene nanocomposites using latex technology. Comp. Sci. Technol. **68**, 2254–2259 (2008)
71. Vigolo, B., Pénicaud, A., Coulon, C., Sauder, C., Pailler, R., Journet, C., Bernier, P., Poulin, P.: Macroscopic fibers and ribbons of oriented carbon nanotubes. Science **290**, 1331–1334 (2000)
72. Xia, H., Wang, Q., Li, K., Wu, G.H.: Preparation of polypropylene/carbon nanotube composite powder with a solid-state mechanochemical pulverization process. J. Appl. Polym. Sci. **93**, 378–386 (2004)
73. Masuda, J., Torkelson, J.M.: Dispersion and major property enhancement in polymer/multiwall carbon nanotube nanocomposites via solid-state shear pulverization followed by melt mixing. Macromolecules **41**, 5974–5977 (2008)
74. Sen, R., Zhao, B., Perea, D.E., Itkis, M.E., Hu, H., Love, J., Bekyarova, E., Haddon, R.C.: Preparation of single-walled carbon nanotube reinforced polystyrene and polyurethane nanofibers and membranes by electrospinning. Nano Lett. **4**, 459–464 (2004)
75. Yuan, J.M., Fan, Z.F., Chen, X.H., Chen, X.H., Wu, Z.J., He, L.P.: Preparation of polystyrene–multiwalled carbon nanotube composites with individual-dispersed nanotubes and strong interfacial adhesion. Polymer **50**, 3285–3291 (2009)
76. Kasaliwal, G., Göldel, A., Pötschke, P.: Influence of processing conditions in small-scale melt mixing and compression molding on the resistivity and morphology of polycarbonate–MWNT composites. J. Appl. Polym. Sci. **112**, 3494–3509 (2009)
77. Mamunya, Y., Boudenne, A., Lebovka, N., Ibos, L., Candau, Y., Lisunova, M.: Electrical and thermophysical behaviour of PVC-MWCNT nanocomposites. Compos. Sci. Technol. **68**, 1981–1988 (2008)
78. Castillo, F.Y., Socher, R., Krause, B., Headrick, R., Grady, B.P., Prada-Silvy, R., Pötschke, P.: Electrical, mechanical, and glass transition behavior of polycarbonate-based nanocomposites with different multi-walled carbon nanotubes. Polymer **52**, 3835–3845 (2011)
79. McNally, T., Pötschke, P., Halley, P., Murphy, M., Martin, D., Bell, S.E.J., Brennan, G.P., Bein, D., Lemoine, P., Quinn, J.P.: Polyethylene multiwalled carbon nanotube composites. Polymer **46**, 8222–8232 (2005)
80. Wei, J.: (ed.) Advances in Chemical Engineering, vol. 25, Academic Press, San Diego, (1999)
81. Kasaliwal, G.R., Göldel, A., Pötschke, P., Heinrich, G.: Influences of polymer matrix viscosity an molecular weight on MWCNT agglomerate dispersion. Polymer **52**, 1027–1036 (2011)

82. Villmow, T., Pötschke, P., Pegel, S., Häussler, L., Kretzschmar, B.: Influence of twin-screw extrusion conditions on the dispersion of multi-walled carbon nanotubes in a poly(lactic acid) matrix. Polymer **49**, 3500–3509 (2008)
83. Sathyanarayana, S., Weiss, P., Hübner, C., Diemert, J., Henning, F.: Twin-screw compounding: Process optimization for multiwalled carbon nanotubes reinforced polypropylene composites. ANTEC Mumbai 2012—Proceedings of the Annual Technical Conference and Exhibition, Society of Plastics Engineers, Mumbai, India, pp. 394–400 (2012) 6–7 Dec
84. Müller, M.T., Krause, B., Kretzschmar, B., Pötschke, P.: Influence of feeding conditions in twin-screw extrusion of PP/MWCNT composites on electrical and mechanical properties. Compos. Sci. Technol. **71**, 1535–1542 (2011)
85. Mikonsaari, I.: (ed.) Direktprozess zur Herstellung von Nanosuspensionen und deren Zudosierung in thermoplastische Matrices zur Herstellung von Nanocomposites. Fraunhofer Verlag, Stuttgart, (2011)
86. Lewis, T.M., Isayev, A.I.: Continuous high power ultrasonic extrusion of PEEK-CNT nanocomposites. ANTEC 2013-Proceedings of the 71st Annual Technical Conference and Exhibition, Society of Plastics Engineers, Cincinnati, OH, USA, pp.422–427 21–25 April (2013)
87. Isayev, A.I., Kumar, R., Lewis, T.M.: Ultrasound assisted twin screw extrusion of polymer–nanocomposites containing carbon nanotubes. Polymer **50**, 250–260 (2009)
88. Weiss, P., Just, D., Diemert, J., Hübner, C.: Use of ultrasound for the dispersion of MWNT in melt compounding. Poster in the proceedings of 3rd Annual Conference of the Innovation Alliance Carbon Nanotubes (Inno.CNT), Ettlingen, Germany, 25–27 Jan (2011)
89. Nguyen, Q.T., Baird, D.G.: An improved technique for exfoliating and dispersing nanoclay particles into polymer matrices using supercritical carbon dioxide. Polymer **48**, 6923–6933 (2007)
90. Zerda, A.S., Caskey, T.C., Lesser, A.J.: Highly concentrated, intercalated silicate nanocomposites: Synthesis and characterization. Macromolecules **36**, 1603–1608 (2003)
91. Manke, C.W., Gulari, E., Mielewski, D.F., Lee, E.C.: System and method of delaminating a layered silicate material by supercritical fluid treatment. United States Patent 6,469,073 (2002)
92. Chen, C., Bortner, M., Quigley, J.P., Baird, D.G.: Using supercritical carbon dioxide in preparing carbon nanotube nanocomposite: Improved dispersion and mechanical properties. Polym. Compos. **33**, 1033–1043 (2012)
93. Müller, M.T., Krause, B., Pötschke, P.: A successful approach to disperse MWCNTs in polyethylene by melt mixing using polyethylene glycol as additive. Polymer **53**, 3079–3083 (2012)
94. Villmow, T., Pegel, S., Pötschke, P., Wagenknecht, U.: Influence of injection molding parameters on the electrical resistivity of polycarbonate filled with multi-walled carbon nanotubes. Compos. Sci. Technol. **68**, 777–789 (2008)
95. Coleman, J.N., Cadek, M., Blake, R., Nicolosi, V., Ryan, K.P., Belton, C., Fonseca, A., Nagy, J.B., Gun'ko, Y.K., Blau, W.J.: High-performance nanotube-reinforced plastics: Understanding the mechanism of strength increase. Adv. Funct. Mater. **14**, 791–798 (2004)
96. Krause, B., Pötschke, P., Häußler, L.: Influence of small scale melt mixing conditions on electrical resistivity of carbon nanotube-polyamide composites. Compos. Sci. Technol. **69**, 1505–1515 (2009)
97. Prashantha, K., Soulestin, J., Lacrampe, M.F., Krawczak, P., Dupin, G., Claes, M.: Masterbatch-based multi-walled carbon nanotube filled polypropylene nanocomposites: Assessment of rheological and mechanical properties. Compos. Sci. Technol. **69**, 1756–1763 (2009)
98. Sathyanarayana, S., Wegrzyn, M., Weiss, P., Giménez, E., Hübner, C., Henning, F.: Influence of short glass fiber addition on the morphology and properties of PC-MWCNT composites. PPS28—Proceedings of the 28th Annual Meeting, Polymer Processing Society Pattaya, Thailand, 11–15 Dec (2012)

99. Krause, B., Boldt, R., Pötschke, P.: A method for determination of length distributions of multiwalled carbon nanotubes before and after melt processing. Carbon **49**, 1243–1247 (2011)
100. Schaefer, D.W., Justice, R.S.: How nano are nanocomposites? Macromolecules **40**, 8501–8517 (2007)
101. Wu, D., Wu, L., Zhou, W., Sun, Y., Zhang, M.: Relations between the aspect ratio of carbon nanotubes and the formation of percolation networks in biodegradable polylactide/carbon nanotube composites. J. Polym. Sci. B **48**, 479–489 (2010)
102. Wang, X., Bradford, P.D., Liu, W., Zhao, H., Inoue, Y., Maria, J.P., Li, Q., Yuan, F.G., Zhu, Y.: Mechanical and electrical property improvement in CNT/nylon composites through drawing and stretching. Compos. Sci. Technol. **71**, 1677–1683 (2011)
103. Liu, T.X., Phang, I.Y., Shen, L., Chow, S.Y., Zhang, W.D.: Morphology and mechanical properties of multiwalled carbon nanotubes reinforced nylon-6 composites. Macromolecules **37**, 7214–7222 (2004)
104. Chen, G., Kim, H., Park, B.H., Yoon, J.: Multi-walled carbon nanotubes reinforced nylon 6 composites. Polymer **47**, 4760–4767 (2006)
105. Bryne, M.T., Gun'ko, Y.K.: Recent advances in research on carbon nanotube-polymer composites. Adv. Mater. **22**, 1672–1688 (2010)
106. Spitalsky, Z., Tasis, D., Papagelis, K., Galiotis, C.: Carbon nanotube–polymer composites: Chemistry, processing, mechanical and electrical properties. Prog. Polym. Sci. **35**, 357–401 (2010)
107. Coleman, J.N., Khan, U., Blau, W.J., Gun'ko, Y.K.: Small but strong: A review of the mechanical properties of carbon nanotube–polymer composites. Carbon **44**, 1624–1652 (2006)
108. Ma, P.C., Siddiqui, N.A., Marom, G., Kim, J.K.: Dispersion and functionalization of carbon nanotubes for polymer-based nanocomposites: A review. Compos. A **41**, 1345–1367 (2010)
109. Li, C., Thostenson, E.T., Chou, T.W.: Dominant role of tunneling resistance in the electrical conductivity of carbon nanotube–based composites. Appl. Phys. Lett. **91**, 223114 (2007). 3pp
110. Munson-McGee, S.H.: Estimation of the critical concentration in an anisotropic percolation network. Phys. Rev. B **43**, 3331–3336 (1991)
111. Celzard, A., McRae, E., Deleuze, C., Dufort, M., Furdin, G., Marêché, J.F.: Critical concentration in percolating systems containing a high-aspect-ratio filler. Phys. Rev. B **53**, 6209–6214 (1996)
112. Bai, J.B., Allaoui, A.: Effect of the length and the aggregate size of MWNTs on the improvement efficiency of the mechanical and electrical properties of nanocomposites-experimental investigation. Compos. A **34**, 689–694 (2003)
113. Martin, C.A., Sandler, J.K.W., Shaffer, M.S.P., Schwarz, M.K., Bauhofer, W., Schulte, K., Windle, A.H.: Formation of percolating networks in multi-wall carbon nanotube/epoxy composites. Compos. Sci. Technol. **64**, 2309–2316 (2004)
114. Krause, B., Villmow, T., Boldt, R., Mende, M., Petzold, G., Pötschke, P.: Influence of dry grinding in a ball mill on the length of multiwalled carbon nanotubes and their dispersion and percolation behaviour in melt mixed polycarbonate composites. Compos. Sci. Technol. **71**, 1145–1153 (2011)
115. Saeed, K., Park, S.Y.: Preparation and properties of multiwalled carbon nanotube/ polycaprolactone composites. J. Appl. Polym. Sci. **104**, 1957–1963 (2007)
116. Mahmoodi, M., Arjmand, M., Sundararaj, U., Park, S.: The electrical conductivity and electromagnetic interference shielding of injection molded multi-walled carbon nanotube/ polystyrene composites. Carbon **50**, 1455–1464 (2012)
117. Bauhofer, W., Kovacs, J.Z.: A review and analysis of electrical percolation in carbon nanotube polymer composites. Compos. Sci. Technol. **69**, 1486–1498 (2009)
118. Krusic, P.J., Wassermann, E., Keizer, P.N., Morton, J.R., Preston, K.F.: Radical reactions of C_{60}. Science **254**, 1183–1185 (1991)

119. Moniruzzaman, M., Winey, K.I.: Polymer nanocomposites containing carbon nanotubes. Macromolecules **39**, 5194–5205 (2006)
120. Ma, P.C., Tang, B.Z., Kim, J.K.: Effect of CNT decoration with silver nanoparticles on electrical conductivity of CNT-polymer composites. Carbon **46**, 1497–1505 (2008)
121. Cai, D., Mo, Song: Latex technology as a simple route to improve the thermal conductivity of a carbon nanotube/polymer composite. Carbon **46**, 2107–2112 (2008)
122. Gojny, F.H., Wichmann, M.H.G., Fiedler, B., Kinloch, I.A., Bauhofer, W., Windle, A.H., Schulte, K.: Evaluation and identification of electrical and thermal conduction mechanisms in carbon nanotube/epoxy composites. Polymer **47**, 2036–2045 (2006)
123. Bryning, M.B., Milkie, D.E., Islam, M.F., Kikkawa, J.M., Yodh, A.G.: Thermal conductivity and interfacial resistance in single-wall carbon nanotube epoxy composites. Appl Phys Lett **87**, 161909 (2005). 1–3
124. Du, D., Guthy, C., Kashiwagi, T., Fischer, J.E., Winey, K.I.: An infiltration method for preparing single wall carbon nanotubes/epoxy composites with improved thermal conductivity. J. Polym. Sci. B **44**, 1513–1519 (2006)
125. Wu, Z., Xue, M., Wang, H., Tian, X., Ding, X., Zheng, K., Cui, P.: Electrical and flame-retardant properties of carbon nanotube/poly(ethylene terephthalate) composites containing bisphenol A bis(diphenyl phosphate). Polymer **54**, 3334–3340 (2013)
126. Kashiwagi, T., Grulke, E., Hilding, J., Harris, R., Awad, W., Douglas, J.: Thermal degradation and flammability properties of poly(propylene)/carbon nanotube composites. Macromol. Rapid Commun. **23**, 761–765 (2002)
127. Wu, Q., Zhu, W., Zhang, C., Liang, Z., Wang, B.: Study of fire retardant behavior of carbon nanotube membranes and carbon nanofiber paper in carbon fiber reinforced epoxy composites. Carbon **48**, 1799–1806 (2010)
128. O'Flaherty, S.M., Hold, S.V., Brennan, M.E., Cadek, M., Drury, A., Coleman, J.N., Blau, W.J.: Nonlinear optical response of multiwalled carbon-nanotube dispersions. J. Opt. Soc. Am. B **20**, 49–58 (2003)
129. Henley, S.J., Hatton, R.A., Chen, G.Y., Gao, C., Zeng, H.L., Kroto, H.W.: Enhancement of polymer luminescence by excitation-energy transfer from multi-walled carbon nanotubes. Small **3**, 1927–1933 (2007)
130. Zhou, Y., Pervin, F., Lewis, L., Jeelani, S.: Fabrication and characterization of carbon/epoxy composites mixed with multi-walled carbon nanotubes. Mat. Sci. Eng. A **475**, 157–165 (2008)
131. Wardle, B.L., Saito, D.S., García, E.J., de Villoria, R.G., Verploegen, E.A.: Fabrication and characterization of ultrahigh-volume- fraction aligned carbon nanotube–polymer composites. Adv. Mater. **20**, 2707–2714 (2008)
132. Wood, C.D., Palmeri, M.J., Putz, K.W., Ho, G., Barto, R., Brinson, L.C.: Nanoscale structure and local mechanical properties of fiber-reinforced composites containing MWCNT-grafted hybrid glass fibers. Compos. Sci. Technol. **72**, 1705–1710 (2012)
133. He, X.D., Zhang, F., Wang, R.G., Liu, W.: Preparation of a carbon nanotube/carbon fiber multi-scale reinforcement by grafting multi-walled carbon nanotubes onto the fibers. Carbon **45**, 2559–2563 (2007)
134. Zhu, J., Imam, A., Crane, R., Lozano, K., Khabashesku, V.N., Barrera, E.V.: Processing a glass fiber reinforced vinyl ester composite with nanotube enhancement of interlaminar shear strength. Compos. Sci. Technol. **67**, 1509–1517 (2007)
135. Bekyarova, E., Thostenson, E.T., Yu, A., Kim, H., Gao, J., Tang, J., Hahn, H.T., Chou, T.W., Itkis, M.E., Haddon, R.C.: Multiscale carbon nanotube-carbon fiber reinforcement for advanced epoxy composites. Langmuir **23**, 3970–3974 (2007)
136. Gao, B., Yue, G.Z., Qiu, Q., Cheng, Y., Shimoda, H., Fleming, L., Zhou, O.: Fabrication and electron field emission properties of carbon nanotube films by electrophoretic deposition. Adv. Mater. **13**, 1770–1773 (2001)
137. Lv, P., Feng, Y.Y., Zhang, P., Chen, H.M., Zhao, N., Feng, W.: Increasing the interfacial strength in carbon fiber/epoxy composites by controlling the orientation and length of carbon nanotubes grown on the fibers. Carbon **49**, 4665–4673 (2011)

138. Zhang, F.H., Wang, R.G., He, X.D., Wang, C., Ren, L.N.: Interfacial shearing strength and reinforcing mechanisms of an epoxy composite reinforced using a carbon nanotube/carbon fiber hybrid. J. Mater. Sci. **44**, 3574–3577 (2009)
139. Gojny, F.H., Wichmann, M.H.G., Fiedler, B., Bauhofer, W., Schulte, K.: Influence of nano-modification on the mechanical and electrical properties of conventional fibre-reinforced composites. Compos. A **36**, 1525–1535 (2005)
140. Loos, M.R., Schulte, K.: Is it worth the effort to reinforce polymers with carbon nanotubes? Macromol. Theory Simul. **20**, 350–362 (2011)
141. Dresselhaus, M.S., Lin, Y.M., Rabin, O., Jorio, A., Souza Filho, A.G., Pimenta, M.A., Saito, R., Samsonidze, G., Dresselhaus, G.: Nanowires and nanotubes. Mat. Sci. Eng. C **23**, 129–140 (2003)
142. Smalley, R.E., Hauge, R.H., Kittrell, W.C., Sivarajan, R., Strano, M.S., Bachilo, S.M., Weisman, R.B.: Method for separating single-walled carbon nanotubes and compositions thereof. United States Patent 7,074,310 (2006)
143. Dresselhaus, M.S., Dresselhaus, G., Saito, R.: Physics of carbon nanotubes. Carbon **33**, 883–891 (1995)
144. Yellampalli S.: (ed.) Carbon Nanotubes—Synthesis, Characterization, Applications. InTech.   http://www.intechopen.com/books/carbon-nanotubes-synthesis-characterization-applications/flame-synthesis-of-carbon-nanotubes (2011)
145. Sahoo, N.G., Rana, S., Cho, J.W., Li, L., Chan, S.W.: Polymer nanocomposites based on functionalized carbon nanotubes. Prog. Polym. Sci. **35**, 837–867 (2010)

Graphite-Based Nanocomposites to Enhance Mechanical Properties

Shanta Desai and James Njuguna

Abstract Carbon based materials such as diamond and graphite are known to mankind for ages. Graphite is highly anisotropic and the properties of single layer of graphite were known for long. In recent years, nanoscale materials using carbon nanotubes have provided opportunities for researchers to engineer new materials with enhanced properties but graphite-based fillers in the polymer nanocomposites has taken forefront in research of many area upon the discovery of graphene, a single layer of graphite by Andre et al. in 2004 due to its extraordinary properties. Due to high surface energy and low density, it is difficult to disperse graphene in polymeric matrix and hence some of the methods identified to homogenously disperse graphite-based fillers are described here such as solution mixing, melt mixing, in situ polymerization and grafting. Nanocomposites prepared using graphite-based reinforcements to enhance mechanical properties in different polymeric matrix is discussed. Finally, applications and challenges of commercialization of these nanocomposites are presented.

1 Background

Nanocomposites attracted researchers since 1987 after a research group in Toyota demonstrated large mechanical property enhancement using reinforcements (montmorillonite) of less than 100 nm in size in a Nylon-6 matrix [52]. Since then, many polymer systems with different nano-reinforcements have been investigated to achieve properties of interest.

S. Desai (✉)
Composites Evolution Ltd, Chesterfield S41 9QG, England
e-mail: sdesai4@gmail.com

J. Njuguna
Institute for Innovation, Design & Sustainability, Robert Gordon University,
Aberdeen AB10 7GJ, UK
e-mail: j.njuguna@rgu.ac.uk

J. Njuguna (ed.), *Structural Nanocomposites*, Engineering Materials,
DOI: 10.1007/978-3-642-40322-4_3, © Springer-Verlag Berlin Heidelberg 2013

In recent years, the development of synthesis for nanoscale materials such as silica, clay, carbon nanotubes has provided opportunities for researchers to engineer new materials with enhanced properties. Amongst these, polymer nanocomposites comprise an emerging class of materials where nanoscale fillers with at least one characteristic length smaller than 100 nm are well-dispersed within a polymer matrix have shown enhancement in properties compared to those with micron-size fillers [6, 18, 24, 33, 38, 46, 55, 62] Exfoliated clay nanocomposites have been studied intensely and some of the materials are now in commercial applications such as automobile exteriors and food packaging [17].

Carbon based materials such as diamond and graphite has been known to mankind for ages. Graphite has high strength and excellent transport (electrical and thermal) properties. Natural graphite flakes based composites with reasonably high thermal conductivity (750 W/m K) have been fabricated by Desai [11]. Graphite is the stiffest material found in nature with Young's modulus 1,060 MPa which is several times that of clay hence, graphite based nanocomposites can offer superior properties compared to that with clay platelets/nanoclay [8]. Also, graphite based nanocomposites have lower cost compared to some other carbon based nanocomposites such as carbon nanotubes (CNT) reinforced nanocomposites.

Since late 1990s research has reported nanocomposites fabricated using intercalated, expanded or exfoliated graphite nanoflakes in polymeric matrix for enhancement of properties and its applications in different areas [5, 7, 12, 33, 40, 49, 61].

However, a new enthusiasm has been developed in the polymer nanocomposites research area upon the discovery of graphene, a single layer of graphite [30]. The properties of single layer of graphite were known for long but It was only recently that Andre K. Geim and Konstantin S. Novoselov showed a method to produce a single layer of graphite/ graphene and were awarded the 2010 Nobel prize in Physics.

Graphene is a monolayer of sp^2-hybridized carbon atoms arranged in a two-dimensional lattice. It has attracted research in many areas due it its extraordinary properties [9, 15, 64]. It has been predicted that a single defect-free graphene platelet could have an intrinsic tensile strength higher than that of any other material [59]. James Hone's group [23] measured mean breaking force of 1,770 nN for free standing monolayer of graphene membrane using nano-indentation in atomic force microscopy. They reported the material to be able to withstand ultra-high strains (~ 25 %) and had Young's modulus of 1.0 TPa.

Graphene sheets that make graphite suffer in composites is due to its strong surface attraction. Unless these sheets are separated and homogenously dispersed within the polymer matrix, the full potential of graphene-based nanocomposites cannot be realised. Hence, pristine graphene materials are unsuitable for intercalation by large species such as polymer chains (graphene has a tendency to agglomerate in a polymeric matrix), the agglomeration can be prevented by different chemical modification methods [16]. First graphite oxide is prepared from natural graphite. After oxidation, a number of methods have been identified to obtain soluble graphene some of them include covalent modification by amidation of the carboxylic groups [31, 54], non-covalent functionalisation of reduced graphene oxide [2, 39, 44], nucleophilic substitution to epoxy groups [4], diazonium salt coupling [27] and reduction of graphite oxide in a stabilisation medium [32].

In this chapter we have attempted to present in brief the preparation routes, studies showing the use of graphite-based reinforcement in nanocomposites to enhance mechanical properties and applications/future of the graphite-based nanocomposites. It is rather impossible to present every area of composites with graphite-based reinforcement as these fillers have been used to enhance different properties based on application due to the multifunctional properties of pristine graphite. Hence, this chapter is delimited to present the use of graphite-based reinforcement in nanocomposites to enhance of mechanical properties.

2 Preparation Routes of Graphite-based Nanocomposites

Since expanded graphite, exfoliated graphite or graphite nanoplatelets found their place in polymer matrix based composites, a number of methods have been identified to prepare graphite reinforced composites. In recent years, polymer based nanocomposites reinforced with expanded graphite (EG) have shown substantial improvements in mechanical, electrical, thermal and barrier properties over the unmodified polymer. As graphite nanosheets could have enormous surface area (up to 2,630 m^2/g) considering both sides of the sheets are accessible, the dispersion of these nanosheets in a matrix plays an important role in the improvement of physical and mechanical properties of resultant composites [51]. This can be achieved by combination of both synthesis and processing techniques that produce exfoliated graphite and good dispersion of the prepared particles in the matrix. The production of exfoliated graphite is well documented in literature and many dispersion routes of EG, graphite nanoplatelets (GNP) and graphite oxide (GO) are reported.

There are different processing techniques that are described here are solution mixing sonication, shear combined sonication and shear mixing, melt mixing, in situ polymerization, grafting (from and to approaches).

2.1 Solution and Shear Mixing

Solution based methods involve mixing of colloidal suspensions of graphite based filler/material with the selected polymer either in solution form or by dissolving the polymer in the suspension by stirring or shear mixing. The composite is then precipitated from the suspension and dried or further processed for testing and application. Composite can also be obtained by casting the suspension into a mold and extracting the solvent. However, in this method there is a danger of aggregation of the filler content which can potentially lower the final properties of the resultant composite [29]. This method is widely reported in the literature and solutions are prepared in water or organic solvents. Using this method a wide range of composites with GNPs/GO are prepared in different polymeric matrix e.g. GO-polystyrene [44], GO-polycarbonate [19] and GO-PMMA [10]. Sonication can be used to mix water based polymers in solution e.g. GO-poly vinyl alcohol (PVA) [56].

2.2 Melt Mixing

In melt mixing method the graphite-based filler usually in a powder form is mixed under high shear with the melted polymer. Although this method is economical, it is difficult to get a homogenous dispersion of the filler using this method [34]. Hence, to improve dispersion filler can be pre-treated with surfactants to improve miscibility with the polymer melt. Some studies have been reported where melt mixing method is used to prepare composites without prior treatment but handling and processing of such melt material was found to be difficult and challenging [22].

2.3 In-Situ Polymerisation

In this method of dispersion, the filler is mixed in a monomer or multiple monomers followed by polymerization in the presence of the dispersed filler. The composite is obtained using solution mixing method that is precipitation, extraction or casting. This method is highly effective in dispersing graphene based fillers and is sometimes referred to as intercalation polymerization. Originally this method was investigated for nanoclay/polymer composites and it is now extended to graphite-based fillers [41, 42]. Recent studies have shown improved dispersion of GO filler in PMMA/GO composites where a microintiator was used to intercalate GO prior to in situ polymerization of methyl methacrylate [21].

2.4 Grafting

Pure carbon materials have relatively lower surface functionality and hence it is difficult to form covalent bonds between the polymer and such surface when used as filler material in composites. However, GO has surface rich in functional groups and these have been exploited to introduce covalent bonding between GO platelets and polymer matrix. There are two ways of forming the bonds (a) grafting-from approach and (b) grafting-to approach.

In grafting-from approach, the alcohol groups present in GO platelet surfaces are covalently bonded to atom transfer radical polymerization (ATRP) initiators via esterification to form chemically modified graphite (CMG). The polymer is grown via ATRP from the surface of GO platelets i.e. CMG platelet is added to a polymer matrix (compatible to ATRP) via solution mixing method and a polymer is grown on GO via ATRP. A composite is then precipitated/extracted or casted. The nanocomposites prepared using this technique is shown to have improved mechanical and thermal properties as compared to neat matrix polymer [14].

A grafting-to approach is a converse of grafting-from approach where functional groups of GO platelet are attached to a polymer via end groups of the polymer e.g. grafting of PVA to GO platelets via carbodiimide-activated esterification [47]. In this technique, it is difficult to precisely determine the reactive site on GO platelet as can be determined in grafting-from approach technique [35].

3 Enhancement of Mechanical Properties Using Graphite-based Reinforcement

The in-plane elastic module of pristine defect free graphene is approximately 1.1 TPa and is the strongest material measured on a micron length scale [23]. However, due to its low wet-ability, it gives rise to poor dispersion in polymer matrix and thereby decreases its mechanical properties of the resulting nanocomposites [52]. The wet-ability can be increased by oxidation and or functionalisation (introducing defects) of graphite/graphene sheets/GNPs that provide improved dispersion in polymer matrix at the risk of lowering conductivity of the resulted nanocomposites.

Miller et al. [28] mitigated the risk of lowering conductivity by using a covalent bonding approach between graphene and epoxy matrix. This approach showed 5-fold increase in electrical conductivity, 30 % improvement in strength and a 50 % improvement in stiffness.

In a study by Ramanathan et al. [37], it was shown that functionalised graphite sheets (FGS) had wrinkled morphology which attributed to superior mechanical and thermal properties in FGS-PMMA nanocomposites. They studied the variation of graphene loading on Young's modulus of the composite and compared it to that of neat PMMA and found that at 0.01 % by weight of graphene increased the modulus of the nanocomposites by over 30 % as compared to neat PMMA. When the graphene loading was increased to 1 % by weight the Young's modulus of the resultant composite increased by 80 % above that of neat PMMA however the ultimate tensile strength of the composite increased only by 20 %.

Liang [25] prepare PVA/GO nanocomposites and these nanocomposites showed enhanced mechanical performance which is attributed to the large aspect ratio of the graphene sheets, the molecular-level dispersion of the graphene sheets in the PVA matrix, and mainly due to strong interfacial adhesion due to hydrogen-bonding between graphene and the PVA matrix. For example, with only 0.7 wt % GO the tensile strength increased by 76 % from 49.9 to 87.6 MPa and the Young's modulus increased by 62 % from 2.13 to 3.45 GPa with respect to the parent polymer.

Zhao et al. [60] used sodium dodecylbenzene sulphonate (SDBS) as a surfactant and prepared nanocomposites with chemically reduced graphite oxide (CRGO) in an aqueous solution containing PVA and the stabilizing SDBS. They studied the effect of filler loading (0–3.0 % volume fraction) on tensile strength and elongation at break (see Fig. 1). 150 % improvement of tensile strength and a nearly 10 times increase of Young's modulus were achieved at a graphene loading of only 1.8 % volume fraction. The value of elongation at break decreased from 220 % for the polymer to 98 % for the composite with the same loading. They found the tensile strength to increase with an increase in volume fraction of the filler however, elongation at break showed an inverse relation with filler volume fraction.

EG/epoxy nanocomposites were prepared using different process methods by Yasmin et al. [58] (see Fig. 2) and mechanical behavior of the prepared nano-composites were compared.

The nanocomposites prepared by different techniques using 1 % by weight EG/Epoxy nanocomposites showed higher elastic modulus (see Fig. 3) but lower tensile strength as compared to neat epoxy when tested under identical conditions. The increase is modulus in generally is found to be due to better dispersion of nanoparticles/fillers and good interfacial adhesion between the particles and the epoxy matrix which restrict the mobility of polymer chains under loading [53]. The stiffening effect can also be attributed to the alignment of the graphite layers and the polymer chains to the loading direction as seen in PP/clay composites where the alignment of clay layers and PP chains with loading direction contributed to the stiffening effect [26].

The modulus of the nanocomposite was found to increase with increase in percentage weight of particle content for a given preparation method. Figure 4 shows comparison of the elastic modulus of EG/epoxy and clay/epoxy nano-composites prepared by shear mixing method with 1 and 2 % by weight particle content.

Rafiee et al. [35, 36] compared mechanical properties of graphene-based epoxy nanocomposites with single-walled nanotubes (SWNT) and multi-walled carbon nanotubes (MWNT) based epoxy nanocomposites with 0.1 ± 0.002 % nanofiller and found that graphene-based epoxy nanocomposites showed superior mechanical properties and fatigue resistance compared to SWNT and MWNT filler based epoxy nanocomposites.

Zhen and Gao [63] prepare nylon-6 – (PA6–) graphene (NG) composites by in situ polymerization of caprolactam in the presence of graphene oxide (GO). Composites of varying weight fractions of graphene were prepared by varying the ratio of caprolactum to GO. Homogenous grafting of up-to 78 % by weight of nylon 6 was seen on graphene sheets. This uniform grafting favoured homogenous dispersion of graphene sheets in PA6 matrix and suppressed the crystallization of PA6 chains. They further prepared NG fibres by melt-spinning process and found that the tensile strength and the Young's modulus increased by 2.1 and 2.4 times respectively with only 0.1 wt % of graphene loading thus showing graphene to be a promising reinforcement in composites to enhance mechanical properties (see Fig. 5).

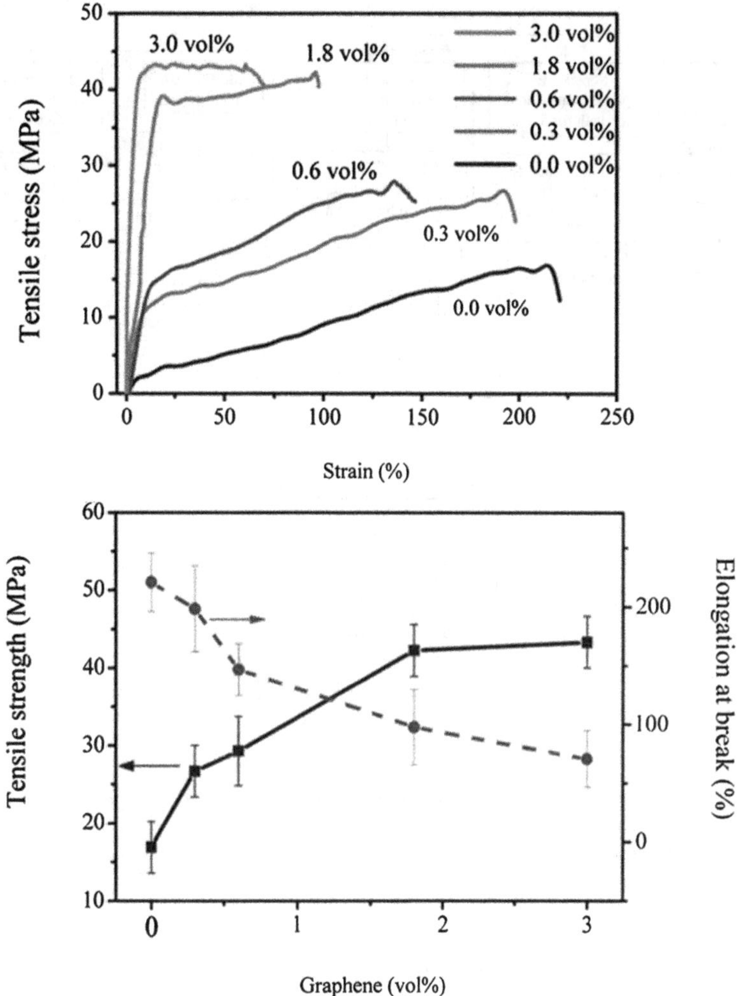

Fig. 1 Tensile strength and elongation at break versus volume percentage of graphene filler (taken from Zhao et al. [60])

Bortz et al. [3] prepared graphene based nanocomposites at concentrations of 0.1, 0.25, 0.5, and 1 wt % with a bisphenol A/F diglycidyl ether blend as the polymer matrix. The resin was added to the graphene oxide/acetone suspension and heated for slow solvent evaporation. It was further heated under vacuum to ensure complete removal of acetone. The dispersion was passed through 3-roll calendar mill with different rotations per minute and the obtained dispersion was diluted with neat resin to obtain the different weight fractions. The test samples were prepared by mixing a harder to the dispersion, degassing it in the mould and casting individual test samples.

                                                                S. Desai and J. Njuguna

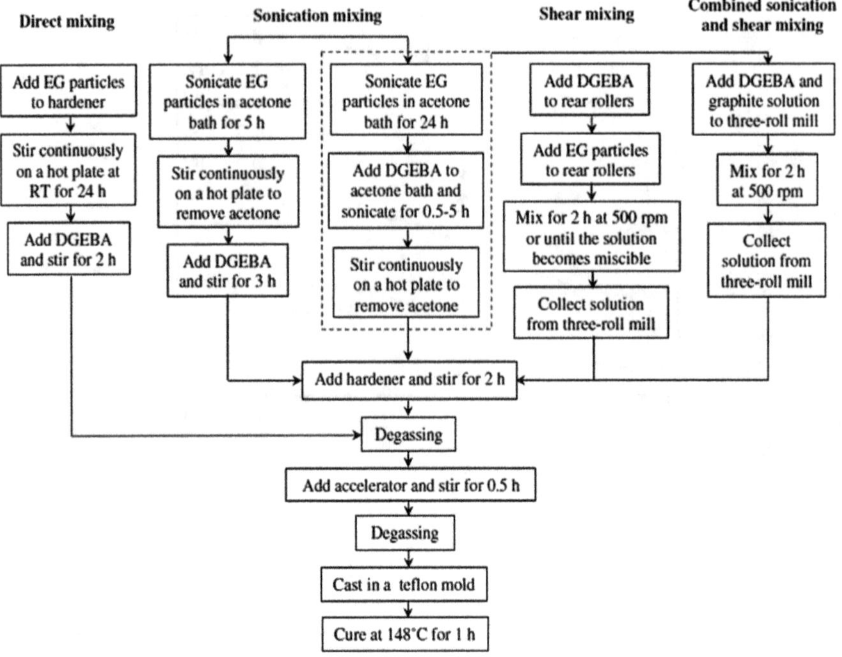

Fig. 2 Processing techniques used to prepare EG-Epoxy nanocomposites (taken from Yasim et al. [58])

Fig. 3 Variation of elastic modulus of 1 % by weight EG/epoxy nanocomposites for different processing techniques (taken from Yasmin et al. [58])

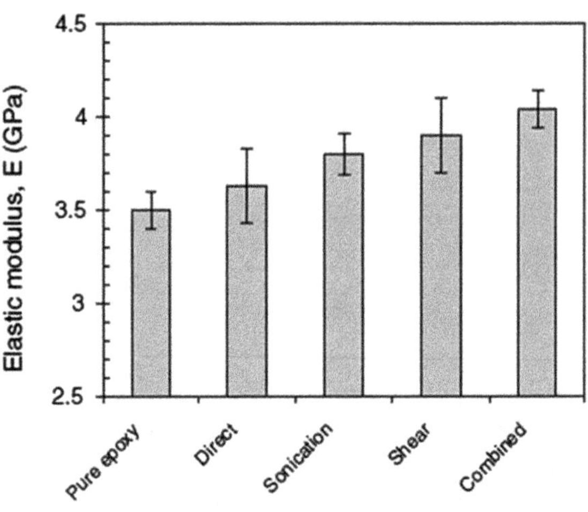

They found that the flexural strength and flexural modulus increased with increase in GO content whereas Tensile modulus was enhanced by 12 % at 0.1 wt % GO but with further increase in GO content a decrease in tensile modulus was observed. At 1 % wt of GO, the tensile modulus was similar to that of

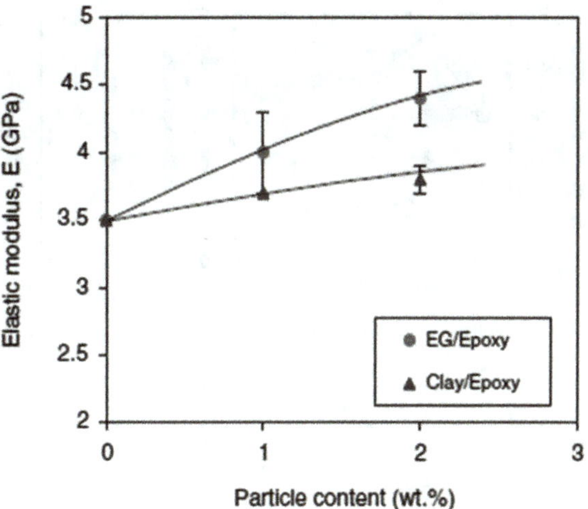

Fig. 4 Variation of elastic modulus as a function of particle content for nanocomposites prepared using shear mixing method (taken from Yasmin et al. [58])

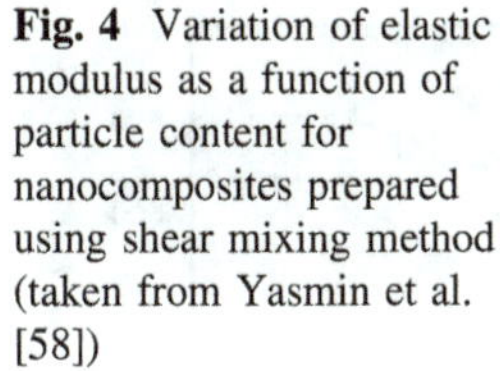

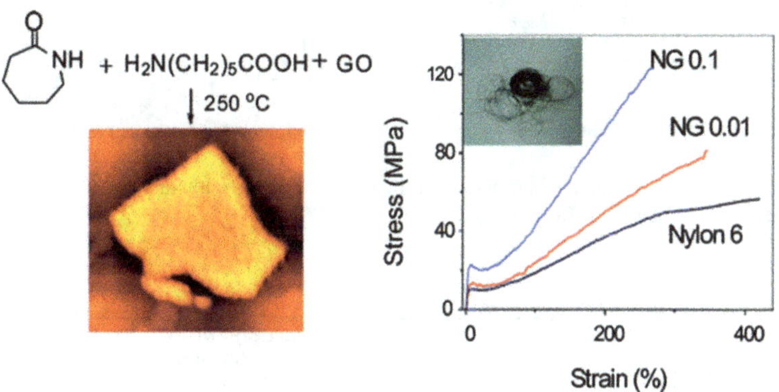

Fig. 5 AFM image showing grafting of polymer on graphene sheet (brush like structure) and graph showing stress versus strain curves for NG fibres and neat Nylon 6 (taken from Zhen and Gao [63])

neat resin. A similar trend was seen in ultimate tensile strength but the maximum enhancement was seen with 0.5 % weight of GO. Enhancement was also seen (see Fig. 6) in critical stress intensity factor (K_{IC}) and critical strain energy release rate (G_{IC}). It can be seen that this enhancement was significant as GO content increased up-to 0.5 % weight however; with further increase in GO content a saturation of toughening effect was observed.

Steurer et al. [45] studied thermally reduced GO (TrGO) or exfoliated GO based polymers in different polymer matrices [poly(styrene-co-acrylonitrile) (SAN) and polycarbonate (PC) shown below Fig. 7]. Since the density of TrGO is very low, solution blending was carried out by pre-mixing TrGO with base polymer in different loadings and then a melt-compound was formed of the

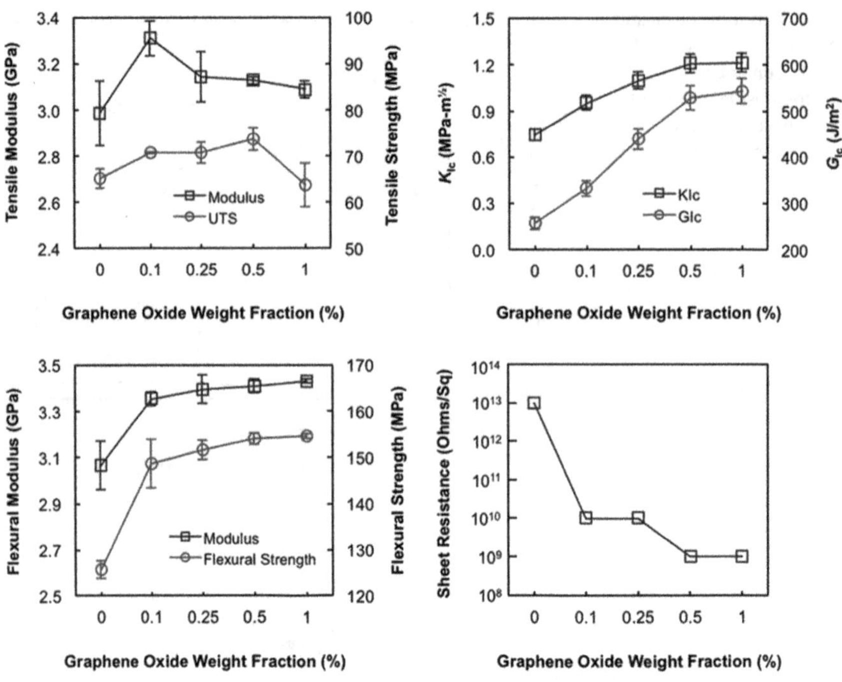

Fig. 6 Behaviour of mechanical properties of the nanocomposite with varying weight fraction of graphene oxide (taken from Bortz et al. [3])

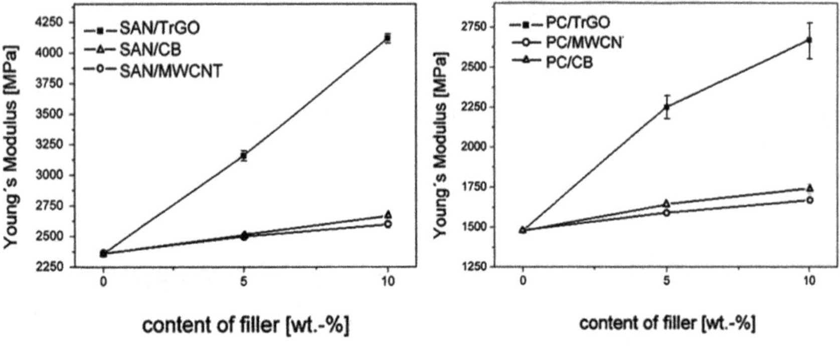

Fig. 7 comparison of Young's modulus of nanocomposites with different reinforcements in SAN and PC matrix (taken from Steurer et al. [45])

additive with the base polymer using a mini-twin screw extruder. The properties of the formed nanocomposites were compared to composites with MWNT and conducting carbon black based polymer nanocomposites.

Jiang and Drzal [22] also prepared nanocomposites with GNP-high density polyethylene (HDPE) using melt-blending technique followed by injection molding and compared with composites filled with carbon fibres (CF), carbon black and glass fibres. They found that GNP-HDPE nanocomposites showed equivalent flexural stiffness and strength compared to HDPE based composites with carbon black and glass fibres but lower than the composites with carbon fibres for a fixed volume fraction of reinforcement.

4 Application and Future

Polymer nanocomposites can have with additives of one-dimensional, such as nanotubes and fibres, two-dimensional, which include layered clay minerals or graphene sheets, or three-dimensional, including spherical particles in a polymer matrix. Graphene has attracted considerable attention because of its unique and outstanding mechanical, electrical and electronic properties, which result in it being one of the most popular candidates for the development of functional and structural graphene-reinforced composites.

The introduction of graphene sheets into polymeric matrices has been proposed as an alternative (Stankovich et al. [44]) or supplementation [57] to more traditional carbon nanotube (CNT) reinforcement. Several studies have shown improved mechanical, electrical and thermal properties for graphene-based polymer composites.

Indeed, graphene-based polymer nanocomposites is a rapidly growing area of nanoengineered materials, providing lighter weight alternatives to CNTs-based nanocomposites with additional functionality at nano-scale [20]. Graphene appears to bond better to the polymers in the epoxy, allowing a more effective coupling of the graphene into the structure of the composite. This property could result in the manufacture of components with high strength to weight ratio for such uses as windmill blades or aircraft components.Conductive graphene/graphite-based nanocomposites have applications in field effect transistors (FETs), Solar cells (and other opto-electronic devices) and memory storage devices. All these applications depend on high conductivity. Photo-voltaics and optoelectronics reply on the fact that monolayers of graphene are about 98 % transparent but still have high electrical conductivity. This makes graphene suitable for photoexcitation and exiciton mobility/diffusion as transparent conducting electrodes [13, 43, 48].

Although we have seen here and in other published research findings that graphene based polymer nanocomposites have tremendous potential for application in automotive, electronics, aerospace and packaging due to its multifunctional properties, however, the development and applicability of graphene-based polymer nanocomposites will be limited by the lack of effective methods for scalable graphene production, difficult manipulation of graphene sheets in processing due to its extremely low bulk density, and the lack of local sites or tensioned bonds on the graphene sheets to anchor functional moieties to make it process friendly and compatible with other materials.

References

1. Ash, B.J., Stone, J., Rogers, D.F., et al.: Investigation into the thermal and mechanical behavior of PMMA/Alumina Nanocomposites. Mater. Res. Soc. Symp. Proc. **661**, 661 (2000)
2. Bai, H., Xu, Y., Zhao, L., Li, C., Shi, G.: Non-covalent functionalization of graphene sheets by sulfonated polyaniline. Chem. Commun. **13**, 1667 (2009)
3. Bortz, D.R., Heras, E.G., Martin-Gullon, I.: Impressive fatigue life and fracture toughness improvements in graphene oxide/epoxy composites. Macromolecules **45**, 238 (2012)
4. Bourlinos A.B., Gournis D., etridis, D., Szabó, T., Szeri, A., Dékány, I.: Graphite oxide: chemical reduction to graphite and surface modification with primary aliphatic amines and amino acids. Langmuir **19**, 6050 (2003)
5. Chen, G.H., Wu, D.J., Weng, W.G., He, B., Yan, W.L.: Preparation of polystyrene/graphite nanosheets composite. Polymer **42**, 4813 (2001)
6. Chen, G., Weng, W., Wu, D., Wu, C.: PMMA/graphite nanosheets composite and its conducting properties. Eur. Polymer J. **39**, 2329 (2003)
7. Chen, X.M., Shen, J.W., Huang, W.Y.: Novel electrically conductive polypropylene/graphite nanocomposites. J. Mater. Sci. Lett. **21**, 213 (2002)
8. Chung D.D.L.: Review: Exfoliation of graphite. J. Mater. Sci. **22**, 4190 (1987)
9. Compton, O.C., Nguyen, S.B.T.: Graphene Oxide, Highly Reduced Graphene Oxide, and Graphene: Versatile Building Blocks for Carbon-Based Materials. Small **6**, 711 (2010)
10. Das, B., Rasad, K.E., Ramamurty, U., Rao, C.N.R.: Nano-indentation studies on polymer matrix composites reinforced by few-layer graphene. Nanotechnology **20**(12), 125705 (2009)
11. Desai, S.: Fabrication and analysis of highly conducting graphite flake composites. PhD Thesis, Institute for Materials Research, University of Leeds, Leeds. United Kingdom (2006)
12. Desai, S., Njuguna, J.: Enhancement of thermal conductivity of materials using different forms of natural graphite Minea A A (ed.) Advances in Industrial Heat Transfer. CRC press 2012, Chapter 6, 201 (2012)
13. Eda, G., Chhowalla, M.: Graphene-based composite thin films for electronics. Nano Lett. **9**(2), 814 (2009)
14. Fang, M., Wang, K.G., Lu, H.B., Yang, Y.L., Nutt, S.: Covalent polymer functionalization of graphene nanosheets and mechanical properties of composites. J. Mater. Chem. **19**(38), 7098 (2009)
15. Geim, A.K., Novoselov, K.S.: The rise of graphene. Nat. Mater. **6**, 183 (2007)
16. Geng, Y., Wang, S.J., Kim, J.-K.: Preparation of graphite nanoplatelets and graphene sheets. J. Colloid Int. Sci. **336**, 592 (2009)
17. Giannelis, E.P.: Polymer-layered silicate nanocomposites: synthesis, roperties and applications. Appl. Organometalic Chem. **12**, 675 (1998)
18. Gonsalves, K.E., Chen, X., Baraton, M.I.: Mechanistic investigation of the preparation of polymer/ceramic nanocomposites. Nanostruct. Mater. **9**, 181 (1997)
19. Higginbotham, A.L., Lomeda, J.R., Morgan, A.B., Tour, J.M.: Graphite oxide flame-retardant polymer nanocomposites. ACS Appl. Mater. Int. **1**(10), 2256 (2009)
20. Salavagione, H.J., Martínez, G., Ellis, G.: Graphene-based polymer nanocomposites, hysics and applications of Graphene—Experiments, Dr. Mikhailov, S. (ed.), (2011). ISBN: 978-953-307-217-3
21. Jang, J.Y., Kim, M.S., Shin, C.M.: Graphite oxide/poly(methyl methacrylate) nanocomposites prepared by a novel method utilizing macroazoinitiator. Compos. Sci. Technol. **69**(2), 186 (2009)
22. Jiang, X., Drzal, L.T.: Multifunctional high density polyethylene nanocomposites produced by incorporation of exfoliated graphite nanoplatelets 1: Morphology and mechanical properties. Polym. Compos. **31**(6), 1091 (2010)
23. Lee, C., Wei, X., Kysar, J.W., Hone, J.: Measurement of the elastic properties and intrinsic strength of monolayer graphene. Science **321**, 385 (2008)

24. LeeBaron, P.C., Wang, Z., Innavaia, T.: Polymer-layered silicate nanocomposites: an overview. App Clay Sci **15**, 11 (1999)
25. Liang, J.: Molecular-level dispersion of graphene into poly(vinyl alcohol) and effective reinforcement of their nanocomposites. Advanced Functional Materials ol. **19**(14), (July 2009) 2297-2302, ISSN: 1616-301X
26. Liu, X., Wu, Q.: PP/clay nanocomposites prepared by grapfting-melt intercalation. Polymer **42**, 10013 (2001)
27. Lomeda, J.R., Doyle, C.D., Kosynkin, D.V., Hwang, W.F., Tour, J.M.: Diazonium functionalization of surfactant-wrapped chemically converted graphene sheets. J. Am. Chem. Soc. **130**, g1620 (2008)
28. Miller, S.G., Bauer, J.I., Maryanski, M.J., Heimann, P.J., Barlow, J.P., Gosau, J.M., Allred, R.E.: Characterization of epoxy functionalized graphite nanoparticles and the physical properties of epoxy matrix nanocomposites. Compos. Sci. Technol. **70**, 1120 (2010)
29. Moniruzzaman, M., Winey, K.I.: Polymer nanocomposites containing carbon nanotubes. Macromolecules **39**(16), 5194 (2006)
30. Mukhopadhyaya, P., Gupta R.K., (2011) Trends and frontiers in Graphene-based polymer nanocomposites. Plast. Eng. 32–42
31. Niyogi, S., Bekyarova, E., Itkis, M.E., McWilliams, J.L., Hamon, M.A., Haddon, R.C.: Solution properties of graphite and graphene. J. Am. Chem. Soc. **128**, 7720 (2006)
32. Park, S., An, J., iner, R.D., Jung, I., Yang D., elamakanni, A., Nguyen S.B.T., Ruoff, R.S.: Aqueous suspension and characterization of chemically modified graphene sheets. Chem. Mater. **20**, 6592 (2008)
33. Pan, Y., Yu, Z., Ou, Y., Hu, G.: A new process of fabricating electrically conducting nylon6/ graphite nanocomposites via intercalation polymerisation. J. Polym. Sci., Part B: Polym. Phys. **38**, 1626 (2000)
34. Paul D.R., Robeson L.M.: Polymer nanotechnology: Nanocomposites. Polymer **49** (15), 3187 (2008)
35. Rafiee, M.A., Rafiee, J., Wang, Z., Song, H., Yu, Z–Z., Koratkar, N.: Enhanced mechanical properties of nanocomposites at low graphene content. ACS Nano **3**(12), 3884 (2009)
36. Rafiee, M.A., Rafiee, J., Srivastava, I., Wang, Z., Song, H., Yu, Z–.Z., Koratkar, N.: Fracture and fatigue in graphene nanocomposites. Small **6**(2), 179 (2010)
37. Ramanthan, T., Stankovich, S., Dikin, D.A., Liu, H., Shen, H., Nguyen, S.T., Brinson, L.C.: Graphitic nanofillers in PMMA nanocomposites—An investigation of particle size and dispersion and their influence on nanocomposite properties. J. Polym. Sci., Part B: Polym. Phys. **45**(15), 2097 (2007)
38. Ramanthan, T., Liu, H., Brinson, L.C.: Functionalized SWNT/polymer nanocomposites for dramatic property improvement. J. Polym. Sci., Part B: Polym. Phys. **43**(17), 2269 (2005)
39. Salavagione, H.J., Gomez, M.A., Martınez, G.: Salavagione, Horacio JPolymeric modification of graphene through esterification of graphite oxide and poly (vinyl alcohol). Macromolecules **42**, g6331 (2009)
40. Shioyama, H.: Polymerization of Isoprene and Styrene in the interlayer spacing of graphite. Carbon **35**, 1664 (1997)
41. Shioyama, H.: The interactions of two chemical species in the interlayer spacing of graphite. Synth. Met. **114**(1), 1 (2000)
42. Sinha, Ray S., Okamoto, M.: Polymer/layered silicate nanocomposites: a review from preparation to processing. Prog. Polym. Sci. **28**(11), 1539 (2003)
43. Spitsina, N.G., Lobach, A.S., Kaplunov, M.G.: Polymer/nanocarbon composite materials for photonics. High Energy Chem. **43**(7), 552 (2009)
44. Stankovich, S., Piner, R.D., Chen, X., Wu, N., Nguyen, S.T., Ruoff, R.S.: Stable aqueous dispersions of graphitic nanoplatelets via the reduction of exfoliated graphite oxide in the presence of poly (sodium 4-styrenesulfonate). J. Mater. Chem. **16** (2), 155 (2006)
45. Steurer, P., Wissert, R., Thomann, R., Mulhaupt, R.: functionalized graphenes and thermoplastic nanocomposites based upon expanded graphite oxide. Macromolecular Rapid Commun **30**, 316 (2009)

46. Sumita, M., Tusukishi, H., Miasaka, K.: Dynamic mechanical properties of polypropylene composites filled with ultrafine particles. J. Appl. Polym. Sci. **29**(5), 1523 (1984)
47. Sun, S.T., Cao, Y.W., Feng, J.C., Wu, P.Y.: Click chemistry as a route for the immobilization of well-defined polystyrene onto graphene sheets. J. Mater. Chem. **20**(27), 5605 (2010)
48. Tien, C.P., Teng, H.: Polymer/graphite oxide composites as high-performance materials for electric double layer capacitors. J. Power Sources **195**(8), 2414 (2010)
49. Uhl, F.M., Wilkie, C.A.: Polystyrene/graphite nanocomposites: effect on thermal stability'. Polym Degrad Stab **76**, 111 (2002)
50. Usuki, A., Kojima, Y., Kawasumi, M., Okada, A., Fukushima, Y., Kurauchi, T., et al.: Synthesis of nylon 6-clay hybrid. J. Mater. Res. **8**(5), 1179–1184 (1993)
51. Viculis, L.M., Mack, J.J., Kaner, R.B.: A chemical route to carbon nanoscrolls. Science **299**(28), 1361 (2003)
52. Wang, S., Zhang, Y., Abidi, N., Cabrales, L.: Wettability and surface free energy of graphene films. Langmuir **25** (18), 11078 (2009)
53. Wei, C.L., Zhang, M.Q., Rang, M.Z., Friedrich, K.: Tensile performance improvement of low nanoparticles filled-polypropylene composites. Compos. Sci. Technol. **62**, 1327 (2002)
54. Worsley, K.A., Ramesh, P., Mandal, S.W., Niyogi, S., Itkis M.E., Haddon, R.C.: Soluble graphene derived from graphite fluoride Chem. Phys. Lett. **445**, 51 (2007)
55. Xu, J., Hu, Y., Song, L., et al.: Increasing the electromagnetic interference shielding effectiveness of carbon fiber polymer-matrix composite by using activated carbon fibers. Carbon **40**(3), 445 (2002)
56. Xu, Y.X., Hong, W.J., Bai, H., Li, C., Shi, G.Q.: Strong and ductile poly(vinyl alcohol)/graphene oxide composite films with a layered structure. Carbon **47**(15), 3538 (2009)
57. Yang, S.Y., Lin, W.N., Huang, Y.L., Tien, H.W., Wang, J.Y., Ma, C.C.M., Li, S.M., Wang, Y.S.: Synergetic effects of graphene platelets and carbon nanotubes on the mechanical and thermal properties of epoxy composites. Carbon **49**(3), 793 (2011)
58. Yasmin, A., Luo, J–J., Danial, I.M.: Processing of expanded graphite reinforced polymer nanocomposites. Compos. Sci. Technol. **66**, 1179 (2006)
59. Zhao, Q.Z., Nardelli, M.B., Bernhole, J.: Ultimate strength of carbon nanotubes: a theoretical study. Physical Review B. **65**(14), 144105 (2002)
60. Zhao, X., Zhang, Q.H., Chen, D.J., Lu, P.: Enhanced mechanical properties of graphene-based poly(vinyl alcohol) composites. Macromolecules **43**, 2357 (2010)
61. Zheng, W., Wong, S.: Electrical conductivity and dielectric properties of PMMA/expanded graphite composites. Compos. Sci. Technol. **63**, 225 (2003)
62. Zheng, W., Lu, X., Wong, S.: Electrical and mechanical properties of expanded graphite-reinforced high-density polyethylene. J. Appl. Polym. Sci. **91**(5), 2781 (2004)
63. Zhen Xu and Chao Gao: *In situ* Polymerization Approach to Graphene-Reinforced Nylon-6 Composites. Macromolecules **43**(16), 6716 (2010)
64. Zhu, Y., Murali, S., Cai, W. et al.: Graphene and graphene oxide: synthesis, roperties, and applications. Adv. Mater. **22** (35), 3906 (2010)

Optimization and Scaling up of the Fabrication Process of Polymer Nanocomposites: Polyamide-6/Montmorillonite Case Study

K. Pielichowski, T. M. Majka, A. Leszczyńska and M. Giacomelli

Abstract Although polyamide (PA) nanocomposites reinforced with montmorillonite (MMT) are processed for more than two decades, the primary technological problem related to optimization of processing conditions to fully exploit properties of these nanomaterials has still to be addressed. The processing of polymer nanocomposites by melt intercalation consists, in principle, of the following stages: preparation and drying of raw materials, preparation of a premix masterbatch, dosing the premix masterbatch into a feeding zone, heating and melting the polyamide-based matrix, an extrusion of a fluid composition followed by a set of auxiliary operations. The process of obtaining polyamide nanocomposites with the desired properties depends on numerous processing parameters that, when varied, affect the quality of manufactured products. Therefore, in this chapter techniques for obtaining polyamide-6/montmorillonite nanocomposites (PA6/MMT NCs), are presented along with discussion of the optimization process for the preparation of PA nanocomposites. Important technological problems arising during the processing are discussed in this chapter, as well as present issues which need to be addressed in scaling up the production from laboratory to industrial scale.

1 Introduction

Nanotechnology has been playing an increasingly important role in science and technology for the last 20 years. Popularity of the nano- concept is reflected in numerous academic and industrial research activities in this area and emergence

K. Pielichowski · T. M. Majka (✉) · A. Leszczyńska
Department of Chemistry and Technology of Polymers, Cracow University of Technology,
ul. Warszawska 24, 31-155 Kraków, Poland
e-mail: tomasz.m.majka@gmail.com

M. Giacomelli
Grado Zero Espace, via Nove, 2/A, Montelupo Fiorentino, Italy

J. Njuguna (ed.), *Structural Nanocomposites*, Engineering Materials,
DOI: 10.1007/978-3-642-40322-4_4, © Springer-Verlag Berlin Heidelberg 2013

need for opening new courses specializing in the field of nanotechnology [1–3]. The new polymer nanomaterials can be produced on the base of both thermoplastic and thermoset polymers [3, 4]. Thermoset nanocomposites find their use mostly in the shipbuilding and coating industry. The ability to easily and repeatedly form thermoplastic nanomaterials is a big advantage, which makes these materials useful in many industries, especially in the automotive or packaging sectors. Thermoplastic polymer nanocomposites can be obtained in several ways described in numerous scientific publications, however, the most common and efficient method is melt intercalation [3–8]. Polymer melt intercalation involves annealing (usually under shear) of a mixture of polymer and layered silicate above the softening point of the polymer [5]. During annealing, polymer chains diffuse from the bulk polymer melt into the galleries between the silicate layers. This method enabled the production of nanocomposites of e.g. polyolefins, polyoxymethylene, polyamide, polycarbonate, poly(ethylene terephthalate), and polystyrene [3–13]. Science is focused both on physical modification of traditional polymers as well as synthesis of new monomers from which new polymers are obtained with interesting new properties that can provide opportunities for novel applications. These new materials may show better properties such as chemical, barrier, physical and mechanical properties [2–5, 11, 13–15]. In melt mixing process crucial role plays the use of different types of fillers which, thanks to their unique characteristics, influence the properties of final products. Intercalation of nanofiller into the polymer matrix, depending on its nature, changes mechanical, optical, biological, surface or thermal properties of the composites. The achievement of significantly better nanocomposites properties primarily depends on nanoadditive surface energy and specific surface area developed in nanocomposite, the degree of organization on the surface, the shape and size of the nanofiller particles and the spatial arrangement of particles in polymer matrix [2–5, 11, 13, 16–20]. 3D structured compounds which are widely used are, e.g. metals and non-metal oxides, carbides, borides, nitrides or carbon black, talc and chalk; compounds having the 2D structure, such as layered silicates or graphene and molybdenium disulfide and compounds of 1D structure as carbon nanotubes and carbon nanofibers [3, 5, 11, 13, 15, 18, 20–23]. Renewable additives are also used such as natural fibers, microfibrillated cellulose [3, 5, 24]. Very often, the polymer matrix is added with a certain percentage of recycled granulate or materials are produced exploiting recycled or biodegradable matrix. Now, these nanobiocomposites are produced also by using advanced methods and technologies [1–6, 19, 20, 25]. Novel nanomaterials are used e.g. in medicine as modern medical supplies, in packaging industry as multi-layer food films (Bayer [26], Clariant [27], Honeywell—Aegies [28], Nanocor Inc. [29]), in automotive industry as interior decorating applications of cars, tanks and car bumpers (GE Plastics [30], UBE—Ecobesta [31], Polykemi [32]), in aeroplane industry [33], in the electrical and electronic industry as insulations for cables and electrical equipment (Hyperion-Catalysis Int. [34], Kabelwerk Eupen [35]), or in coating materials, including marine paints and varnishes (Polymeric Supply Company [36]).

2 Processing of Polymer Nanocomposites

Polymer nanocomposites are two- or multi- phase materials containing continuous and dispersed phase(s), usually obtained by modification of the traditional polymers by introducing and dispersing the nano-sized additives in the polymer matrix [3, 5, 9, 37–41]. The polymer matrix is a thermoplastic or thermoset polymer, and the second component (the filler) may be selected on the basis of its physical structure (including particles size and shape) and chemical nature [3, 5, 9, 41]. Until now, various polymer/ceramic, polymer/metal, and polymer/non-metal composites were obtained and characterized. The well-known methods for preparation of polymer nanocomposites include:

- Intercalation of polymer or prepolymer from solution,
- *In situ* intercalative polymerization,
- Sol–gel technology,
- Melt intercalation.

Intercalation of the polymer or prepolymer from solution is a technique involving the preparation of polymer solution in the solvent, that is modified by addition of cation exchanged layered silicate. The polymer become adsorbed onto the nanoparticles. After removal of the solvent by e.g. evaporation under vacuum nanofiller remains intercalated/exfoliated and the polymer nanocomposite is obtained. The advantage of this method is that intercalated nanocomposites can be synthesized based on polymers with low or even no polarity, but the major technological drawback is the excessive use of the solvent [3, 5, 9, 41].

In the first step of intercalative *in situ* polymerization in the presence of layered silicates, the filler should be swollen by liquid monomer or pre-polymer solution. The monomer migrates into the galleries of layered silicate so that the polymerization reaction can occur between the intercalated sheets. In the second step, the intercalated substrate is subjected to polymerization, which occurs between the layers of the silicate. Depending on the type of monomer or prepolymer the polymerization can take place under the influence of the added initiator such as azobisisobutyronitrile (AIBN) or catalyst such as metallocene ($[Zr(n-C_2H_5)_2\ Me(THF)]^+$) or Ziegler–Natta catalyst, or it may be initiated by temperature or irradiation. In the *in situ* intercalative polymerization, the crucial issue is the good choice of a suitable catalyst, which affects homogeneity of the dispersion [3, 5, 9, 41].

The sol–gel method involves the synthesis of the silicate mineral in the polymer matrix, using aqueous solutions or gels containing the polymer and the silicate building blocks. As precursors for the clay silica sol, magnesium hydroxide sol and lithium fluoride were used. During this process the polymer assists a nucleation and growth of inorganic host crystals and gets trapped in layers when they are increased. However, synthesis of minerals requires high temperatures which degrade the polymer matrix. Another problem is the tendency of silicate layers to aggregate [3, 5, 9, 41, 42].

Quite efficient way to obtain polymer nanocomposites with layered silicates is melt intercalation. It involves dispersing of appropriately modified layered silicate in the polymer melt. Shear forces cause delamination of nanosilicate layers. Moreover, when the silicate is compatible with the polymer matrix, the polymer can enter the space inside the layers and form either an intercalated or an exfoliated nanocomposite. This process requires the use of a single or, better, a twin-screw extruder, having a suitable profile. Intercalation in melt allowed to obtain nanocomposites based on PE, PP, PA, PS, PVA, PVC, PET and the liquid crystalline polymers on lab and industrial scale [1–3, 5, 7–9, 41].

Not all of the presented methods applied for polymer nanocomposites preparation are used on an industrial scale due to technical and economical reasons. The most common method in technological practice is the melt intercalation which requires the use of processing machines that are usually already available in polymer processing companies. Production losses during the use of this technique are small and the technique is quite versatile, i.e. it allows the manufacture of many nanocomposite products using a variety of thermoplastic polymer matrix mixed with various types of additives. Other advantages include high production speed, high quality of products while assuring their reproducibility. On the other hand, there are some technological problems related to the processing of different types of polymer matrices in the presence of a nanofiller, especially degradation of low molecular weight compounds that are used as nanofiller's modifiers (e.g. to enhance organophilization) under processing temperature and shearing forces action [2, 3, 6–9, 12, 13, 20, 25, 39, 41, 43].

3 Fabrication and Characterization of PA6/ Montmorillonite Nanocomposites

Polymer/ceramic nanocomposites are one of the most interesting classes of materials that have been developed in recent years. Under the concept of layered silicates the natural 2:1 clay minerals are enclosed. In this group montmorillonite, which is the main component of bentonite rocks occurring mainly in Europe, North America and Asia, was found very attractive for the preparation of polymeric nanocomposites and, without a doubt, the most commonly used for this type of applications [3, 5, 17, 20, 41].

3.1 Characteristics of Montmorillonite

Montmorillonite (MMT) is a mineral mined from the ground and respectively purified, processed and modified to be compatible with the polymer matrix. Montmorillonite is used as nanoadditive, due to its specific structure. Silicates are based on the structure of SiO_4 whose corners are occupied by oxygen atoms and the silicon atom is in the center of the tetrahedron [3, 5, 17, 20, 41]. Montmorillonite (Fig. 1) is composed of 2:1 packages, containing one layer of octahedral with aluminum—oxygen—hydroxyl sheet, which is closed between two layers of tetrahedral with silicon—oxygen sheets. MMT crystals posses permanent negative layer charge because of the isomorphous substitution in either the octahedral sheet, where species such as Mg^{2+}, Fe^{2+}, or Mn^{2+} replace Al^{3+}, or in the tetrahedral sheet, typically because of substitution of Al^{3+} or occasionally Fe^{3+} cations for Si^{4+}. This charge must be balanced by the alkaline cations such as H^+, Li^+, Na^+, K^+ or Ca^{2+}.

Oxygens anions placed in tetrahedra vertices are directed to the center of structure and with hydroxyl groups, they are surrounded by other cations of aluminum, iron or magnesium, thereby forming together octahedrons [3, 5, 17, 20, 41, 44–54]. The two tetrahedral layers are connected by strong atomic-ion bonds with one octahedral layer, creating a package, which is a characteristic structural unit of montmorillonites, joined in an infinite two-dimensional nets. Between these packages are metal cations, especially sodium or calcium and water molecules, connecting them together [3, 5, 15, 17, 20, 41, 44–54]. Polymer–clay interactions have been actively studied in the sixties and seventies of the last century but since 1990s MMT is consciously applied as a nanofiller for polymer composite materials. It creates so called gallery or interlayer of cations balancing electric charge [55–57]. The thickness of the layer packet is about 1 nm and lateral dimensions range from about 300 Å to several microns, depending on the composition and structure of silicate and method of preparation [55–60]. Due to the nature of the galleries there are different types of montmorillonite, for instance sodium montmorillonite (Na^+—MMT), calcium montmorillonite (Ca^{2+}—MMT) or hydrogen montmorillonite (H^+—MMT).

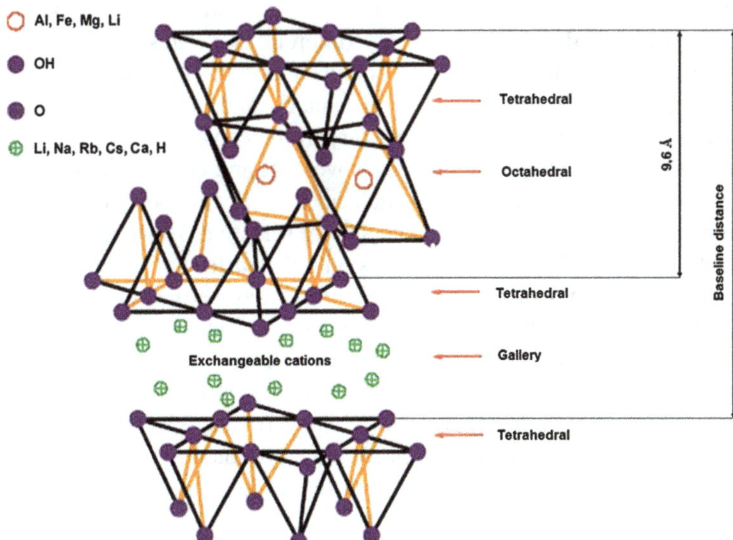

Fig. 1 Basic structure of montmorillonite [5]

3.2 Interactions Between Polyamide-6 and Montmorillonite

Layered silicates are hydrophilic and incompatible with most polymers (due to their hydrophobicity), and in particular with non-polar polymers, such as large-scale produced polyolefins. They are miscible in their original form only with hydrophilic polymers such as poly(ethylene oxide) or poly(vinyl alcohol). Therefore, for the use of layered silicates as nanofillers it is necessary to modify them to achieve better compatibility with polymer matrix, and, in consequence better properties [61–68]. This process involves the exchange of the metal cations, located in the interlayer space of MMT, with an organic cations, particularly a quaternary ammonium cations. After such an exchange, the modified montmorillonite, in which the interlayer distance is increased (hence facilitating the penetration of the monomer or polymer molecules in the interlamellar spaces), has been achieved. Besides, the mineral becomes more hydrophobic, thus increasing its compatibility with most of polymers [69–75]. From the point of view of processing of polymer nanocomposites, it is possible to obtain three different structures of polymer/layered silicate composite:

- The microcomposite structure—where the polymer and organoclay are present as two separated phases;
- The intercalation structure—regular introduction of polymer chains between the layers of clay;
- The exfoliation structure—silicate layers with a thickness of 1 nm are randomly dispersed in polymer matrix.

The components of the nanocomposite have different specific properties, which cause that next to the nanoparticle surface an adsorption layer of polymer is formed. This adsorbed polymer layer has different properties from that of pure polymer in melt. The thickness of the boundary polymeric layer increases with increasing interactions between the polymer matrix and the nanofiller and ranges usually between 2 and 9 nm. Adsorption of macromolecules on the surface of solid nanofiller reduces the possibility of their movement and conformational changes. If adsorption is selective, then macromolecules of lower molecular weight are much more easily adsorbed. The phenomenon of selective adsorption may include an interaction of suitable functional groups with the surface of nanofiller. Content of the nanofiller also affects rate of crystallization, degree of crystallinity and the nature of the crystalline phase [57–59, 72, 73]. Small content causes that the phase separation boundaries in nanocomposite are increased several times, compared to the conventional composite. In nanocomposites there are strong interactions between the polymer matrix and surface of nanoparticles, which, along with high specific surface area of nanoparticles, bring more pronounced changes in polymer properties, in comparison with conventional fillers. To improve the properties of polyamide composites as structural materials two parameters have to be considered: the size of the contact surface of the dispersed phase (filler) and the interaction between the continuous phase and the dispersed one [55–60, 69–75].

If the polymer is not able to intercalate between silicate layers followed by dispersing nanofiller therein, a conventional phase separated microcomposite is obtained. Weak physical attractions between the organic and inorganic components in the nanocomposite systems lead to relatively poor mechanical properties. Furthermore, agglomeration of particles tends to reduce the strength of composite materials. Therefore, the type of surfactant used for surface modification of MMT, the length of hydrocarbon chain, and packing density of surfactant on the clay layer play an important role in structure formation [55–60, 69–76]. The nanocomposites with uniformly distributed silicate plates having irregular spatial arrangement in the polymer matrix, without preserving its original parallel packing, are referred as exfoliated nanocomposites and were reported as having the best physical and mechanical properties [77, 78]. For instance, nanocomposites based on organoclay having two alkyl chains were found to have better properties than those based on organoclays with only one alkyl chain. This was attributed to better affinity of polymer to alkyl chains than silicate surface. Therefore, in this case the increased amount of the alkyl chains should lead to a better dispersion of organoclay. The low packing density of the inner layer of organic modifier causes clasp adsorbed monolayers. If the packing density is increased, the chains adopt a more extended conformation. Exfoliated or delaminated structure increases interaction between polymer matrix and clay, changing the properties of the final nanocomposite radically. In technological practice it is not easy to achieve complete exfoliation of clay—this is due to the fact that the silicate layers are highly anisotropic and most of the polymer chains in systems are physically bound to the surface of the layered silicate [55–60, 69–75].

4 Optimization

Preparation of polyamide nanocomposites reinforced by montmorillonite requires an individual approach. Numerous factors influence the process of preparation of nanocomposites with desired properties. Even a small change in one parameter of processing significantly affects the quality of final materials. In the 'Laboratory of Processing of Polymer Nanocomposites' in the Department of Chemistry and Technology of Polymers at the Cracow University of Technology optimization procedures and guidelines of fabrication conditions of polymer nanocomposites have been intensively developed in recent years. The processing of polymer nanocomposites by melt intercalation consists of the following steps: the selection of the suitable raw materials, preparation of a premix masterbatch, dosing the premix masterbatch into a filling zone, heating and melting the polymer matrix, an extrusion of a fluid composition solidification of the nanocomposite in water bath, and finally pelletizing and drying. The optimization process for the preparation of polyamide-6/montmorillonite nanocomposites in the laboratory scale has been performed based on a mini-technological line shown in Fig. 1.

This mini-technological line includes: a twin screw co-rotating extruder Thermo Scientific Rheomex PTW 16/25 XL, cooling tank and granulator ZAMAK G-16/325 and volumetric feeder produced by Brabender, which is compatible with extruder. Based on the results obtained from studies using this mini-line, some suggestions and recommendations related to optimized fabrication of polyamide-6/ montmorillonite nanocomposites were formulated.

4.1 Selection of Raw Materials

The proper selection of raw materials is a crucial step that governs the quality of the final product. Manufacturers involved in the production of fillers offer a large variety of modified nanoadditives. It is possible in most cases to choose the type of filler dedicated to a given polymer matrix and specific recommendations for processing of such system are given, too. Depending on the type of organic modifier used, fillers are more or less compatible with a specific polymer matrix. However, the thermal stability of modifier limits the temperature of processing of polymer matrix. Usually, the lower the molecular weight of the modifier, the faster the decomposition at high temperatures, deteriorating the compatibility and initiating the decomposition reactions of the polymer matrix. An useful example could be a layered silicate—montmorillonite, which is usually modified with quaternary ammonium salts, mainly dialkyl-dihydrogenated tallow ammonium salt (DADHT) (in our work denoted as 72T) or dialkyl-benzyl-hydrogenated tallow ammonium salt (DABHT) (in our work denotated as 43B). The flashpoint of dimethyl-dihydrogenated tallow ammonium (DMDHT) salt is 25 °C and boiling point is 135 °C while, on the other hand, the flashpoint of dimethyl-benzyl-hydrogenated tallow ammonium salt (DMBHT) is 110 °C. These temperatures are about 100 °C lower than e.g. polyamide-6 processing temperatures. Despite this fact, montmorillonite modified by DMDHT salt and other alkyl derivatives is recommended as nanoadditive for polyamide matrices, but appropriate processing conditions (described in the following sub-chapters) must be selected. The content of a modifier may also influence the thermal stability and other properties of MMT-based systems— investigations on polyamide-6/MMT nanocomposites, in which montmorillonite was modified by different DMDHT salt content, revealed that the addition of MMT with low content of DMDHT increased thermal stability of the nanocomposite, and the addition of MMT with high content of DMDHT increased strength properties of the nanomaterials [79–81] From technological point of view, montmorillonites as minerals differ in elemental composition which is associated with mining place. Differences are also due to performance of the crushing, grinding and cleaning operations. As a consequence, the montmorillonites can differ in the degree of purity and the number of exchangeable cations in their galleries. Generally, similar types of the quaternary ammonium salts are used to modify MMT, but they usually differ in the percentage share of long hydrocarbon chains originating from natural allows. It can be understood that the modifier is in fact a mixture of quaternary ammonium salts with a tallow chain composed of e.g. 20 % of C16, 50 % of C18 and 30 % of C20.

Selection of raw materials includes proper choice of polymer matrix. Each of the polymer types has characteristic technological features and specifications that must be considered during processing. The molecular weight of polymer determines its viscosity in the melted phase, and, in consequence the processing conditions, e.g. rotational speed of screws. For instance, the use of polyamide-6 with low molecular weight (in the order of 40,000) allows its processing at temperatures around 220–235 °C, whereby the higher molecular weight polymer has to be processed at temperatures higher by about 10–15 °C, which in turn significantly promotes degradation of low molecular weight organic modifiers. Therefore, when using a polyamide matrix with high molecular weight it would be logical to use the filler with lower content of modifier.

4.2 Drying of Raw Materials

A very important requirement related to the processing of some polymeric materials, especially polyamides, is to provide a dry masterbatch into the feeder or directly into filling zone. Polyamide must be stored in a dry place at temperature close to this in the workplace. If the material is stored at a temperature lower than the ambient temperature, before the opening of the container or big bag, the polyamide must be brought to a room temperature of the processing machinery to avoid condensation of water from the air on cold granules. Polyamide, delivered by the manufacturer in the form of granules, should be dried under appropriate conditions, even despite the assurances that it is ready for immediate use. The moisture content in the polyamide is a particularly important factor that has a direct influence on the melt viscosity, stability, mechanical properties, and, finally, the appearance of the product. At processing conditions water or its vapour take part in chemical reaction with the polyamide (hydrolysis) causing a reduction of polymer molecular weight. It may also interact with volatile low molecular weight degradation products of quaternary ammonium salts leading to different secondary products. Therefore, manufacturers of polyamide suggest that the moisture content in the material shall not exceed 0.2 wt%, and in some cases even 0.1 wt%. On the other hand, drying is a highly energy-consuming process, and, therefore, a compromise between the speed of drying, the drying temperature and the moisture content in the polymer matrix should be reached. Such measurements have been conducted in our laboratory and the results indicate that the polyamide-6 should be vacuum-dried in the temperature range 75–80 °C for 3–24 h, with the recommendation for 75 °C/6 h [82]. Apart from polymer drying, the nanoadditive—montmorillonite—should be dried as well, since it contains small amount of water adsorbed on the crystal structures. Usually montmorillonite is dried using vacuum ovens at a temperature of 80–100 for 3 h.

4.3 Preparation of a Masterbatch

Mass content of a nanofiller in the polymer matrix plays a fundamental role as it directly influences the structure and morphology of resultant polymer/MMT nanocomposite. The high degree of intercalation of montmorillonite between polymer chains significantly improves the mechanical and barrier properties of the nanocomposite, but may result in a reduction of thermal stability of nanomaterials [83, 84]. Larger enhancement effects were reported when greater surface area of filler contact with the matrix, and there are strong interactions between the two phases. Therefore, nanocomposites should display a high surface of contact area between clay layers and polymer, which considerably improves the properties of the products even when the silicate content in the matrix is at low level [2–5, 11, 13]. For instance, concentrations of filler in the range of 2.5–5 wt% caused an increase in the strength modulus. Moreover, complete exfoliation had a positive influence on the stiffness of the nanocomposites [3–7]. This effect was explained by the fact that the number of planes in a single MMT stack is equal to the number of planes in exfoliated single plane system. In polymer nanocomposite when the number of planes per one stack increases, the number of MMT planes being in contact with polymer decreased, then the strengthening efficiency is reduced. Nanocomposites containing more than 5 wt% of layered silicates are characterized by higher brittleness, due to collapse of the bonds between the silicate and the polymer matrix. It leads to spatial defects of nanocomposites, often in the form of visible cracks on the surface [3–7, 85, 86]. When the planar bonds between matrix and clay stack breaks, the microgaps are formed, usually around the large silicate heterogeneities [6, 7, 85–88]. The morphology plays an important role in the development of strengthening mechanism because the distribution of silicate intercalated regions is equally important to improve the stiffness of the nanocomposites [5–7, 85, 86, 89]. It has been found that the presence of larger content of layer silicate than 5 wt% reduces also the thermal stability of polymer nanocomposites. The nanofiller is able to accumulate heat that, when emitted later on, may accelerate the polymer decomposition [90]. Moreover, organically modified clays (e.g. by alkylammonium cations) can catalyze degradation of the polymer matrix [3–7, 85–89]. The appropriate amount of silicate allows to obtain a suitable distribution of clay platelets in the polymer matrix which improves the barrier properties by creating a labyrinth path for the evaporating gases [3–7, 85–89].

Carefully selected raw materials should be mixed in a specially designed tumble mixer, and the mixing time is 5–15 min. This period of time is sufficient to allow the nanofiller to equally cover the polymer granules. The mixing time for a given polymer/MMT system is dependent on the amount of raw material and the mixer type. The raw materials should be mixed at not too high rotational speed to avoid dust formation that could be dangerous to the workers exposed to its action [91].

4.4 Dosing of a Masterbatch

Results of tests performed using the processing line presented in Fig. 2 show that the dosing rate of the masterbatch into the feed zone has a significant influence on the whole fabrication procedure. In the laboratory conditions, feeders are often used to dose at an appropriate speed raw materials into the feed zone of the extruder. The control is done using an integrated software applied to control the extruder that is able to detect the feeder. The feeder (Fig. 3) is provided with two spring screws for dispensing the masterbatch granules.

The rotational speed of the screws is adjusted to the corresponding rotational speed of extruder and is given as a percentage of extruder screws speed. Before performing any operation, the feeder should be calibrated. Such a calibration has been successfully performed using the 0.2, 0.3, 0.5, and 1.0 % of extruder's screws speed which was set on 50, 100, 150, 200, 240 rpm. Information on the stable work of the feeder at a speed of 0.3 % in relation to the rotational speed of the extruder screws was used to optimize the extrusion process, and further tests were performed in order to check the properties of the polyamide nanocomposite with montmorillonite when using various screws speeds. Nanocomposites of PA6 containing 3 wt% of organically modified MMT were prepared at four different rotational speeds 100, 150, 200 and 240 rpm and then tested in terms of mechanical and thermal properties. The values of modulus of elasticity measured for neat PA6 and PA6/MMT nanocomposites during static bending tests showed clear dependence on the rotational speed of screw applied in the production step and the behaviour of unmodified PA6 was different from that observed for nanocomposites. The modulus of elasticity (E) of PA6 decreased with increasing screw speed showing a minimum at speed of 200 rpm while the E values measured for nanocomposites displayed a maximum for the same rotational speed (Fig. 4).

It is worth to notice that in industrial practice increasing the capacity of production line by increasing the rotational speed of extruder could be done at the expense of mechanical performance of pristine polymer. This unfavourable effect

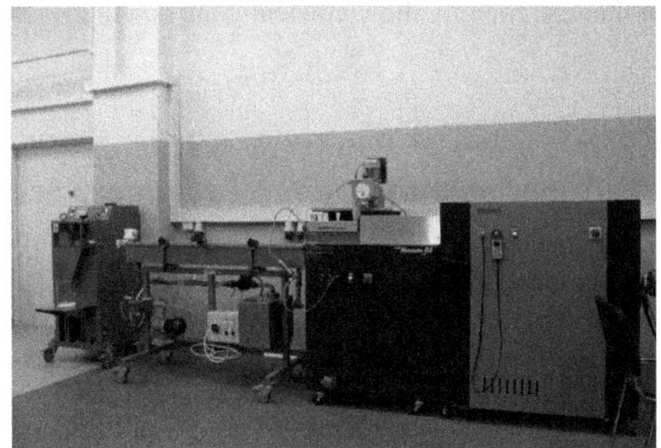

Fig. 2 Mini-technological line located in the 'Laboratory of Processing of Polymer Nanocomposites' at the Cracow University of Technology

Fig. 3 Feeder for dosing a masterbatch

is not expected to occur for PA6/MMT nanocomposites since those materials exhibited maximum enhancement when produced at high rotational speed.

An important issue is the choice of the heating zone, to which the masterbatch is dispensed. For instance, fibrous fillers are usually dosed into the successive zones whereby the polymer is added to the first feed zone—this is due to the fact that the fibers may be disintegrated by screws if they were supposed to follow the whole extrusion route. Powder fillers are rather dosed together with the granulate in the first zone to allow efficient compounding.

4.5 Extrusion of Polyamide-6/Montmorillonite Nanocomposites in the Laboratory Scale

Optimization of the extrusion process is the most important part of the optimization of whole technology for obtaining the polyamide-6/montmorillonite nanocomposites, which includes selection of the type of extruder, configuration of the extruder screws, selection of heating zones temperature, selection of the length of screws, selection of the temperature zones of the die, selection of screws speed

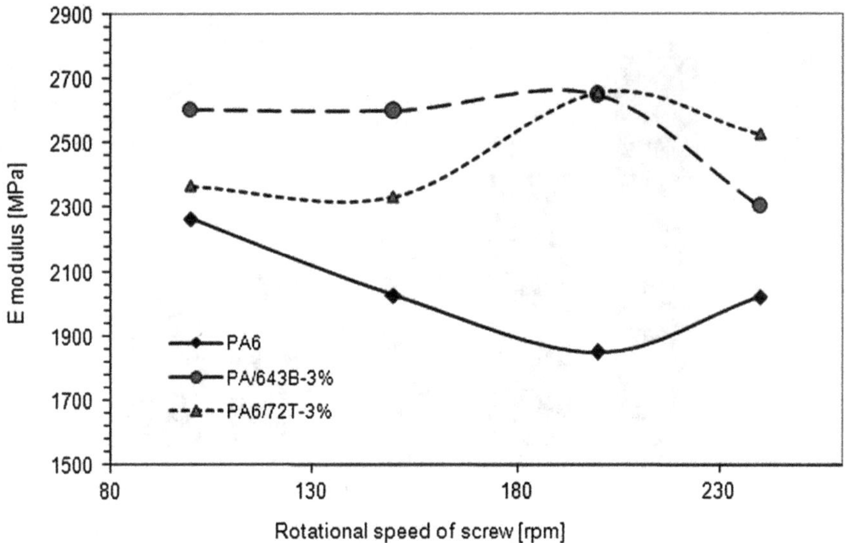

Fig. 4 Modulus of elasticity for PA6 and its nanocomposites versus extruder screw speed

and residence time of the material in cylinder. Depending on the type of extruder different melt homogenization can be achieved. Co-rotating twin screw extruders are ensuring a high homogenization of the composition and these extruders are usually used for the preparation of polymer nanocomposites.

In order to achieve the appropriate homogenization optimization of the extrusion process is necessary. Homogenization process can be divided into thermal and mechanical one. Thermal homogenization occurs due to temperature fluctuations along the extruder barrels, as well as in the cross-section. The so called 'active' part of the screw is responsible for the transfer of a particle (in relation to other particles) along the axis of the screw, whereby the passive part is responsible for the transport of melt from the feed zone to the die. In modern processing equipment, different segments of the screws are responsible for different processes. Some of them are responsible for melting of the pellets, some other for appropriate dispersion of the filler in the matrix, and there are also parts designed for venting. The way the individual segments are placed along the screw affects the mechanical and thermal properties of the obtained products. Sometimes, wrong configuration of screws does not allow to achieve the selected rotational speed, and therefore the residence time of the material in the extruder gets changed. The residence time of polyamide-6/montmorillonite nanocomposites in the extruder with different screws speeds is shown in Fig. 5.

Using the rotational speed of 50 rpm, 1 g of sample remained inside the extruder for 93.3 s; with 100 rpm—63.0 s; with 150 rpm—48.3 s; with 200 rpm—43.0 s; and when using 240 rpm it remained for only 37.3 s. Certainly, an increase of the screws speed reduces the residence time of the composition inside the cylinder, but, on the other hand, higher speed provides high shear forces

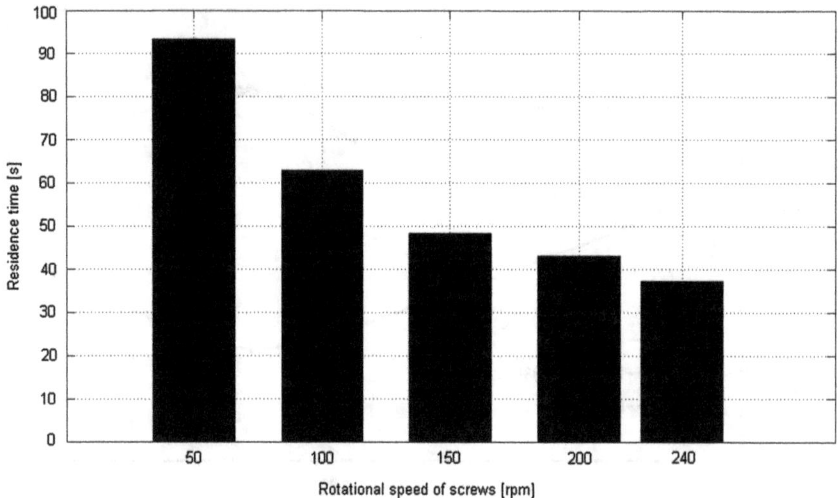

Fig. 5 The residence time of polyamide-6/montmorillonite nanocomposites in extruder with different screws speeds

that facilitate the exfoliation of clay layers, though, excessive shearing may cause decomposition of the material. Therefore, a compromise should be sought between the (short) residence time of the material inside the extruder and the screws rotation speed to achieve proper homogenization. Concerning the screws dimensions, too long screws not only lengthen the residence time inside the machine but also increase the consumption of raw materials. In the laboratory scale, an efficient extruder {with a length of screws equal to 25 L/D [L—length, D—diameter of the screw (16 mm)]; 6 heating zones} is able to produce (at a speed of 240 rpm and the dosing speed of 0.3 %) 300 g of the composition during 30 min of processing, provided an appropriate configuration of the screws is secured (Fig. 6).

Abrasive wear of screws occurs primarily at the edges of the thread and on the surface contact of the cylinder when clay minerals are dosed with a polymer as a premix masterbatch. Over the time, diameter of screw core gets slightly reduced due to wear in the transition and feeding zone. The viscosity of the molten material has a significant influence on the pressure inside the cylinder. If the polymer contains an (nano)additive, its viscosity increases and liquidity decreases. This means that the torque on screws is increased and in some cases the increase of zones temperature was forced. The viscosity of the liquid material is primarily a function of the molecular weight—the higher the molecular weight the highest the viscosity and thus higher temperatures are required to reduce the processing torque since too high torque can damage the laboratory equipment. On the other hand, high temperatures have negative impact on polymer matrix and organic modifiers of the filler. Therefore, it is necessary to identify the proper temperatures of the heating zones and the die, as it was done for polyamide-6/montmorillonite system (Table 1).

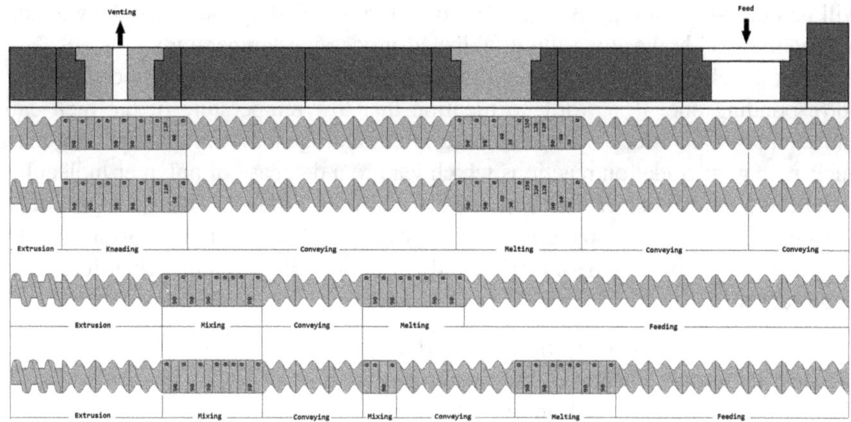

Fig. 6 Examples of different configurations of screws for polyamide-6 composites

Table 1 The selected processing conditions for polyamide-6/montmorillonite nanocomposites using mini-technological line

Twin co-rotating screw extruder

Flow rate (%)	Rotational speed (1/min)	Heating zones						
0.3	240	1	2	3	4	5	6	Die
Temperature (°C)		245	245	245	250	255	250	260
Atmospheric venting		–	–	–	–	Yes	–	–
Length of the zones (mm)		80	60	60	64	60	76	23
L/D		5.00	3.75	3.75	4.00	3.75	4.75	–
Cooling tank								
Length of cooling surface (mm)		1,500						
Tank volume (dm3)		27						
Height of bath (mm)		1,081						
Water temperature (°C)		18						
Granulators								
Size of pellets (mm)		1						
Rotational speed (1/s)		12						

The melt viscosity decreases with an increase in shear rate; for this reason, by increasing the speed of extrusion, one can also reduce the viscosity of the melt. It is generally agreed that energy delivered by the screw should constitute about 80 % of the energy required to raise the temperature evenly. It follows that it is necessary to take into account the construction and operating characteristics of the screw. Attention should also be paid to the location of the heating elements in small laboratory equipment. If the extruder heaters are located at the bottom of the cylinder, substantial temperature gradient may occur even though the sensors will show that the desired temperature has been reached. The upper part of the cylinder will be under-heated for long time. For this purpose, first cylinder may be warmed up about 20 °C higher than the established processing temperature.

After reaching a given value, the temperature should be reduced to the corresponding one. It is worth to mention that low temperature of the melt can cause undesirable effects such as formation of heterogeneities in the melt that may nucleate e.g. degradation reactions which lead to a decrease of polymer molecular weight and worsening of the properties of the final product. There are generally two degradation routes operating at processing conditions: (1) hydrolysis due to presence of moisture in the material, and (2) thermal degradation. Therefore, it is important to:

- Assure a moisture content in the material as low as possible;
- Select the proper content of raw materials;
- Set the appropriate dose rate of materials;
- Select proper length of screws, as well as their configuration and the rotational speed;
- Set the proper temperature profile of particular heating zones and the die.

4.6 Cooling of Polyamide-6/Montmorillonite Nanocomposites

After extrusion process the nanocomposite strip should be cooled down. In polymer compounding lines this process is done using a cooling bath. Polyamides are *semi*-crystalline polymers and depending on cooling mode different values of crystallinity degree can be achieved. The degree of crystallization has a significant influence on the product properties, including mechanical properties, moisture absorption and chemical resistance. The neat polyamide-6 displays a polymorphic behavior showing the γ- and α-crystalline phase. Under the processing conditions where PA6 was subjected to high shearing and cooling rates as well as stretching of the extruded polymer string while solidifying it showed mainly the γ crystalline phase that reflected a strong diffraction peak with a maximum at 21.5° covering the small double-peak of the α-modification on WAXD diffractograms. Such crystal morphology could origin from the processing condition which favored the formation of γ crystals. The main diffraction peak observed on WAXD diffractogram of PA6/MMT nanocomposite occurred at the same value of $2\theta°$ as that observed for neat PA6. Therefore, no changes in the crystallographic structure of PA6 were induced in this material by the presence of MMT layers and processing conditions were principal factors controlling the crystalline morphology of PA6 in as obtained nanocomposites.

It has been observed that the water temperature in cooling bath increased by 2–4 °C due to collecting the heat if bath was not cooled. The increase was moderate due to small mass of polymeric material extruded on laboratory scale, but it is recommended to control the water temperature in cooling bath at industrial scale and provide constant parameters of PA6 crystallization.

4.7 Pelletising of Polyamide-6/Montmorillonite Nanocomposites

From the cooling bath, the nanocomposite strip is directly brought to the granulator, where the successive step of cutting it into the granules takes place. It is important that the pellets present dimensional consistency. In the processing line settings it is possible to fix the drawing speed and rotation speed of a cutting edges to produce pellets with a given size. In practice, to avoid decontaminations it is important to clean regularly the pellets' container in the granulator and the cutting knives.

4.8 Drying of Polyamide-6/Montmorillonite Nanocomposite Pellets

Drying of pellets of polyamide-6/montmorillonite nanocomposite is necessary because the material during cooling is in direct contact with water. The content of moisture in polyamide nanocomposites has to be carefully controlled as it strongly influences the material's behaviour in next processing steps. The moisture content changes rheological behavior of molten mass, and it can lead to formation of inhomogeneities, such as interior bubbles, during injection molding. Besides, it could disturb maintaining of the dimensional stability of the final product within an assumed tolerance. Therefore, nanocomposite pellets should be dried following the protocol established for pristine polyamide-6, i.e. for 6 h at 75 °C using a vacuum dryer.

4.9 Injection Molding of Polyamide-6/Montmorillonite Nanocomposite Pellets

Optimization of injection molding process of PA6/MMT nanocomposites was done using a laboratory scale injection molding machine presented in Fig. 7. This machine is equipped with a cylinder and one-slot injection molds which are heated electrically. Instead of screw plasticizing the polymer matrix (as usual in industrial-size machines) here is a plunger, which has several important functions in the process.

During operation, the cylinder, the form and the plunger are heated to an appropriate processing temperature. The plunger is connected with a pneumatic system of injection molding machine driven by compressed air coming from the compressor. Nanocomposite pellets are fed into the cylinder and the tip of plunger is placed in the cylinder. After plastification, the material is injected. When the cylinder comes into contact with the mold, the plunger is pushed into the cylinder transferring thus the melt to the mold. After the composition is injected, the

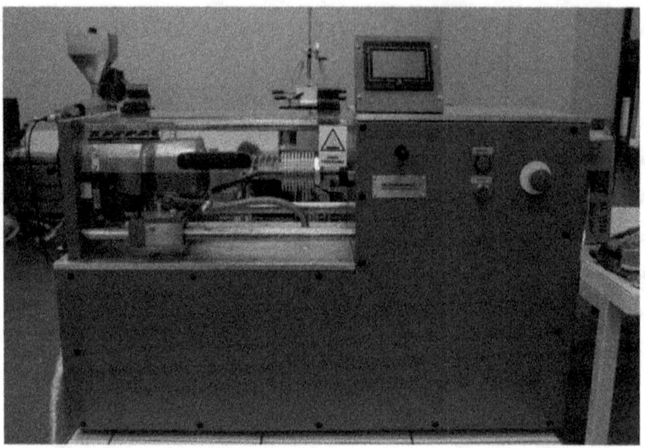

Fig. 7 Laboratory-scale injection molding machine

pneumatic system moves the plunger and cylinder to its original position. During this time, the injected bar is cooled in the mold. Then the sample is removed from the mold, already ready for testing/further use.

Before proceeding to injection molding on laboratory scale, preliminary tests should be performed to find out:

- The number of profiles that can be obtained with a single filling of cylinder;
- The optimum temperature of the cylinder;
- Residence time;
- The optimum temperature of the mold;
- The total duration of the process.

The temperature in the cylinder is the temperature at which the polymer (in this case polyamide-6) undergoes melting. The melting point of the polymer matrix can be determined using e.g. differential scanning calorimetry (DSC). Initially, in our experiments the cylinder temperature was set at 232 °C. Next, it was investigated how much time it takes to melt the polyamide at the given cylinder temperature—if this time is longer than 3.5 min, it would require to increase the cylinder's temperature. A longer residence time of the heated material within the cylinder may lead to oxidative degradation of the polymer. Since PA6/MMT nanocomposites displayed rather narrow processing window the optimal cylinder temperature can be defined as this one that allows the polyamide melting within 2–3.5 min without decomposition. The residence time of nanocomposite materials in the cylinder shall be kept within this time limits to avoid decomposition of organic modifier, too. The next parameter to be optimized is the temperature of the mold. The mold temperature influences the mechanisms of crystallization of polyamide-6/montmorillonite nanocomposites. In the process of crystallization two basic processes occur: the formation of the nucleating sites and growth of the crystals. It is possible to distinguish between homogeneous and heterogeneous nucleation depending on the processing conditions. In the first case, in the molten polymer material certain groups of atoms present close spatial arrangement; when these clusters achieve the necessary critical size, they can act as (homogeneous) nucleating agents; this requires the use of large undercooling, which can be achieved by lowering the temperature of the mold. In the second case, the formation of nucleating agents occurs on the surface of the solid phase in contact with the molten polymer. The nucleation occurs on the fine particles of montmorillonite dispersed in the polymer matrix. Under such conditions, the crystallization takes place at a much lower supercooling than the pure polyamide-6 [92–94]. According to this theory, during the injection molding of profiles it can also be expected that the crystalline phase will quickly appear next to the walls of the mold. In this zone, due to the large drop in temperature, and thus a considerable supercooling, a number of nucleating agents may be formed. In the process of injection molding of the tested materials the temperature of the mold was 80 °C. On the other hand, residence time of the material in the mold (Fig. 8) is in the range of 10 s when using a single mould form of laboratory injection molding machine.

The sample after removal from the mold is subjected to further, slow cooling under air. Different temperature of the mold causes different absorption of the heat, lengthening or reducing the time of cooling to room temperature, and, hence, influencing the mechanical properties of the obtained products. Both for pure polyamide matrix and polyamide-based nanocomposites the larger undercooling favors the crystallization, increasing thus the impact strength of materials.

Venting takes place on the mold surface and in this micro-scale there is no need or possibility to construct the mold with heated ingots or with venting parts. It is important to keep the injection pressure at constant level so the plunger is inserted to the cylinder always with the same force. In our experiments the injection pressure was 10 Bar; higher pressure may cause distortions, spoiling of the plunger and cylinder, as well as may be dangerous for the operator. Operator plays an important role in this technological process, performing the processing and controlling the final product's quality. The operator can visually detect defects such as scorch marks due to local overheating, pollutions, voids, etc., and try to avoid them by changing raw materials. Processing parameters or replacing the mold.

Fig. 8 A single mould form

5 Technological Problems Associated with the Fabrication Process of Polymer Nanocomposites

Each technological process, including fabrication of polymer nanocomposites by processing methods, reveals some problems which must be eliminated during the development stages of this technology. It is not an easy task and usually requires previous experience with the materials involved, as well as skills to predict (as much as possible) the behavior of the composition during numerous operations throughout the process. Sometimes, small oversights can cause large changes at the technology output, leading to major worsening of the final product properties. Some technological problems have already been presented in the previous sub-chapters, but the full description should include also other issues. One of them is proper storage of raw materials. For polyamides, even if the polymer granulate is packed in laminated bags to protect it against absorption of moisture, after a long storage time a certain amount of water is absorbed. The bags should be kept closed at the ambient temperature with no sunlight access. The hopper or feeder should be loaded with quantity of material which is sufficient for at least 1 h of work, and closed by a tight cover. Unused pellets should not come into direct contact with atmospheric air and sunlight, and later on brought back to the bag, which should be then sealed. At the stage of preparation and dosing of the masterbatch for the feed zone, there are sometimes problems with deposition of nanofiller powder on the walls of the container and feeder. If containers, feeders or drum mixers have a rough surface, the nanofiller may enter into micro-spatial surface defects. Therefore, it is recommended to use devices with a smooth surface. Another technological problem—adhesion of the nanofiller to the granules—may be solved by wetting the pellets with an easily volatile solvent and then mixing them with an appropriate amount of additive. The solvent improves the adhesion of the nanofiller to the surface of the granules and, after evaporation, it leaves nanofiller well-dispersed on polymeric granules. However, it seems that losses during stirring and dosing of the masterbatch to the extruder on laboratory scale are hard to avoid—in our studies the total amount of losses (from weighing of the raw materials to producing pellets) was ca. 19 % for composition mass of 350 g with 3 wt% of MMT (Table 2).

Those losses are mainly caused by the adhering of nanofiller and pellets on the walls of feeder, and in the areas which are not or hardly available for masterbatch dosing screws. This is the case when the dosing screws are located in such a way

Table 2 Total amount of materials losses for PA6/MMT nanocomposites fabrication on lab scale

Polymer weight (g)	MMT weight (g)	wt% of filler (%)	Masterbatch weight (g)	The losses during extrusion (g)	Total amount of losses (g)	Total amount of losses (%)
339,519	10,503	3.09	**350,022**	36,929	66,976	**19.13**

that the filler is not properly dosed since pellets settle on the bottom. It should be noted that the remaining unused materials can be returned to the feeding zone, but then the properties of products may change.

The losses during injection molding of nanocomposites pellets, using laboratory injection molding machine, are estimated at 2–5 %. Breaks occurred during extrusion are due to sudden machinery stops, when the torque increases rapidly. Each modern extruder has safety lock that enables the work in some restricted processing conditions when e.g. too high torque may damage the screws. Therefore, for example, a quite widely used safety lock automatically stops the machinery whenever the torque reaches a certain value (e.g. 100 nm). To reduce the torque higher processing temperatures may be applied, but this may cause decomposition of polymer matrix and/or organic modifier. For instance, during extrusion of polyamide-6/montmorillonite nanocomposites under the conditions presented in Table 3 a torque of 80–100 nm was achieved. The obtained nanocomposites were partially degraded, showing a dark-yellow appearance.

This technological problem can be solved by applying an increased speed of the screws. The residence time of material in the extruder is then short enough to prevent an undesirable thermal degradation of material. The examples of torque behavior during PA 6/MMT extrusion processes are shown in Fig. 9a and b.

The improper extrusion conditions were too low temperature and screws speed (150 rpm) were applied resulted in continuous shutdown of the machine (Fig. 9a). Nanocomposite material during the machine downtime was still present and heated inside the cylinder, resulting in degradation of polymer matrix. After increasing the temperature and rotational speed (to 240 rpm), the extrusion process stabilized, as shown in Fig. 9b. Visible torque oscillations in the range of 20 nm result from irregular dosing of raw materials to the feed zone. Introduction of each portion of

Table 3 Wrong conditions set during the extrusion of polyamide-6/montmorillonite nanocomposites

Twin co-rotating screw extruder							
Flow rate (%)　Rotational speed (1/min)	Heating zones						
0.2–0.5　　　100–170	1	2	3	4	5	6	Die
Temperature (°C)	250	250	260	260	270	270	275
Atmospheric venting	–	–	–	–	Yes	–	–
Length of the zones (mm)	80	60	60	64	60	76	23
L/D	5.00	3.75	3.75	4.00	3.75	4.75	–
Cooling tank							
Length of cooling surface (mm)	1,500						
Tank volume (dm3)	27						
Height of bath (mm)	1,081						
Water temperature (°C)	18						
Granulator							
Size of pellets (mm)	4–6						
Rotational speed (1/s)	1–7						

raw materials resulted in the torque increase. Other factors that affect the quality and characteristics of the process are (1) venting and (2) cooling of the feeding zone. Venting in one of the last heating zones of the extruder should be applied, otherwise even for the materials properly dried, bubbles can appear in the melt. They arise because the low molecular weight compounds present in the composition degrade during processing without the possibility to evaporate. The use of

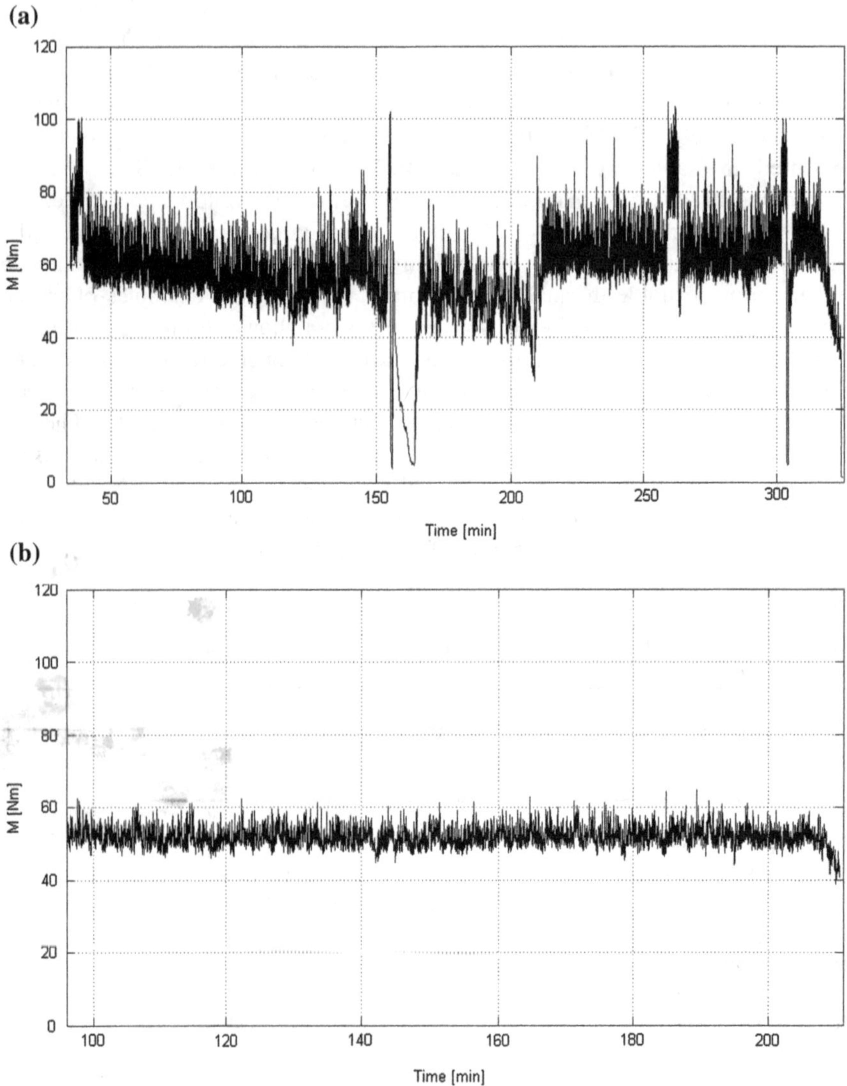

Fig. 9 The examples of the behavior of the torque during extrusion process: **a** wrong extrusion conditions, **b** proper extrusion conditions

venting not only decreases the pressure inside the cylinder but also efficiently eliminates the bubbles present in melt state, making their evaporation easier. Cooling of the feeding zone is necessary to compensate the large amounts of heat coming from adjacent heating zones and caused by abrasion of granules by screws. An overheating may cause degradation of polymer nanocomposite material and eventually a breakdown of the machine. Therefore, efficient cooling of first feed zone with compressed air (preferred) or water should be applied.

6 Scaling up of the Fabrication Process of Polymer Nanocomposites

As polymer nanocomposites become more and more popular in many real-world applications, development of effective and versatile industrial-scale nanocomposite processing methods is of great interest nowadays. For instance, the commercial success of nano-enabled products for the automotive market is expected for numerous applications such as external body parts, interior and under-bonnet parts, coating and fuel system components, etc. Scaling up will primarily increase capacity of production but this process requires several steps to be performed properly:

- Design of micro-scale technologies,
- Implementation of lab-scale experiments,
- Design, implementation and optimalisation of macro-scale technologies,
- Technical verification of the technology,
- Troubleshooting.

Scaling up is a complex issue that needs to take into account various factors. This can be confirmed by the results achieved within different EC funded projects aimed at scaling up production of polymer nanocomposites based on thermoset and thermoplastic matrices from laboratory scale to industrial scale [95]. As an example one could name the "Polyfire" project, carried out at Sheffield Hallam University [96] which deals with layered silicates-based fillers that have already been proven to have flame-retarding properties when added to certain polymers including polypropylene, polystyrene and polyamide, among others. The scaling up production goes from tests performed in the laboratory scale to the macro scale experiments [97].

However, there are still some limitations and challenges for scaling up nanocomposites production, such as:

- Processing limitations—extruders and injection molding machines have usually a different configuration in the micro (lab) and macro (industry) versions. Processing on a large scale requires a new configuration of the screws, cylinders, dies and molds scaled to larger dimensions. This requires changes in the conditions of nanocomposite processing, e.g. the length and configuration of the screws, rotational speed of the screws, temperatures of the heating zones, the number of heating zones, the cooling mode of zones, type and size of molds used for injection, etc.

- Processing costs—the production of nanocomposites on a commercial scale at viable prices requires proper optimization of the technology in terms of material and energy balance, as well as equipment depreciation costs. The performance-to-cost ratio is the key to success in commercialization of the new nanoproduct.
- Environmental and safety limitations—emphasis should be put on careful assessment of health and safety as well as environmental issues associated with the nanomaterials, which is imperative before introducing the technology in the industry.
- Compatibility, consistency and reliability in volume production—it is possible to get good compatibility, consistency and reliability in volume production of polymeric nanocomposites. However, particle size distribution control may cause problems because nanofillers show tendency to agglomerate authorization of the final products for common use—commercializing selected end-use products may take a longer time, mainly due to stringent approval of e.g. European Commission or US Environment Protection Agency. Generally, the products containing new nanomaterials require a long series of positive tests to fulfill all the formal requirements.

7 Conclusions

Polymer nanocomposites represent a new class of multiphase materials containing dispersion of nano-sized fillers such as nanoclays, nanotubes, nanofibers etc. within the polymer matrix. These multifunctional nanocomposites not only exhibit excellent mechanical properties, but they also show outstanding combinations of optical, electrical, thermal, magnetic and other physico-chemical properties. It is believed that the molecular level interactions between the nanoparticles and polymer matrices along with the presence of very high nanoparticle-polymer interfacial areas play a major role in influencing the physical and mechanical properties of nanocomposites [2–5, 11, 13–16, 18–20]. Among nanofillers, nanoclays offer a reduction in relative heat release, excellent dispersion and ex-foliation, excellent flame retardant synergy, and reduced weight [3, 5, 17, 20, 41, 44–54]. The use of thermoplastics-based nanocomposites fabricated by processing methods has been steadily growing, especially in automotive applications, largely due to the material's low cost, high performance, low density, longer shelf life, easy dispersion and ability to recycling [3, 5, 11, 13, 15, 17, 18, 20–22]. A market research report states that world nanocomposites market is forecast to reach 1.3 billion pounds by the year 2015, and growth in the nanocomposites market will be driven by robust demand in automotive and aerospace markets. By 2020, the automotive sector is expected to become the third-largest market for polymer nanocomposite applications, with over 15 % of the market [98–100].

However, the scaling up of the fabrication process of polymer nanocomposites still requires increased attation. Techniques for preparation of polymer/montmorillonite nanocomposites on industrial scale need to be optimized, taking into account important technological problems occurring during the processing both on equipment and materials side.

Acknowledgments This work was partially funded by the European Commission (FP7 Project CP-FP; Project Reference: 228536-2) and by the National Science Centre in Poland under contract UMO-2011/01/M/ST8/06834.

References

1. Fengge, G.: Clay/polymer composites: The story. Rev. Feature Mater. Today **7**, 50–55 (2004). doi:10.1016/S1369-7021(04)00509-7
2. Schmidt, D., Shah, D., Giannelis, E.P.: New advances in polymer/layered silicate nanocomposites. Curr. Opin. Solid State Mater. Sci. **6**, 205–212 (2002)
3. Ray, S.S., Okamoto, M.: Polymer/layered silicate nanocomposites: A review from preparation to processing. Prog. Polym. Sci. **28**, 1539–1641 (2003)
4. Azeez, A.A., Rhee, K.Y., Park, S.J., Hui, D.: Epoxy clay nanocomposites—processing, properties and applications: A review. Compos.: Part B **45**, 308–320 (2013)
5. Pavlidou, S., Papaspyrides, C.D.: A review on polymer–layered silicate nanocomposites. Prog. Polym. Sci. **33**, 1119–1198 (2008)
6. Pfaendner, R.: Nanocomposites: Industrial opportunity or challenge? Polym. Degrad. Stab. **95**, 369–373 (2010)
7. Lepoittevina, B., Pantoustier, N., Devalckenaere, M., Alexandre, M., Calberg, C., Jerome, R., Henrist, C., Rulmont, A., Dubois, P.: Polymer/layered silicate nanocomposites by combined intercalative polymerization and melt intercalation: A masterbatch process. Polymer **44**, 2033–2040 (2003)
8. Majka, T.M., Leszczyńska, A., Pielichowski, K.: Comparison of rheological properties of polyamide-6 and its nanocomposites with montmorillonite, obtained by melt intercalation. Czasopismo Techniczne in print (2013)
9. Meneghetti, P., Qutubuddin, S.: Application of mean-field model of polymer melt intercalation in organo-silicates for nanocomposites. J. Colloid Interface Sci. **288**, 387–389 (2005)
10. Huang, Z.M., Zhang, Y.Z., Kotaki, M., Ramakrishna, S.: A review on polymer nanofibers by electrospinning and their applications in nanocomposites. Compos. Sci. Technol. **63**, 2223–2253 (2003)
11. Raman, N., Sudharsan, S., Pothiraj, K.: Synthesis and structural reactivity of inorganic–organic hybrid nanocomposites: A review. J. Saudi Chem. Soc. **16**, 339–352 (2012)
12. Nguyen, T.R.: Polymer-based nanocomposites for organic optoelectronic devices: A review. Surf. Coat. Technol. **206**, 742–752 (2011)
13. Zhang, S., Sun, D., Fu, Y., Du, H.: Recent advances of superhard nanocomposite coatings: A review. Surf. Coat. Technol. **167**, 113–119 (2003)
14. Silva, F., Njuguna, J., Sachse, S., Pielichowski, K., Leszczynska, A., Giacomelli, M.: The influence of multiscale fillers reinforcement into impact resistance and energy absorption properties of polyamide 6 and polypropylene nanocomposite structures. Mater Des in print (2013)
15. Njuguna, J., Pielichowski, K., Desai, S.: Nanofiller-reinforced polymer nanocomposites. Polym. Adv. Technol. **19**, 947 (2008)

16. Shirazi, S.M.: Investigation of physical and chemical properties of polypropylene hybrid nanocomposites. Mater. Des. **34**, 474–478 (2012)
17. Beyer, B.: Nanocomposites: A new class of flame retardants for polymers. Plast. Addit. Compound. **4**, 22–28 (2002)
18. Krishnamoorti, R., Yurekli, K.: Rheology of polymer layered silicate nanocomposites. Curr. Opin. Colloid Interface Sci. **6**, 464–470 (2001)
19. Faucheu, J., Gauthier, C., Chazeau, L., Cavaille, J.Y., Mellon, V., Lami, M.B.: Miniemulsion polymerization for synthesis of structured clay/polymer nanocomposites: Short review and recent advances. Polymer **51**, 6–17 (2010)
20. Choudalakis, G., Gotsis, A.D.: Permeability of polymer/clay nanocomposites: A review. Eur. Polym. J. **45**, 967–984 (2009)
21. Shokuhfar, A., Shahabadi, A.Z., Atai, A.A., Nejad, S.E., Termeh, M.: Predictive modeling of creep in polymer/layered silicate nanocomposites. Polym. Test. **31**, 345–354 (2012)
22. Ma, P.C., Siddiqui, N.A., Marom, G., Kim, J.K.: Dispersion and functionalization of carbon nanotubes for polymer-based nanocomposites: A review. Compos.: Part A **41**, 1345–1367 (2010)
23. Hu, K.H., Wang, J., Schraubed, S., Xua, J.Z., Hua, X.G., Stenglerd, R.: Tribological properties of MoS_2 nano-balls as filler in polyoxymethylene-based composite layer of three-layer self-lubrication bearing materials. Wear **266**, 1198–1207 (2009)
24. Siró, I., Plackett, D.: Microfibrillated cellulose and new nanocomposite materials: A review. Cellulose **17**, 459–494 (2010)
25. Boccaccini, A.R., Erol, M., Stark, W.J., Mohn, D., Hong, Z., Mano, J.F.: Polymer/bioactive glass nanocomposites for biomedical applications: A review. Compos. Sci. Technol. **70**, 1764–1776 (2010)
26. http://www.bayer.com/en/Products-from-A-to-Z.aspx
27. http://www.clariant.com
28. http://www51.honeywell.com/sm/aegis/
29. http://www.nanocor.com
30. http://www.geindustrial.com/cwc/markets/industrial_sub10.htm
31. http://www.ube-ind.co.jp/english/index.htm
32. http://www.polykemi.se/
33. Njuguna, J., Pielichowski, K., Fan, J.: Polymer nanocomposites for aerospace applications. In: Gao, F. (ed.) Advances in Polymer Nanocomposites: Types and Applications. Woodhead Publishing Ltd., Cambridge (2012)
34. http://hyperioncatalysis.com/
35. http://www.eupen.com/
36. http://www.hielscher.com
37. Kouini, B., Serier, A.: Properties of polypropylene/polyamide nanocomposites prepared by melt processing with a PP-g-MAH compatibilizer. Mater. Des. **34**, 313–318 (2012)
38. Rijswijk, K., Lindstedt, S., Vlasveld, D.P.N., Bersee, H.E.N., Beukers, A.: Reactive processing of anionic polyamide-6 for application in fiber composites: A comparitive study with melt processed polyamides and nanocomposites. Polym. Test. **25**, 873–887 (2006)
39. Lin, J.C., Nien, M.H., Yu, F.M.: Morphological structure, processing and properties of propylene polymer matrix nanocomposites. Compos. Struct. **71**, 78–82 (2005)
40. Chandra, A., Turng, L.S., Gopalan, P., Rowell, R.M., Gong, S.: Polymer/layered silicate nanocomposites by combined intercalative polymerization and melt intercalation: A masterbatch process. Polymer **44**, 2033–2040 (2003)
41. Hussain, F., Hojjati, M., Okamoto, M., Gorga, R.E.: Review article: Polymer-matrix nanocomposites, processing, manufacturing, and application: An overview. J. Compos. Mater. **40**, 1511–1575 (2006)
42. Abend, S., Lagaly, G.: Sol–gel transitions of sodium montmorillonite dispersions. Appl. Clay Sci. **16**, 201–227 (2000)

43. Pavlos CM.: Manufacturing processes for advanced materials: Nanocomposites. AeonClad coatings LLC In: http://www.nist.gov/tip/wp/pswp/upload/273_manufacturing_processes_for_advanced_materials.pdf
44. Pagacz, J., Pielichowski, K.: Modification of layered silicates for applications in nanotechnology. Czasopismo Techniczne z 1-Ch: 133–147 (2007)
45. Pospíšil, M., Kalendová, A., Capková, P., Šimoník, J., Valášková, M.: Structure analysis of intercalated layer silicates: Combination of molecular simulations and experiment. J. Colloid Interface Sci. **277**, 154–161 (2004)
46. He, H., Ma, Y., Zhu, J., Yuan, P., Qing, Y.: Organoclays prepared from montmorillonites with different cation exchange capacity and surfactant configuration. Appl. Clay Sci. **48**, 67–72 (2010)
47. Elban, W.L., Howarter, J.A., Richardson, M.C., Stutzman, P.E., Forster, A.M., Nolte, A.J., Holmes, G.A.: Influence of solvent washing on interlayer structure of alkylammonium montmorillonites. Appl. Clay Sci. **61**, 29–36 (2012)
48. Jiang, J.Q., Cooper, C., Ouki, S.: Comparison of modified montmorillonite adsorbents Part I: Preparation, characterization and phenol adsorption. Chemosphere **47**, 711–716 (2002)
49. Vazquez, A., López, M., Kortaberria, G., Martín, L., Mondragon, I.: Modification of montmorillonite with cationic surfactants. Thermal and chemical analysis including CEC determination. Appl. Clay Sci. **41**, 24–36 (2008)
50. Paiva, L.B., Morales, A.R., Díaz, R.F.V.: Organoclays: Properties, preparation and applications. Appl. Clay Sci. **42**, 8–24 (2008)
51. Mittal, V.: Modification of montmorillonites with thermally stable phosphonium cations and comparison with alkylammonium montmorillonites. Appl. Clay Sci. **56**, 103–109 (2012)
52. Chefetz, B., Eldad, S., Polubesova, T.: Interactions of aromatic acids with montmorillonite: Ca_2+- and Fe_3+-saturated clays versus Fe_3+–Ca_2+-clay system. Geoderma **160**, 608–613 (2011)
53. Kozak, M., Domka, L.: Adsorption of the quaternary ammonium salts on montmorillonite. J. Phys. Chem. Solid. **65**, 441–445 (2004)
54. Radojevic, Z., Mitrovic, A.: Study of montmorillonite and cationic activators system rheological characteristic change mechanism. J. Eur. Ceram. Soc. **27**, 1691–1695 (2007)
55. Garcıa-Lopez, D., Gobernado-Mitre, I., Fernandez, J.F., Merino, J.C., Pastor, J.M.: Influence of clay modification process in PA6-layered silicate nanocomposite properties. Polymer **46**, 2758–2765 (2005)
56. Davis, R.D., Gilman, J.W., VanderHart, D.L.: Processing degradation of polyamide 6/montmorillonite clay nanocomposites and clay organic modifier. Polym. Degrad. Stab. **79**, 111–121 (2003)
57. Fornes, T.D., Yoona, P.J., Hunterb, D.L., Keskkula, H., Paul, D.R.: Effect of organoclay structure on nylon 6 nanocomposite morphology and properties. Polymer **43**, 5915–5933 (2002)
58. Liu, X., Wu, Q.: Non-isothermal crystallization behaviors of polyamide 6/clay nanocomposites. Eur. Polym. J. **38**, 1383–1389 (2002)
59. Winberg, P., Eldrup, M., Pedersen, N.J., Es, M.A., Maurer, F.H.J.: Free volume sizes in intercalated polyamide 6/clay nanocomposites. Polymer **46**, 8239–8249 (2005)
60. Picard, E., Vermogen, A., Gerard, J.F., Espuche, E.: Barrier properties of nylon 6-montmorillonite nanocomposite membranes prepared by melt blending: Influence of the clay content and dispersion state consequences on modeling. J. Membr. Sci. **292**, 133–144 (2007)
61. Muzny, C.D., Butler, B.D., Hanley, H.J.M., Tsvetkov, F., Peiffer, D.G.: Clay pellet dispersion in polymer matrix. Mater. Lett. **28**, 379–384 (1996)
62. Farahani, R.D., Ahmad Ramazani, S.A.: Melt preparation and investigation of properties of toughened polyamide 66 with SEBS-g-MA and their nanocomposites. Mater. Des. **29**, 105–111 (2008)
63. Pramoda, K.P., Liu, T., Liu, Z., He, C., Sue, H.J.: Thermal degradation behavior of polyamide 6/clay nanocomposites. Polym. Degrad. Stab. **81**, 47–56 (2003)

64. Zulfiqar, S., Ahmad, Z., Ishaq, M., Sarwar, M.I.: Aromatic–aliphatic polyamide/montmorillonite clay nanocomposite materials: Synthesis, nanostructure and properties. Mater. Sci. Eng., A **525**, 30–36 (2009)
65. Zulfiqar, S., Sarwar, M.I.: Investigating the structure–property relationship of aromatic–aliphatic polyamide/layered silicate hybrid films. Springer Ser. Solid-State Sci. **11**, 1246–1251 (2009)
66. Yu, S., Zhao, J., Chen, G., Juay, Y.K., Yong, M.S.: The characteristics of polyamide layered-silicate nanocomposites. J. Mater. Process. Technol. **192–193**, 410–414 (2007)
67. Jadav, G.L., Singh, P.S.: Synthesis of novel silica-polyamide nanocomposite membrane with enhanced properties. J. Membr. Sci **32**, 257–267 (2009)
68. Chavarria, F., Paul, D.R.: Comparison of nanocomposites based on nylon 6 and nylon 66. Polymer **45**, 8501–8515 (2004)
69. Zulfiqar, S., Kausar, A., Rizwan, M., Sarwar, M.I.: Probing the role of surface treated montmorillonite on the properties of semi-aromatic polyamide/clay nanocomposites. Appl. Surf. Sci. **255**, 2080–2086 (2008)
70. Zulfiqar, S., Lieberwirth, I., Ahmad, Z., Sarwar, M.I.: Influence of oligomerically modified reactive montmorillonite on thermal and mechanical properties of aromatic polyamide–clay nanocomposites. Acta Mater. **56**, 4905–4912 (2008)
71. Monticelli, O., Musina, Z., Frache, A., Bellucci, F., Camino, G., Russo, S.: Influence of compatibilizer degradation on formation and properties of PA6/organoclay nanocomposites. Poly. Degrad. Stab. **92**, 370–378 (2007)
72. Fornes, T.D., Paul, D.R.: Crystallization behavior of nylon 6 nanocomposites. Polymer **44**, 3945–3961 (2003)
73. Zammarano, M., Bellayer, S., Gilman, J.W., Franceschi, M., Beyer, F.L., Harris, R.L., Meriani, S.: Delamination of organo-modified layered double hydroxides in polyamide 6 by melt processing. Polymer **47**, 652–662 (2006)
74. Liu, T.X., Liu, Z.H., Ma, K.X., Shen, L., Zeng, K.Y., He, C.B.: Morphology, thermal and mechanical behavior of polyamide 6/layered-silicate nanocomposites. Compos. Sci. Technol. **63**, 331–337 (2003)
75. Incarnato, L., Scarfato, P., Russo, G.M., Di Maio, L., Iannelli, P., Acierno, D.: Preparation and characterization of new melt compounded copolyamide nanocomposites. Polymer **44**, 4625–4634 (2003)
76. Pielichowski, K., Leszczyńska, A.: Polyoxymethylene-based nanocomposites with montmorillonite: An introductory study. Polimery **2**, 143 (2006)
77. Leszczyńska, A., Njuguna, J., Pielichowski, K., Banerjee, J.R.: Polymer/montmorillonite nanocomposites with improved thermal properties part I. Factors influencing thermal stability and mechanisms of thermal stability improvement. Thermochim. Acta **453**, 75 (2007)
78. Leszczyńska, A., Njuguna, J., Pielichowski, K., Banerjee, J.R.: Polymer/montmorillonite nanocomposites with improved thermal properties part II. Thermal stability of montmorillonite nanocomposites based on different polymeric matrixes. Thermochim Acta **454**, 1 (2007)
79. http://www.lookchem.com/cas-710/71011-26-2.html
80. http://datasheets.scbt.com/sc-257099.pdf
81. http://cameochemicals.noaa.gov/chemical/20490
82. Majka, T.M., Leszczyńska, A., Pielichowski, K., Dworakowska, S.: Optymalizacja procesu suszenia poliamidu-6 w warunkach laboratoryjnych III Ogólnopolska Sesja Kół Naukowych w Tarnobrzegu w Państwowej Wyższej Szkole Zawodowej Tarnobrzeg in Print (2012)
83. Pielichowski, K., Leszczyńska, A., Njuguna, J.: Mechanism of thermal stability enhancement in polymer nanocomposites. In: Mittal, V. (ed.) Optimization of Polymer Nanocomposite Properties. Wiley, Weinheim (2010)
84. Pielichowski, K., Leszczyńska, A., Njuguna, J.: Mechanisms of thermal degradation of layered silicates modified with ammonium and other thermally stable salts. In: Mittal, V.

(ed.) Thermally Stable and Flame Retardant Polymer Nanocomposites. Cambridge University Press, Cambridge (2011)

85. Crosby, A.J., Lee, J.Y.: Polymer nanocomposites: The "nano" effect on mechanical properties. Polym. Rev. **47**, 217–229 (2007)
86. Batista, C.A., Archer, L.A.: Polymer/Silica Nanocomposites. Department of Chemical Engineering, New York (2005)
87. Picu, C.R., Sarvestani, A., Ozmusul, M.S.: Atomistically informed continuum model of polymer-based nanocomposites. Mater. Res. Soc. **740**, I8.1.1–I8.1.6 (2003)
88. Sperling, L.H.: Introduction to Physical Polimer Science. Willey Interscience Bethlehem Pennsylvania (2006)
89. Toth, R., Coslanich, A., Ferrone, M., Fermeglia, M., Pricl, S., Miertus, S., Chiellini, E.: Computer simulation of polypropylene/organoclay nanocomposites: Characterization of atomic scale structure and prediction of binding energy. Polymer **45**, 8075–8083 (2004)
90. Leszczynska, A., Pielichowski, K.: Application of thermal analysis methods for characterization of polymer/montmorillonite nanocomposites. J. Therm. Anal. Calorim. **93**, 677 (2008)
91. Sachse, S., Silva, F., Zhu, H., Irfan, A., Leszczyńska, A., Pielichowski, K., Ermini, V., Blazquez, M., Kuzmenko, O., Njuguna, J.: The effect of nanoclay on dust generation during drilling of PA6 nanocomposites. J. Nanomaterials in print (2012)
92. Alexandre, M., Dubois, P.: Polymer-layered silicate nanocomposites: Preparation, properties and uses of a new class of materials. Mater. Sci. Eng., A **28**, 1–63 (2000)
93. Yamakawa, R.S., Razzino, C.A., Correa, C.A., Hage, E.: Influence of notching and molding conditions on determination of EWF parameters in polyamide 6. Polym. Test. **23**, 195–202 (2004)
94. Yebra-Rodriguez, A., Alvarez-Lloret, P., Yebra, A., Cardell, C., Rodriguez-Navarro, A.B.: Influence of processing conditions on the optical and crystallographic properties of injection molded polyamide-6 and polyamide-6/montmorillonite nanocomposites. Appl. Clay Sci. **51**, 414–418 (2011)
95. Muller, J., Grosse, S., Kummer, S., Masarati, E., Consalvi, M., Fisher, D.: Scale-up of an on line process monitoring system to an industrial extruder to determine the concentration and dispersion of polymer composites. J. Nanostruct. Polym. Nanocomposites 8 (2012)
96. http://www.polyfireproject.eu/
97. Giacomelli, M., Pielichowski, K., Leszczyńska, A.: Thermoplastic polymer nanocomposites with montmorillonite-lab vs industrial scale fabrication. IOP Conf. Ser. Mater. Sci. Eng. **40**, 1–6 (2012)
98. Nanocomposites-A Global Strategic Business Report, Mar 2011, published by Electronics.ca, http://www.electronics.ca/presscenter/articles/1404/1/Global-Nanocomposites-Market-to-Reach-13-Billion-Pounds–by-2015/Page1.html
99. Nanocomposites, Nanoparticles, Nanoclays, and Nanotubes, June 2006, published by BCC Research, http://www.bccresearch.com/report/NAN021C.html
100. Potential Market for Carbon Nanomaterials' Applications, 28 Feb 2011, published by Frost & Sullivan www.nist.gov/cnst/upload/Valenti-NIST.pdf

Influences of Morphology and Doping on the Photoactivity of TiO$_2$ Nanostructures

Nasser A. M. Barakat and Muzafar A. Kanjwal

Abstract Compared to the bulk, the nanoscale provides special characteristics for the functional materials. Moreover, the nanostructural morphology has also distinct influences. Enormous efforts have been devoted to the research of TiO$_2$ material, which has led to many promising applications. Beside the morphology impact, doping of titanium dioxide nanostructures by pristine metal nanoparticles (e.g. Ag, Pt, … etc.) revealed distinct improvement in the photocatalytic activity. Although the doping process can remarkably improve the photoactivity, it has also noticeable influences on the crystal structure. In this chapter, the important parameters affecting the photocatalytic activity of TiO$_2$ are discussed; morphology and silver doping. Also, effect of sliver-doping on the crystal structure and the nanofibrous morphology is investigated. Moreover, the influence of the temperature on the photodegradation process using Ag-doped TiO$_2$ nanostructures will be addressed. Two morphologies were introduced; nanoparticles and nanofibers. The nanofibers were synthesized by electrospinning of a sol–gel consisting of titanium isopropoxide, silver nitrate and poly(vinyl acetate). The silver nitrate amount was changed to produce nanofibers having different silver contents. The nanoparticles were prepared from the same sol-gels, however instead of spinning the gels were dried, grinded and sintered. The experimental and analytical studies indicate that doping by silver reveals to form anatase and rutile when the silver nitrate content in the mother solution was more than 3 wt%. The rutile phase content is directly proportional with the AgNO$_3$ concentration. Negative impact of the silver-doping

N. A. M. Barakat (✉)
Organic Materials and Fiber Engineering Department, Chonbuk National University, Jeonju 561-756, Republic of Korea
e-mail: nasser@jbnu.ac.kr

N. A. M. Barakat
Chemical Engineering Department, Minia University, El-Minia 61111, Egypt

M. A. Kanjwal
Technical University of Denmark, DTU Food, Soltofts plads, B 227 2800 Kgs Lyngby, Denmark

J. Njuguna (ed.), *Structural Nanocomposites*, Engineering Materials,
DOI: 10.1007/978-3-642-40322-4_5, © Springer-Verlag Berlin Heidelberg 2013

on the nanofibrous morphology was observed as increase the silver content caused to decrease the aspect ratio, i.e. producing nanorods rather nanofibers. However, silver-doping leads to modify the surface roughness. In contrast to the known influence of the temperature on the chemical reactions, in case of the nanofibrous morphology of Ag-doped TiO_2, the temperature has negative impact on the photoactivity.

1 Introduction

In chemistry, photocatalysis is the acceleration of a photoreaction in the presence of a catalyst. In catalyzed photolysis, light is absorbed by an adsorbed substrate. In photogenerated catalysis, the photocatalytic activity depends on the ability of the catalyst to create electron–hole pairs, which generate free radicals (e.g. hydroxyl radicals: •OH) able to undergo secondary reactions. Its practical application was made possible by the discovery of water electrolysis by means of titanium dioxide. The commercially used process is called the advanced oxidation process (AOP). There are several ways the AOP can be carried out; these may (but do not necessarily) involve TiO_2 or even the use of UV light. Generally the defining factor is the production and use of the hydroxyl radical. Generally, the semiconductors are used as photocatalysts. Like other solids, semiconductor materials have electronic band structure determined by the crystal properties of the material. The actual energy distribution among the electrons is described by the Fermi energy and the temperature of the electrons. At absolute zero temperature, all of the electrons have energy below the Fermi energy; but at non-zero temperatures the energy levels are filled following a Boltzmann distribution.

In semiconductors the Fermi energy lies in the middle of a forbidden band or band gap between two allowed bands called the valence band and the conduction band. The valence band, immediately below the forbidden band, is normally very nearly completely occupied. The conduction band, above the Fermi level, is normally nearly completely empty. Because the valence band is so nearly full, its electrons are not mobile, and cannot flow as electrical current.

However, if an electron in the valence band acquires enough energy to reach the conduction band, it can flow freely among the nearly empty conduction band energy states. Furthermore it will also leave behind an electron hole that can flow as current exactly like a physical charged particle. Carrier generation describes processes by which electrons gain energy and move from the valence band to the conduction band, producing two mobile carriers; while recombination describes processes by which a conduction band electron loses energy and re-occupies the energy state of an electron hole in the valence band.

In a material at thermal equilibrium generation and recombination are balanced, so that the net charge carrier density remains constant. The equilibrium carrier density that results from the balance of these interactions is predicted by thermodynamics. The resulting probability of occupation of energy states in each energy band is given by Fermi–Dirac statistics.

Fast electrons/holes recombination is the main dilemma facing the semiconductors photocatalysts in general and titanium oxide in particular. Incorporation of noble metal nanoparticles (NPs) into the titania dielectric matrix is a recent strategy to overcome this problem and simultaneously improve the photocatalytic activity of titanium oxide [1, 2]. Doping with foreign metal nanoparticles can produce high Schottky barrier that facilitates electron capture [3]. The capture of electrons postulate to produce a longer electron–hole pair separation lifetime, and therefore hinder the recombination of electron/hole pairs and enhance the transfer of holes and possibly electrons to O_2 adsorbed on the TiO_2 surface. Afterwards, excited electrons migrate to the metal, where they become trapped and the electron/hole pair recombination is suppressed. Therefore, many investigations have reported the enhancement of photoactivities in both liquid and gas phases [4, 5]. Moreover, this incorporation provides an absorption feature due to the surface plasmon resonance (SPR) occurring over the visible range of the spectrum [6]. Particularly, silver and gold nanoparticles are more familiar because of their color varieties in the visible region, which is attributed to oscillations of the electrons at the surface of the nanoparticles [7]. Silver is the most common metal used to modify titania, because its d-s band gap is in the UV region and does not damp out the plasmon mode as strongly as gold [8, 9]. Titania-modified-silver particles have raised extensive interest due to their applications in photocatalytic degradation [10]. The researchers have intensively studied how silver provides titania distinct photocatalytic activity, the mechanism has been well explained in previous reports [11–13]. Moreover, some researchers have concluded that incorporation of silver in titanium dioxide leads to increases in the total surface area of the prepared titanium dioxide [14, 15]. This can be considered an additional benefit of utilizing silver-loaded titanium oxide in the field of photocatalytic degradation. Therefore, many researchers have practically demonstrated that the degradation rates of the dyes can be enhanced by the deposition of silver on titanium dioxide [16–18].

In the last decades, nanostructural materials have been intensively investigated because of their high surface area, which strongly affects their physiochemical properties. Different shapes have been introduced. Of the reported nanostructure shapes, special attention has been paid to one-dimensional forms such as nanorods, nanowires, and nanofibers. This is due to their potential applications in the nanodevices [19–21]. Nanofibers have received special consideration due to their high axial ratio, good mechanical properties, and their manageable and novel physical properties.

Compared to nanoparticles, nanofibers have small surface area which might be considered a negative impact upon using as catalyst in the chemical reactions. Therefore, if the morphology has no impact, the nanoparticles will have higher catalytic activity. Moreover, for the same surface activity, the temperature should have the same influence on both formulations. In most of the chemical reactions, positive effect of the temperature on the reaction rates is a prevailing impression. Temperature is a measure of the kinetic energy of a system, so higher temperature implies higher average kinetic energy of molecules and more collisions per unit time. This hypothesis might be true in the normal cases, however the nanostructures usually have unexpected behaviors compared to the bulk scales.

In this chapter, the aforementioned facts have been investigated for the Ag-doped TiO_2 nanophotocatalyst. Two nano-formulations have been utilized; nanofibers and nanoparticles. Silver-grafted titanium oxide nanofibers have been synthesized using the electrospinning of silver nitrate/titanium isopropoxide/poly(vinyl acetate) sol–gel. However, the nanoparticulate form has been obtained by calcination of a ground powder prepared from the same electrospun sol–gels. Metal-doping may also have an influence on the titania crystal structure leading to distinct improving in the application fields [22, 23]. Therefore, effect of the silver-doping on the crystal structure, the morphology and the photocatalytic activity of TiO_2 nanofibers will be discussed in this chapter.

2 Nanofibers

Nanofibers are an exciting new class of material used for several value added applications such as medical, filtration, barrier, wipes, personal care, composite, garments, insulation, and energy storage. Special properties of nanofibers make them suitable for a wide range of applications from medical to consumer products and industrial to high-tech applications for aerospace, capacitors, transistors, drug delivery systems, battery separators, energy storage, fuel cells, and information technology [24, 25].

If the diameters of polymer fiber materials are shrunk from micrometers to submicrons or down to nanometers, there appear several amazing characteristics such as very large surface area to volume ratio (this ratio for a nanofiber can be as large as 10^3 times of that of a microfiber), flexibility in surface functionalities, and superior mechanical performance (e.g. stiffness and tensile strength) compared with any other known form of the material. These outstanding properties make the polymer nanofibers to be optimal candidates for many important applications [26].

Generally, polymeric nanofibers are produced by an electrospinning process (Fig. 1). Electrospinning is a process that spins fibers of diameters ranging from 10 nm to several hundred nanometers. This method has been known since 1934 when the first patent on electrospinning was filed. Fiber properties depend on field uniformity, polymer viscosity, electric field strength and DCD (distance between nozzle and collector). Advancements in microscopy such as scanning electron microscopy has enabled us to better understand the structure and morphology of nanofibers. At present the production rate of this process is low and measured in grams per hour.

Another technique for producing nanofibers is spinning bi-component fibers such as Islands-In-The-Sea fibers in 1–3 denier filaments with from 240 to possibly

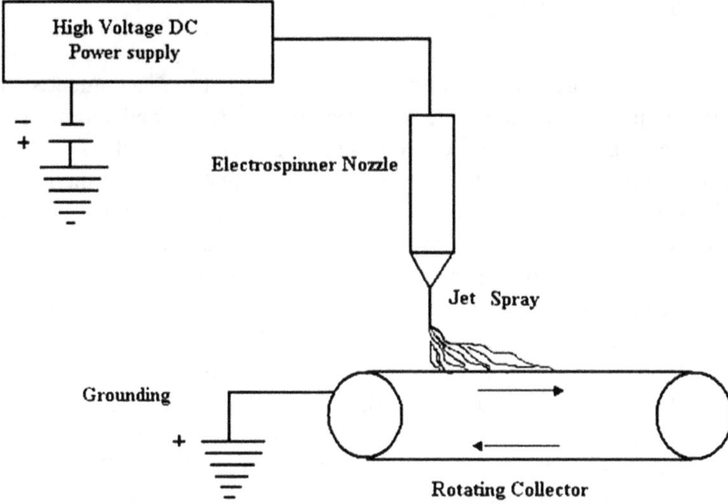

Fig. 1 Schematic representation of electrospinning process

as much as 1,120 filaments surrounded by dissolvable polymer. Dissolving the polymer leaves the matrix of nanofibers, which can be further separated by stretching or mechanical agitation.

The most often used fibers in this technique are nylon, polystyrene, polyacrylonitrile, polycarbonate, PEO, PET and water-soluble polymers. The polymer ratio is generally 80 % islands and 20 % sea. The resulting nanofibers after dissolving the sea polymer component have a diameter of approximately 300 nm. Compared to electrospinning, nanofibers produced with this technique will have a very narrow diameter range but are coarser.

In the electrospinning process, the solution is held at the tip of a capillary tube by virtue of its surface tension. The electrical potential applied provides a charge to the polymer solution. Mutual charge repulsion in the polymer solution induces a force that is directly opposite to the surface tension of the polymer solution. An increase in the electrical potential initially leads to the elongation of the hemispherical surface of the solution at the tip of the capillary tube to form a conical shape known as the Taylor cone [27]. A further increase causes the electric potential to reach a critical value, at which it overcomes the surface tension forces to cause the formation of a jet that is ejected from the tip of the Taylor cone. The charged jet undergoes bending instabilities and gradually thins up in air primarily due to elongation and solvent evaporation [28]. The charged jet eventually forms randomly oriented nanofibers that can be collected on a stationary or rotating grounded metallic collector of the electrospinning set up [29].

2.1 Metallic Nanofibers

In recent years, ceramic material with effective control at the nanometer scale has been actively pursued as a problem of materials science and engineering. In particular, the preparation of 1D ceramic nanostructures such as fibers, wires, rods, belts, tubes, spirals, and rings has received great interest owing to their potential applications in many vital areas of technologically such as electronics, photonics, and mechanics [30, 31]. A large number of chemical and/or physical methods, mostly based on the bottom-up and template directed routes, have been demonstrated for producing 1D ceramic nanostructures with various compositions by controlling the nucleation and growth processes. Top-down approaches such as photolithography, soft lithography and electrospinning have also been exploited to fabricate 1D ceramic nanostructures [30, 32–35]. Moreover, electrospinning of vital materials are fascinating in the field of implantology. Electrospinning of metallic nanofibers has been largely limited due to availability of proper solvent and desired viscosity for bending instability during electrospinning process. Ceramics are usually considered to be not directly spinnable although it is possible (at least, in principle) to electrospun ceramic nanofibers from their melts at extremely high temperatures. Like conventional spinning processes for generating ceramic fibers. Preparation of ceramic fibers by electrospinning has to rely on the use of spinnable precursors or by using a blending template containing colloidal sol [36]. The fabrication of ceramic nanofibers consists of three major steps:

- Preparation of an inorganic sol or a solution containing a matrix polymer together with an alkoxide, salt, or polymer precursor.
- Electrospinning of the solution to generate composite nanofibers consisting of the matrix polymer and precursor. The spinning experiments are usually performed in a well-controlled environment at room temperature.
- Calcination, sintering, or chemical conversion of the precursor into the desired ceramic at an elevated temperature, with concomitant removal of all organic components from the precursor fibers.

As compared with the bulk metals, these nanofibers have been approved to be an excellent templates for the bottom based applications. This potentiality is the basis for superior properties like less weight, high area to volume ratio, improved durability and bioactive interfaces [37–41].

2.2 Modification of Titania Nanofibers

Currently, the heterogeneous photocatalysis has arised as an alternative economical and harmless technology for removal of organic impurities. In such processes, the illuminated semiconductors absorb light and generate electronic species which leads to complete oxidation of organic components in waste waters. Titanium dioxide is considered an interesting material because of its optical, electrical and photochemical properties [42, 43]. However, for practical applications the photocatalytic activity and photoresponse behavior of titanium dioxide are very much needed to be further improved [5, 44]. Recently studies have focussed to modify the surface of titanium dioxide to improve its efficiency for photocatalytic activity and to use in solar cells [5, 45]. Incorporation of noble metal nanoparticles into the titania dielectric matrix is a modern strategy to improve the photocatalytic character of titanium dioxide. This incorporation is providing an absorption feature due to surface plasmon resonance (SPR) occurring over the visible range of the spectrum [6]. In particular, silver and gold metals are the most popular materials do have strong SPR character [46, 47]. Moreover, the nanoscale noble metals are usually classical high-performance heterogeneous catalysts [43, 48]. Silver is the most common metal used to modify titania that is because of its d-s band gap is in the UV region and does not damp out the plasmon mode as strongly as for gold [8, 9]. The titania modified-silver particles have raised an extensive interest due to their applications in photocatalytic degradation [10]. The researchers have intensively studied how silver provides titania distinct photocatalytic activity; the mechanism was well explained [11–13, 49–52]. Many formulations have been introduced for titania/silver composite, for instance, nanoparticles [53], and thin films [54–56].

As it is well known, the surface to volume ratio is an important feature for any catalyst. Accordingly, 1D nanostructures are expected to strongly modify the photocatalytic activity. Among all 1D nanostructure, nanofibers do have especial interest due to large surface area to volume ratio. In the field of metal oxides nanofibers; electrospinning is the most widespread.

3 Experimental Section

3.1 Materials

Silver nitrate (99.8 assay), methylene blue dihydrate dye (95.0 assay), N,N-dimethylformamide (DMF, 99.5 assay), and rhodamine B dye were obtained from Showa, Co. Japan. Titanium (1V) isopropoxide (Ti(Iso), 98.0 assay) was purchased from Junsei Co. Ltd., Japan. Poly (vinyl acetate) (PVAc, MW = 500,000 g/mol) was obtained from Aldrich, USA. These materials were used without any further purification.

3.2 Preparation of Nanofibers and Nanoparticles of Ag/TiO₂ Composite

3.2.1 Nanofibers

The electrospinning process was utilized to prepare the silver-grafted titania Nanofibers (NFs). Typically, a sol–gel was prepared by mixing titanium iso-propoxide (Ti(Iso)) and poly(vinyl acetate) (PVAc, 14 wt% in DMF) with a weight ratio of 2:3, respectively, and then few drops of acetic acid were added until the solution became transparent. The mixing process was carried out at 25 °C using magnetic stirrer rotating at 150 rpm. To prepare sol–gels containing different contents of silver, silver nitrate solutions in DMF were mixed with proper quantities of the prepared Ti(Iso)/PVAc solution to prepare final solutions containing 0.5, 1.0, 1.5, 2.0 and 2.5 wt% $AgNO_3$. Afterwards, these solutions were homogeneously mixed under stirring conditions for 10 min at 25 °C and moderate stirring speed. A high voltage power supply (CPS-60 K02V1, Chungpa EMT Co., Republic of Korea) was used as the source of the electric field. The sol–gel was supplied through a plastic syringe attached to a capillary tip. A copper wire originating from the positive electrode (anode) connected with a graphite pin was inserted into the sol–gel and the negative electrode (cathode) was attached to a metallic collector covered with polyethylene sheet. Briefly, the solution was electrospun at 6 kV and 15 cm working distance (the distance between the needle tip and the collector). The electrospinning process was carried out at 25 °C in 40 % relative humidity atmosphere. The formed nanofiber mats were initially dried for 24 h at 80 °C in a vacuum and then calcined in air atmosphere at 700 °C for one hour with a heating rate of 5 °C/min.

3.2.2 Nanoparticles

Silver nitrate/Ti(Iso)/PVAc solutions having the aforementioned compositions and preparation procedure were utilized to prepare nanoparticles containing different silver contents. The process parameters (temperature and stirring) were not changed. Instead of spinning, the solution was vacuously dried at 80 °C for 48 h to completely remove the solvent. The obtained solid materials were finely grinded and sintered in air at 700 °C for 1 h.

3.3 Photocatalytic Degradation

The photocatalytic degradation of the selected dyes in the presence of Ag-TiO₂ nanofibers and nanoparticles was carried out in a simple photo reactor. The reactor was made of glass (1,000 ml capacity, 23 cm height and 15 cm diameter), covered with aluminum foil, and equipped with ultra-violet lamp emitting radiations at 365 nm. The initial dye solution and the photocatalyst were placed in the reactor and continuously stirred to ensure proper mixing during the photocatalytic reaction. Typically, 100 ml of dye solution (10 ppm, concentration) and 50 mg of catalyst were used. At specific time intervals, a 2 ml sample was withdrawn from the reactor and centrifuged to separate the residual catalyst, and then the absorbance intensity was measured at 664 and 554 nm for methylene blue and rhodamine B dyes, respectively.

3.4 Characterization

Surface morphology of nanofibers was studied by JEOL JSM-5900 scanning electron microscope (JEOL Ltd, Japan) and field-emission scanning electron microscope (FESEM, Hitachi S-7400, Japan). The phase and crystallinity were characterized by using Rigaku X-ray diffractometer (Rigaku Co, Japan) with Cu Kα ($\lambda = 1.54056$ Å) radiation over a range of 2θ angles from 20° to 100°. High resolution image and selected area electron diffraction pattern were observed by JEOL JEM-2200FS transmission electron microscope (TEM) (JEOL Ltd., Japan). The concentration of the dyes during the photodegradation study was investigated by spectroscopic analysis using HP 8453 UV–visible spectroscopy system (Germany). The spectra obtained were analyzed by HP ChemiStation software 5890 series.

4 Results and Discussion

4.1 The Photocatalyst Characterization

The polymer is an essential constituent in sol–gels to achieve the electrospinning process [19, 57–59]. Metal alkoxides are the best candidates to form the gel structure due to their affinity to hydrolysis and condensation in the polymer matrix. Because of the well polycondansation property [60, 61], titanium isopropoxide is a famous precursor to prepare TiO_2 nanofibers by using the electrospinning process and PVAc as a gelling polymer [62]. Addition of silver nitrate does not affect the nanofibrous morphology [5]. Figure 2a displays the obtained PVAc/Ti(Iso)/AgNO$_3$ (the later was 2.0 wt% in the electrospun solution) nanofibers. As shown, good morphology was obtained. It is noteworthy mentioning that all the formulations produced good morphology nanofibers; only slight decrease in the average fiber diameter with increasing the silver nitrate content was observed due to increase the electrical conductivity of the electrospun solutions. Calcination did not affect the nanofibrous morphology for all formulations. Figure 2b represents the SEM image of the obtained powder from calcination of the electrospun mat containing 2.0 wt% silver nitrate. As shown, good nanofibers were obtained. Increase the roughness of the outer surface with increasing the silver nitrate content in the original electrospun sol–gel was a noticeable observation. Figure 2c and d display the FE SEM images of the pristine and Ag-containing ($AgNO_3$ = 2.0 wt%) TiO_2 nanofibers, respectively. Big difference between the surfaces of the investigated nanofibers can be observed. It is noteworthy mentioning that the surface roughness was dependent on the silver content, moreover increase the silver nitrate content more than the maximum value used in this study (i.e. 2.5 wt%) led to destroy the nanofibrous morphology and produce nanorods instead. Concerning the nanoparticulate shape, addition of silver nitrate improves the spherical morphology of the nanoparticles as shown in Fig. 2e and f which demonstrate the SEM images of the pristine and Ag-doped (2.0 wt%) TiO_2 nanoparticles.

To properly investigate the effect of addition of silver nitrate on the surface of the silver-doped nanofibers, the surface area has been measured by using Brunauer–Emmett–Teller (BET) technique (Micromeritics, Norcross, GA). The obtained results indicated that the average surface area of the silver-free nanofibers was about 21.3102 ± 0.1351 m^2/g, while, it was 38.8100 ± 0.1324 m^2/g for the nanofibers obtained from calcination of silver nitrate (2 wt%)/Ti(Iso)/PVAc electrospun nanofiber mats. As can be concluded from these results, the surface area of the silver containing nanofibers was duplicated compared to the pristine ones which strongly enhance the photocatalytic activity. For the nanoparticles, it is clear from the SEM images that addition of silver has negative influence on the surface area as the average diameter decreases with silver addition (Fig. 2e and f).

XRD is a reliable and widespread identification technique especially for crystalline materials. Figure 3 represents the XRD pattern of the nanofibers obtained from two sol–gels, Ti(Iso)/PVAc and silver nitrate (2.0 wt%)/Ti(Iso)/PVAc. As shown in both spectra, the results confirm formation of pure anatase titanium dioxide. The strong diffraction peaks at 2θ values of 25.09°, 37.65°, 38.44°, 47.89°, 53.89°, 55.07°, 62.40°, 68.70°, 70.04°, and 75.00° correspond to the crystal planes (101), (004), (112), (200), (105), (211), (204), (220), (220), and (215), respectively indicate formation of anatase titanium dioxide [JCPDS card no 21-1272]. In the case of the nanofibers obtained from calcination of silver nitrate/ Ti(Iso)/PVAc mats (Fig. 2, spectra A), in addition to the titanium dioxide peaks, extra peaks at 2θ values of 38.11°, 44.29°, 64.43° and 77.48° corresponding to the crystal planes (111), (200), (220), and (311), respectively confirm presence of silver metal [JCPDS card no 04-0783]. To make it easy, in Fig. 3, we have marked the peaks corresponding to titanium oxide and silver as A and S, respectively. It is noteworthy to mention that the morphology did not affect the XRD data (i.e. the same results have been obtained for the nanoparticles). Therefore, the XRD pattern of the nanoparticles was not added to simplify and clarify the figure.

Transmission electron microscope analysis is used to investigate the crystal structure. Figure 3 shows the HR TEM of the obtained Ag-doped TiO_2 nanofibers; A and nanoparticles; B. As shown in these figures, there are some black dots in

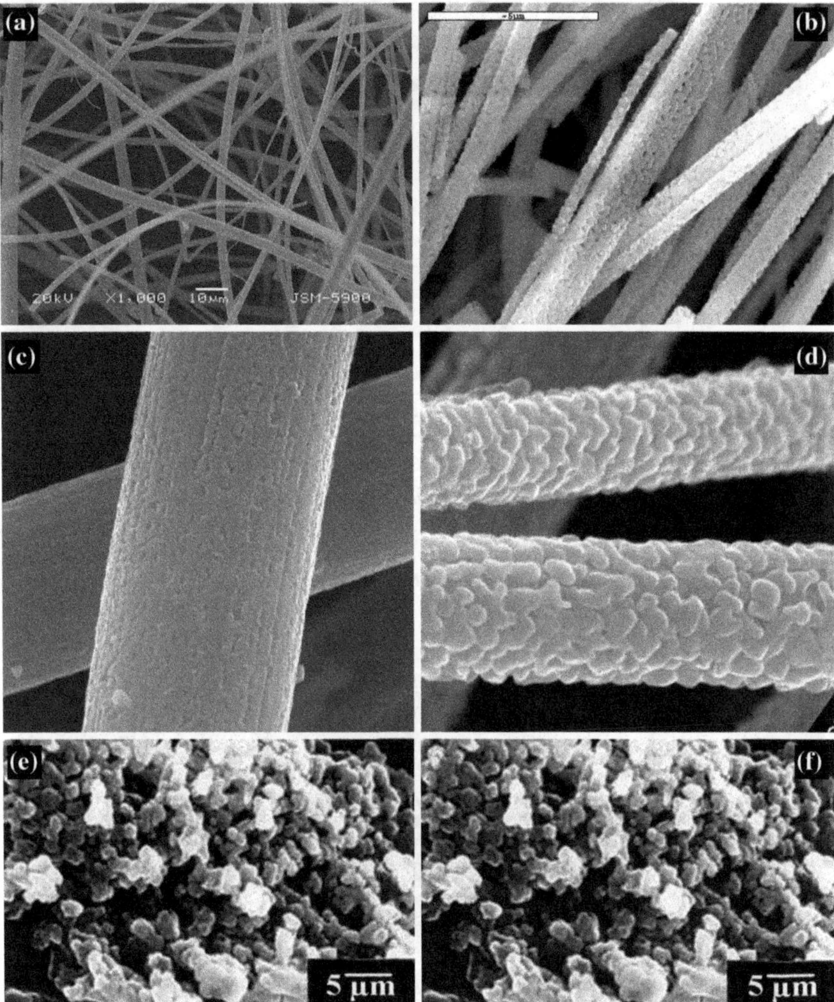

Fig. 2 SEM image of the electrospun PVAc/Ti(Iso)/AgNO$_3$ (AgNO$_3$ = 0.2) nanofibers; **a** the corresponding sintered nanofibers; **b** FE SEM images of the pristine; **c** and Ag-doped TiO2; **d** nanofibers, and SEM images of the pristine; **e** and Ag-doped TiO$_2$ (AgNO$_3$ = 0.2); **f** nanoparticles

both formulations which can be considered as the Ag nanoparticles as these dots have different crystal structures compared to the TiO$_2$ matrices. Both formulations have good crystallinity as shown in Fig. 4c and d which represent the SAED patterns of the Ag-doped TiO$_2$ nanofibers and nanoparticles, respectively. Moreover, the Fast Fourier Transform (FFT) images for both formulations reveal good crystallinity as shown in the corresponding insets.

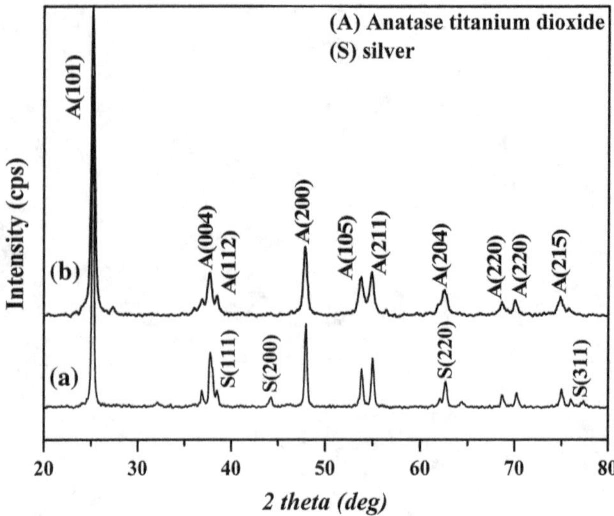

Fig. 3 XRD results for the obtained powder after calcination of PVAc/Ti(Iso) and PVAc/Ti(Iso)/ AgNO$_3$ (AgNO$_3$ = 0.2) electrospun nanofibers in air at 700 °C

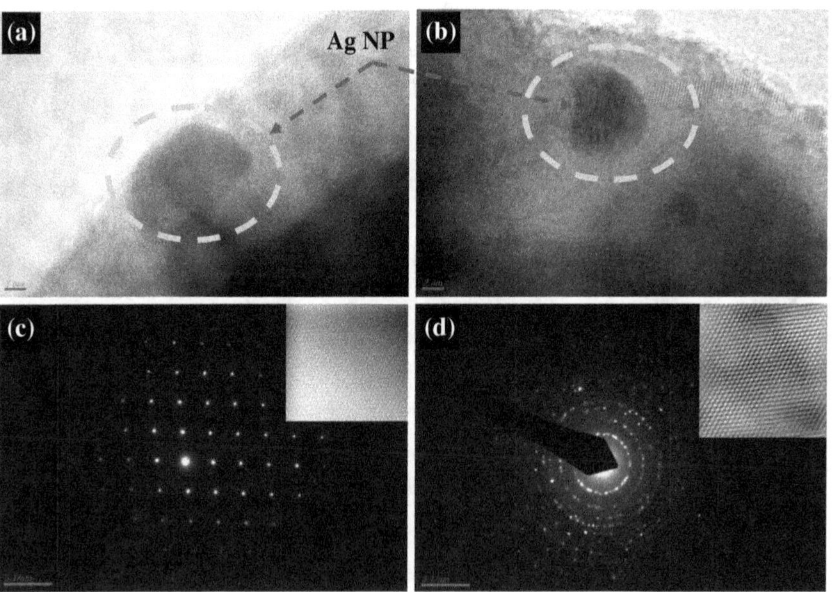

Fig. 4 HR-TEM image of the Ag-doped TiO$_2$ nanofibers; **a** and nanoparticles; **b** and SAED patterns of the marked areas; **c** and **d** the insets show the FFT images of the marked areas

In order to affirm that the silver is present as particles in the TiO_2 nanofibers and simultaneously investigate the nanoparticles appeared in the TEM images, EDX analysis has been invoked. The analysis has been performed in two different points as shown in Fig. 5. As shown in this figure and the corresponding insets, there is big difference in the silver weight percentages between the two chosen points. Therefore, one can say that the point in Fig. 5a represents a point containing silver nanoparticles since the silver content at that point is almost 21 %. However, Fig. 5b might represent almost silver-free point. Taking into consideration that the electron beam in the EDX analysis has specific width more than the silver nanoparticle size, the small weight percentage of silver at this point might be explained as a small part of bordering silver nanoparticles. Actually, these results match all the previous reports about silver-doped titania nanostructures which concluded that the silver exists as small nanoparticles in/on the synthesized nanostructures. Moreover, according to the atomic size, silver-titania system cannot form solid solution alloy, so, theoretically silver has to be in the form of nanoparticles within the titania matrix. Well distribution of silver nanoparticles can be obtained by good mixing of the original $AgNo_3$-Ti(Iso)-PVAc mixture, so, in this study, the electrospinning process was achieved after obtaining well mixed solution.

To explain the phase changes during the calcination process of both formulations (i.e. Ti(Iso)/PVAc and silver nitrate-Ti(Iso)/PVAc), we have utilized thermal gravimetric analysis (TGA). Figure 6 shows the obtained TGA results in all cases,

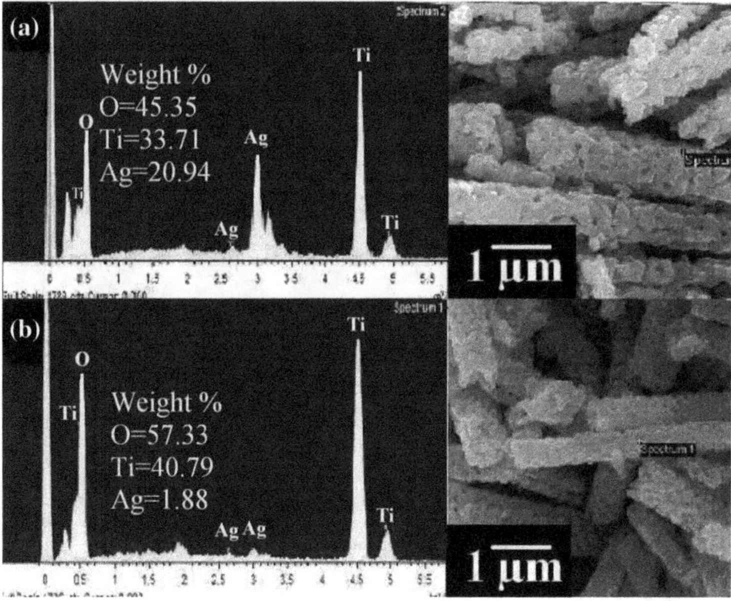

Fig. 5 EDX analysis at two different points, the insets represent the corresponding elemental weight percentages

along with the first derivatives curves were plotted to precisely extract the useful information. Figure 6a represents TGA results for Ti(Iso)/PVAc nanofiber mats, as shown in this figure; two peaks are observed in the first derivative plot. According to the thermal properties of titanium isopropoxide, the first peak (at ∼325 °C) can be explained as decomposition of this organometallic compound into titanium

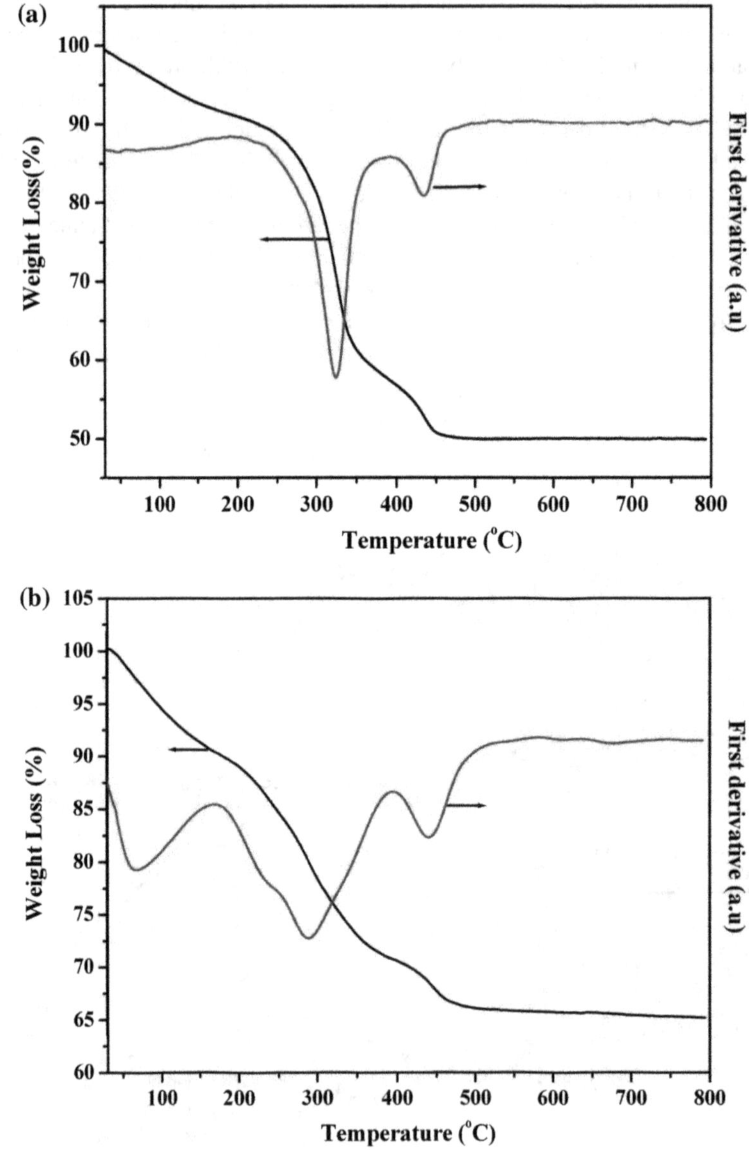

Fig. 6 TGA results in oxygen atmosphere for Ti(Iso)/PVAc (**a**) and silver nitrate-Ti(Iso)/PVAc nanofiber mats (**b**)

dioxide. Because of the calcination process has been performed in air atmosphere, the second peak (at ~ 434 °C) can be explained as complete elimination of PVAc that also matches our previous work in the same polymer [5]. However, in case of silver nitrate- Ti(Iso)/PVAc (Fig. 6b), there are observable change; as shown in this figure, the rate of degradation is comparatively low compared with the first case. Also, there are two main peaks were obtained, a broad peak at ~ 286 °C and another one at 445 °C. According to the thermal properties of silver nitrate [59], and Ti(Iso) (as seen in Fig. 6a), the first peak can be investigated as simultaneous decomposition of silver nitrate and Ti(Iso). However, the second peak represents the decomposition of PVAc polymer. There is another peak at low temperature (~ 60 °C) it might be explained as evaporation of the combined moisture with the sample. Another observation can be noticed by comprising the thermal decomposition of the two formulations is increase the final inorganic residuals in case of the silver nitrate-Ti(Iso)/PVAc than the silver-free case, this increase represents the silver NPs.

4.2 Effect of Silver Doping on the Photoactivity of the Titania

To investigate the influence of the silver doping on the photoactivity of titanium oxide, photodegradation of some organic dyes was performed as explained in the experimental section.

4.2.1 Methylene Blue Dihydrate dye

To safely estimate the dye concentration in the distilled water, the absorbance intensities of the utilized methylene dye have been measured for many dye/distilled water solutions. The UV absorbance spectra for different concentrations of dye solutions (starting from 1 to 8.75 mg/l) were measured within the range of 500–800 nm. Figure 7a reveals the obtained results. As shown in this figure, the absorbance curves have maximum value at almost 664 nm. Moreover, the maximum measured absorbance intensities were linearly increased with increasing of the dye concentration as shown in Fig. 7b which represents the relationship between the dye concentration and the measured absorbance at 664 nm. As shown in this figure, the absorbance varies linearly with the dye concentration in good model. Statistical analyses of this curve indicated high accuracy of the exploited linear model since the coefficient of determination; R^2 of this model was 0.9989 which reveals excellent precision and reproducibility of this calibration curve.

Theoretically, as mentioned in the introduction section some studies have explained the role of silver in enhancing the photocatalytic activity of titania. Generally, these studies have drawn this conclusion. The Ag nanoparticles

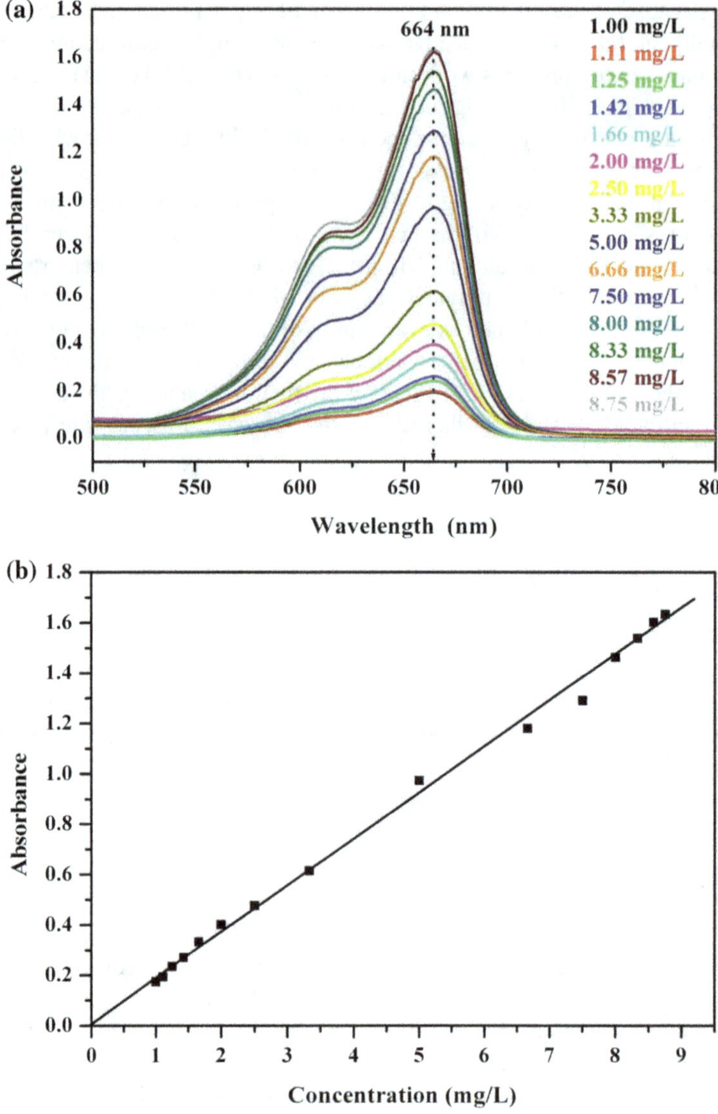

Fig. 7 Absorbance spectra at different concentrations of methylene blue dye solutions within a range of 500–800 nm (**a**), and the relationship between the absorbance intensity and concentration of the dye at wavelength of 664 nm (**b**)

deposited on TiO_2 surface act as electron acceptors, enhancing the charge separation of electrons and holes and consequently the transfer of the trapped electron to the adsorbed O_2 acting as an electron acceptor. The sufficient dye molecules are adsorbed on the surface of Ag-TiO_2 than on the TiO_2 surface, increasing the photoexcited electron transfer from the sensitised dye molecule to the conduction band of TiO_2 and consequently increasing the electron transfer to the adsorbed O_2.

Experimentally, the pure titania and silver-doped titania nanofibers have been invoked as photocatalysts in degradation of methylene blue dye, Fig. 8 shows the obtained results. As shown in this figure, incorporating of silver in titanium oxide nanofibers reveals to significant increase in the degradation rate of this dye. As can be observed in this figure, within 10 min more than 92 % of the dye has been degraded, moreover, the dye completely eliminated after almost two hours. This interesting result comes from exploiting the advantage of the nanofibrous shape as well as the benefit of incorporating of silver in titanium oxide. However, in case of pure titanium oxide nanofibers, almost 50 % of the dye has been oxidized after 30 min and the dye was almost disappeared from the solution after 6 h. Actually, even in case of pure titanium nanofibers, the obtained results are satisfactory compared with the other photocatalyst [63].

4.2.2 Methyl Red Dye

By the same aforementioned fashion, the wavelength corresponding to maximum absorbance intensity in case of methyl red dye has been estimated. Figure 9a shows the relationships between the absorbance intensity and the wavelength for many dye aqueous solutions (from 1 to 8.75 ppm) within wavelength range of 300–600 nm. As shown in the figure, the wavelength matching maximum

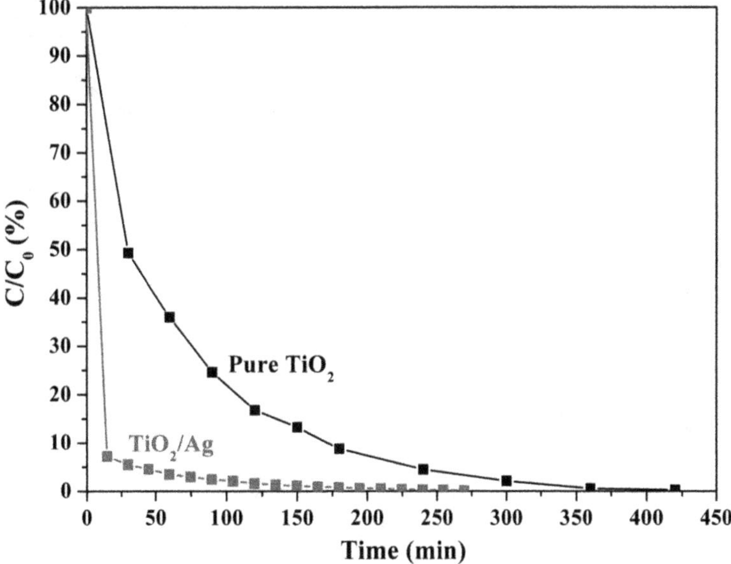

Fig. 8 Photocatalytic degradation of methylene blue dye in presence of pure TiO_2 and Ag-TiO_2 nanofibers

absorbance intensity for all dye concentrations is 428 nm. Moreover, there is very good linearity between the dye concentration and the absorbance intensity at this wavelength as shown in Fig. 9b. A single calibration model was adequate to represent all the data points in Fig. 9b with very low mathematical errors ($R^2 = 0.999$). Therefore, one can confidently apply in this calibration curve (Fig. 9b) to estimate the concentration of any unknown methyl red dye aqueous solution after measuring the absorbance intensity for this solution at 428 nm.

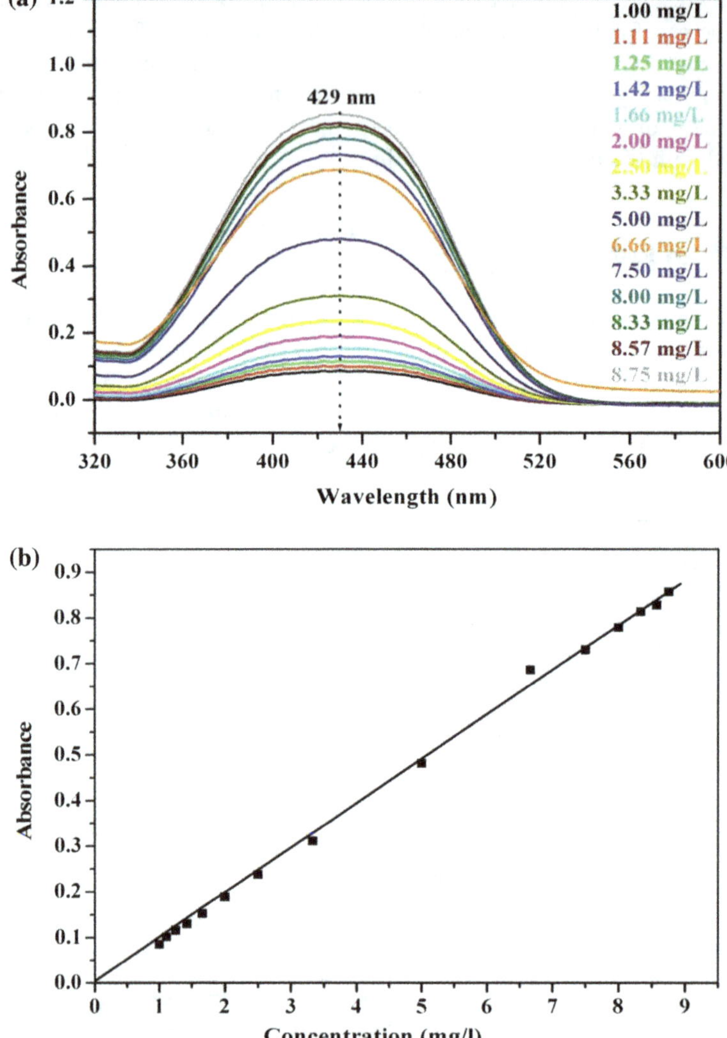

Fig. 9 Absorbance spectra at different concentrations of methyl red dye solutions within a range of 300–600 nm (**a**), and the relationship between the absorbance intensity and concentration of the dye at wavelength of 428 nm (**b**)

It is worth noting that methyl red decay is only due to photocatalysis as already published results showed practically no dye removal by only UV radiation within 300 min [42]. Accordingly, it is difficult to completely remove this dye within relatively short reaction time; e.g. several hours [42]. Figure 10 reveals the obtained results after utilizing the prepared TiO_2 and Ag/TiO_2 nanofibers as photocatalysts to oxidize the methyl red dye. As shown in this figure, the behavior of this dye is completely different than methylene blue, in contrast to that dye; the same removal percentage (i.e. 92 %) was obtained after almost 3 h. Actually, this result is also very satisfactory since according to our best knowledge, no one has reached to such decay percentage using any photocatalyst. Moreover, this result affirms the role of silver in enhancement of the behavior of titanium oxide as photocatalyst since the nanofibrous shape was not enough to eliminate this dye as shown in Fig. 10. Almost 50 % of the dye is remaining after 7 h, this decay efficiency is not rejected compared with the other catalysts [64].

4.3 Effect of Morphology

The effect of the fibrous morphology has been investigated by studying the rates of degradation of methylene blue dye using the two formulations as photocatalyst. The obtained results have strongly recommended utilizing this composite in nanofibrous form. Figure 11 represents the effect of nanostructure shape on the

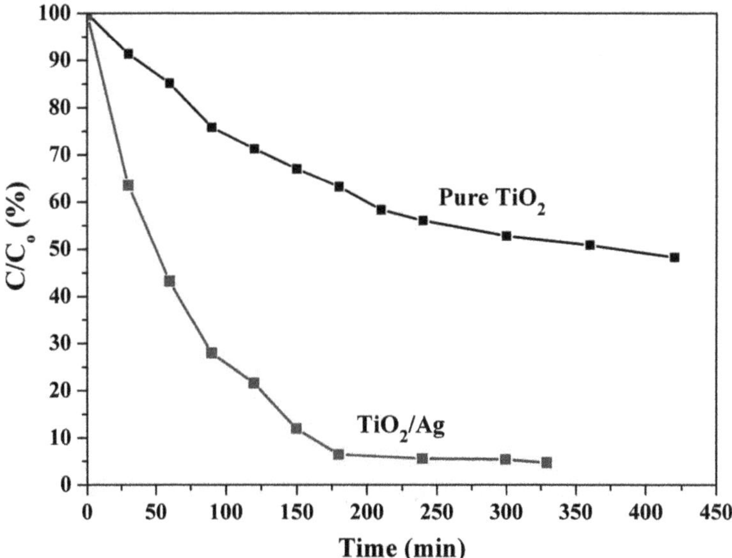

Fig. 10 Photocatalytic degradation of methyl red dye in presence of pure TiO_2 and Ag-TiO_2 nanofibers

photocatalytic activity of pure titanium oxide. As shown in this figure, the nano-fibrous morphology does have great effect. Titanium oxide nanoparticles could not catalyze the photocatalytic reaction to remove the dye even after almost 12 h, 65 % of the original dye was remaining in the solution after such long reaction time. However, the high surface area to volume ration which is the mean feature of nanofibers was the mean reason to get such interesting result compared with nanoparticles. As shown in Fig. 6, all the dye was oxidized and eliminated within 6 h.

Theoretically, some studies have explained the role of silver in enhancing the photocatalytic activity of titania. Generally, these studies have drawn this conclusion: The Ag nanoparticles deposited on TiO_2 surface act as electron acceptors, enhancing the charge separation of electrons and holes and consequently the transfer of the trapped electron to the adsorbed O_2 acting as an electron acceptor. The sufficient dye molecules are adsorbed on the surface of Ag-TiO_2 than on the TiO_2 surface, increasing the photoexcited electron transfer from the sensitized dye molecule to the conduction band of TiO_2 and consequently increasing the electron transfer to the adsorbed O_2.

Experimentally, the silver-doped titania nanofibers and nanoparticles have been invoked as photocatalyst in degradation of methylene blue dye, Fig. 12 shows the obtained results. As shown in this figure, incorporating of silver in titanium oxide nanofibers reveals to significant increase in the degradation rate of this dye. As can be observed in this figure, within 10 min more than 92 % of the dye has been

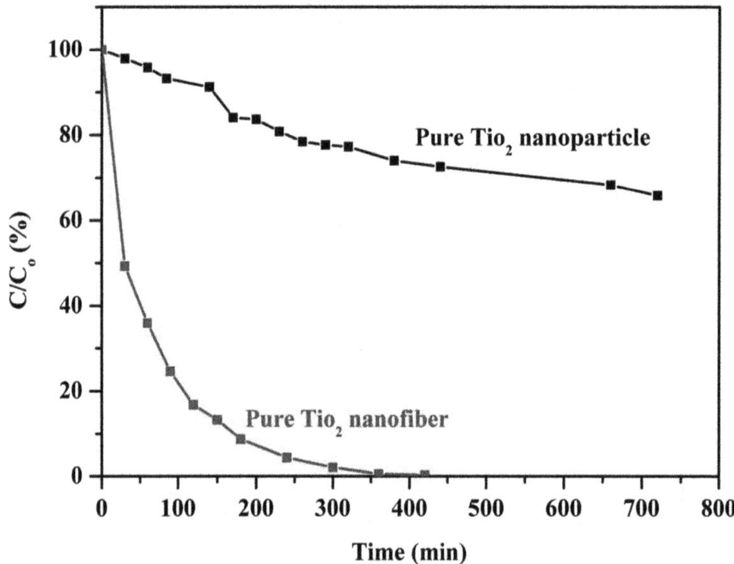

Fig. 11 Photocatalytic degradation of methylene blue dye in presence of pure TiO_2 nanoparticles

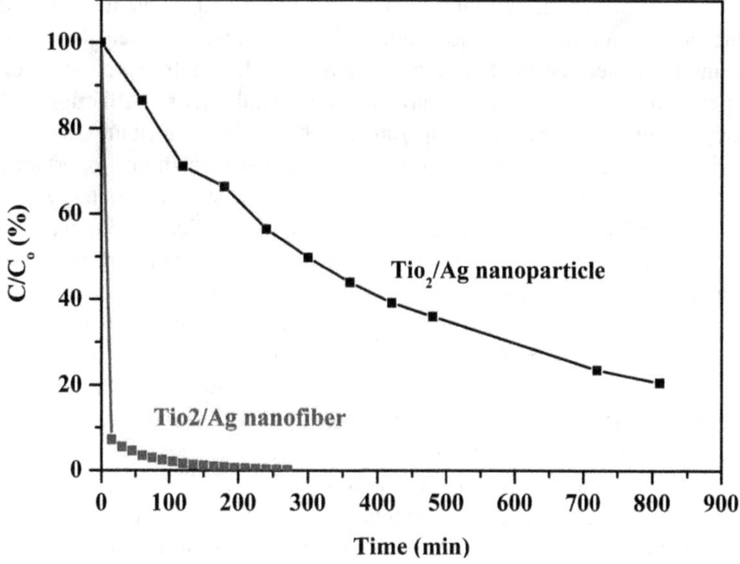

Fig. 12 Photocatalytic degradation of methylene blue dye in presence of Ag-doped TiO$_2$ nanofibers

degraded, moreover, the dye completely eliminated after almost two hours. This fantastic result comes from exploiting the advantage of the nanofibrous shape as well as the benefit of incorporating of silver in titanium oxide. However, in case of Ag-doped titanium oxide nanoparticles, almost 80 % of the dye has been oxidized after 12 h and the dye did not completely disappeared from the solution.

The results obtained from Figs. 11 and 12 support the pervious works reported the advantages of doping of titanium oxide by silver nanoparticles and simultaneously strongly recommend utilizing this interesting photocatalyst in nanofibrous shape.

4.4 Effect of Temperature

Photocatalysis is the acceleration of a photoreaction in the presence of a catalyst. In catalyzed photolysis, light is absorbed by an adsorbed substrate. In photogenerated catalysis, the photocatalytic activity depends on the ability of the catalyst to create electron–hole pairs, which generate free radicals (hydroxyl radicals: •OH) able to undergo secondary reactions. Its comprehension has been made possible ever since the discovery of water electrolysis by means of the titanium dioxide.

Figures 13, 14, 15, 16 and 17 represent the photodegrdation of rhodamine B using Ag-doped TiO$_2$ nanoparticles obtained from sol-gels contain 0.5, 1.0, 1.5, 2.0 and 2.5 wt% AgNO$_3$, respectively at different temperatures (5, 15, 25, 45 and

55 °C). As shown in these figures, increase the silver content in the doped nanoparticles enhances the photoactivity of the utilized photocatalyst. As it is known, increase the reaction temperature improves the photodegration efficiency. Molecules at a higher temperature have more thermal energy. Although collision frequency is greater at higher temperatures, this alone contributes only a very small proportion to the increase in rate of reaction. Much more important is the fact that the proportion of reactant molecules with sufficient energy to react (energy greater than activation energy: E > Ea) is significantly higher. Accordingly, the dye photodegradation modified linearly with temperature increase. The best temperature was found to be 55 °C, at this temperature the maximum decolorization efficiencies at the utilized reaction time (3 h) were 65, 74, 84, 86 and 100 % for the photocatalyst obtained from sol-gels contain 0.5, 1.0, 1.5, 2.0 and 2.5 wt% $AgNO_3$, respectively. It is noteworthy mentioning that the investigated photocatalyst has good performance only under the UV irradiation as normal light has no impact as shown in Fig. 17.

Table 1 shows the rate of the rhodamine B photodegradation (mg/min) at different temperatures and silver contents. As shown in the table, the maximum rate is not always obtained at the maximum temperature; it depends on the silver content. At the lowest silver content (i.e. 0.5 wt%), the highest rate was obtained at 15 °C. Increasing the silver content shifts the maximum rate to be at 25 °C for the samples obtained from sol-gels having 1.0 and 1.5 wt% silver nitrate. However, at the higher silver contents the best rate was obtained at 45 °C. Overall, the

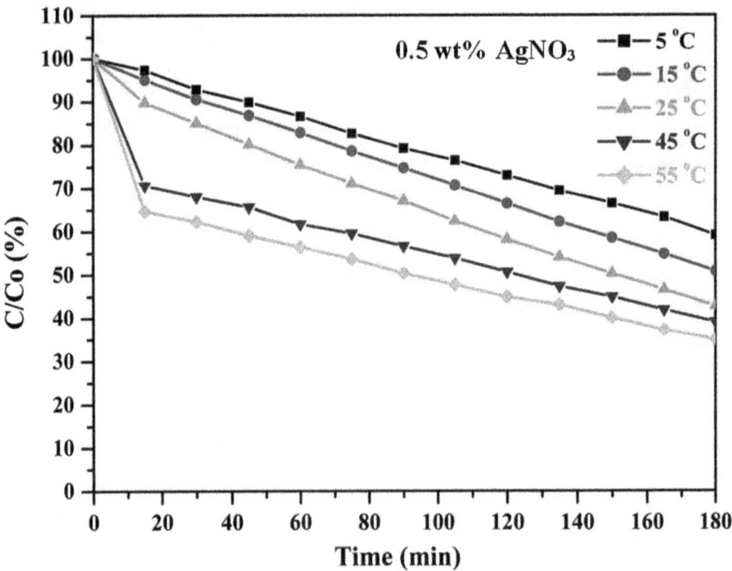

Fig. 13 Effect of temperature on the degradation rate of rhodamine B dye using Ag-doped TiO_2 nanoparticles having 0.5 wt% $AgNO_3$ under UV irradiation

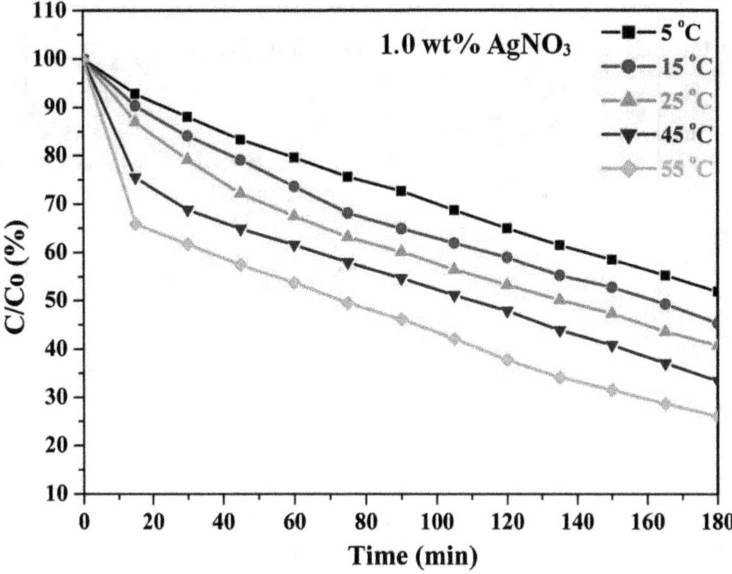

Fig. 14 Effect of temperature on the degradation rate of rhodamine B dye using Ag-doped TiO_2 nanoparticles having 1.0 wt% $AgNO_3$ under UV irradiation

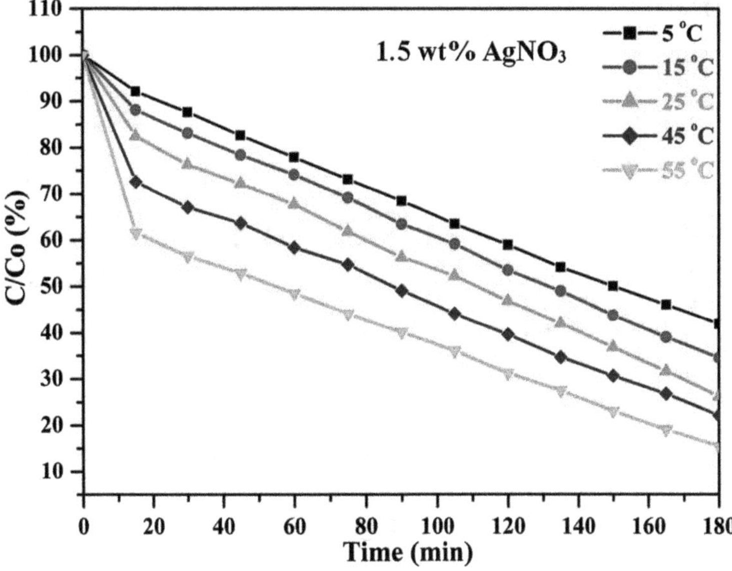

Fig. 15 Effect of temperature on the degradation rate of rhodamine B dye using Ag-doped TiO_2 nanoparticles having 1.5 wt% $AgNO_3$ under UV irradiation

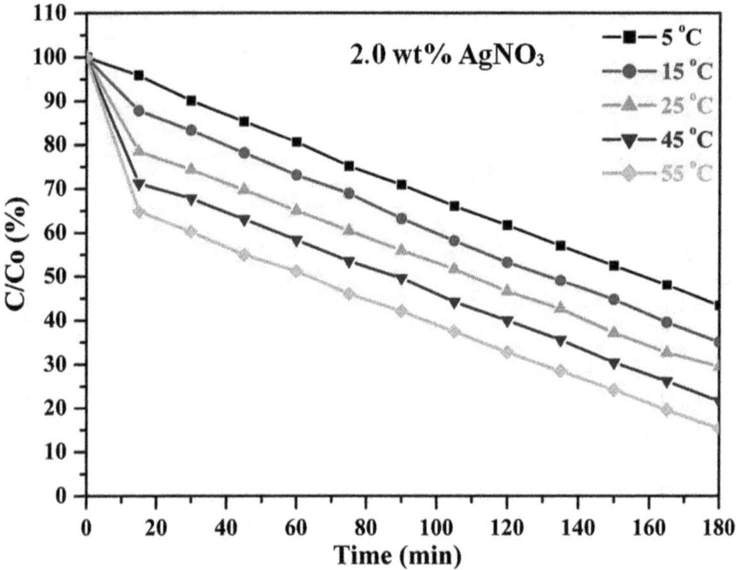

Fig. 16 Effect of temperature on the degradation rate of rhodamine B dye using Ag-doped TiO$_2$ nanoparticles having 2.0 wt% AgNO$_3$ under UV irradiation

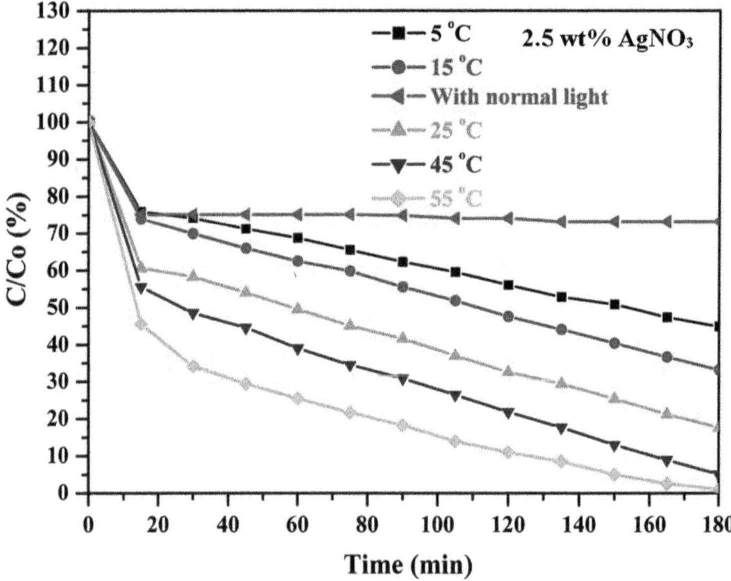

Fig. 17 Effect of temperature on the degradation rate of rhodamine B dye using Ag-doped TiO$_2$ nanoparticles having 2.5 wt% AgNO$_3$ under UV irradiation, and also at room temperature under the normal light

maximum degradation rate is corresponding to the samples obtained from the sol-gels having 1.0 and 1.5 wt% silver nitrate at the room temperature (0.0356 mg/min). As aforementioned, besides enhancing the activation energy, increase the temperature leads to increase the kinetic energy of the dyes molecules which modifies the collisions with the catalyst surface. If the catalyst surface is very active so numerous free radicals will be present in the thin film surrounding the catalyst nanoparticles, so fast cracking reactions of the dye molecules will take place, otherwise the molecules will escape without degradation. Therefore, for the low silver content NPs, the surface is not so active so relatively low temperature (15 °C) is preferred to provide suitable time for dye molecules to react with the available •OH radicals. Increase the temperature leads to escape the molecules. More increase in the silver content results in more activation of the NPs surfaces, so low contact time is enough to degrade the dye molecules in the thin layer surrounding the catalyst NPs. Accordingly, the maximum degradation rates were obtained at 45 °C for the samples having high silver contents (last two columns in Table 1). At 55 °C, the kinetic energy is high enough to liberate the dye molecules from the active zones before degradation so low degradation rate was obtained. However, the number of the degraded dye molecules is high in the beginning of the reaction (the first 10 min) which makes the quantitative degradation is more at 55 °C. For instance after 10 min, at 55 °C, the eliminated dyes were 35, 38 and 55 % for the nanoparticles obtained from sol-gels containing 0.5, 1.5, and 2.5 wt% silver nitrate, respectively. It is noteworthy to mention that the fast initial degradation in the first 10 min (Figs. 13, 14, 15, 16, 17) made the total elimination of the dyes linearly dependent on the temperature. However, after this initial declination, the photodegrdation rate becomes temperature independent as shown in Table 1.

In contrast to the nanoparticles, influence of temperature was different in case of Ag-doped TiO_2 nanofibers. Figures 18, 19, 20, 21 and 22 represent the photodegrdation of rhodamine B using Ag-doped TiO_2 nanofibers obtained from electrospun nanofiber mats contain 0.5, 1.0, 1.5, 2.0 and 2.5 wt% $AgNO_3$, respectively at different temperatures. As shown in these figures, on contrary to the nanoparticulate morphology, the photodegrdation efficiency was temperature-dependent only at low silver content (0.5 wt%, Fig. 18) as the best degradation was obtained at the highest temperature. Moreover, the amount of the eliminated dyes was

Table 1 Rate of rhodamine B dye removal (mg/min) using Ag-doped TiO_2 nanoparticles

Temperature (°C)	$AgNO_3$ content (wt%)				
	0.5	1.0	1.5	2.0	2.5
5	0.0250	0.0269	0.0313	0.0356	0.0200
15	**0.0313**	0.0288	0.0331	0.0325	0.0269
25	0.0294	**0.0356**	**0.0356**	0.0300	0.0269
45	0.0200	0.0263	0.0325	**0.0319**	**0.0300**
55	0.0181	0.0256	0.0294	0.0306	0.0250
Activation energy	3,625.32	1,885.29	1,989.84	1,800.2	315.337

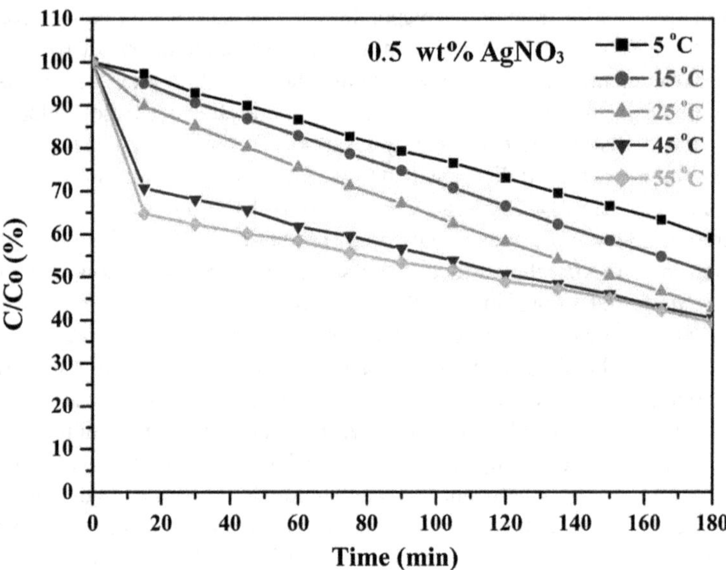

Fig. 18 Effect of temperature on the degradation rate of rhodamine B dye using Ag-doped TiO_2 nanofibers having 0.5 wt% $AgNO_3$ under UV irradiation

gradually and regularly increases with the temperature increase. For the nanofibers obtained from electrospun mats having 1.0 wt% silver nitrate, the optimum reaction temperature was 45 °C as shown in Fig. 19. However, for the remaining formulations the optimum temperature was 25 °C; Figs. 20, 21 and 22. For all formulations, 5 °C reaction temperature is the worst conditions as the lowest degradation was obtained at this temperature. Moreover, contrary to the nanoparticulate morphology, starting from silver nitrate content of 1.0 wt% in the electrospun nanofibers mats, the photodegradation cannot be mathematically correlated with the reaction temperatures as shown in Figs. 19, 20, 21 and 22. These obtained results can be explained by the higher activity of the nanofibrous Ag-doped TiO_2 compared to the nanoparticulate morphology especially at high silver contents. Beside the desorption process and kinetic energy, it is possible also that the temperature affects the e–h pairs transfer and lifetime.

Figure 23 represents a conceptual illustration to explain the influence of the temperature on the photodegradation of the rhodamine B molecules using Ag-doped TiO_2 nanoparticles and nanofibers. In case of nanoparticles, the surface activity is not high so increase the temperature has the normal impact on the chemical reactions and enhances the degradation process in general by modifying the activation energy. However, in case of the nanofibers having high silver content, the surface is very active so instant degradation for the dye molecules takes place at the thin film surrounding the nanofibers. Therefore, increase the temperature which leads to increase the molecules kinetic energy has positive impact in the beginning as it helps to move the molecules to the active zones.

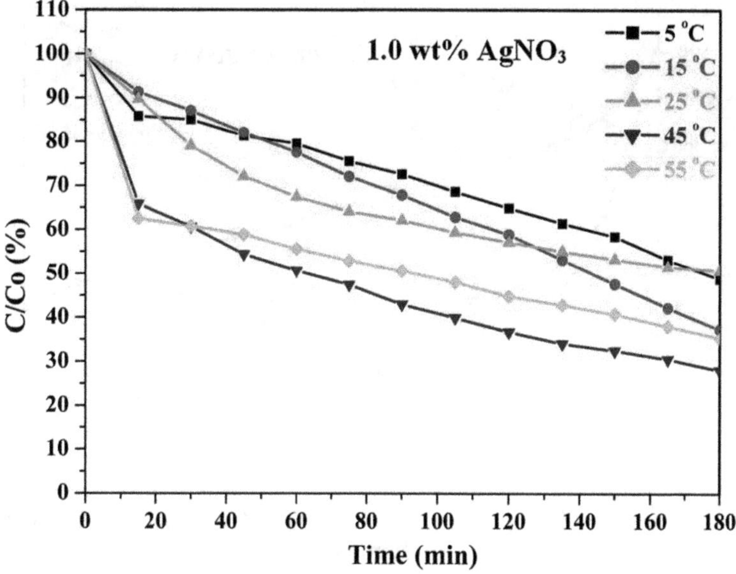

Fig. 19 Effect of temperature on the degradation rate of rhodamine B dye using Ag-doped TiO_2 nanofibers having 1.0 wt% $AgNO_3$ under UV irradiation

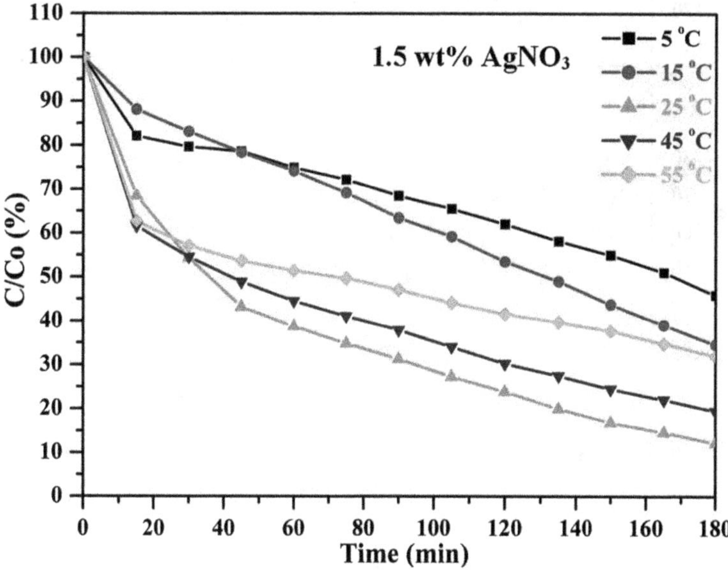

Fig. 20 Effect of temperature on the degradation rate of rhodamine B dye using Ag-doped TiO_2 nanofibers having 1.5 wt% $AgNO_3$ under UV irradiation

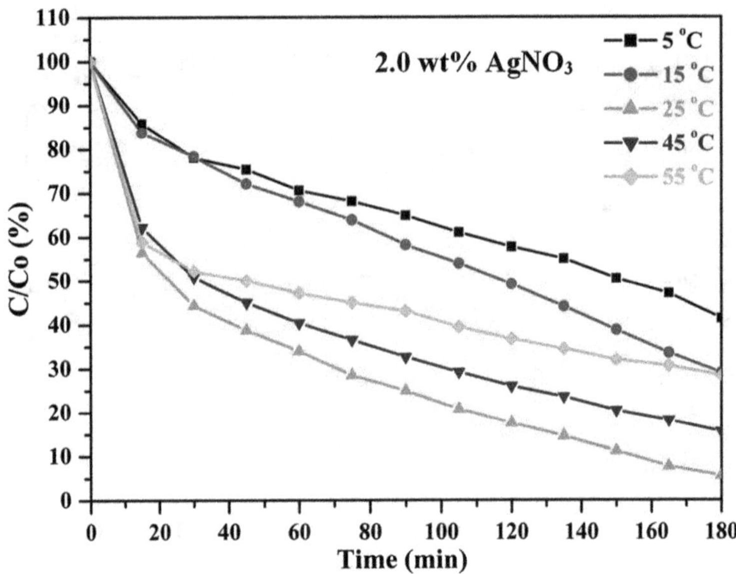

Fig. 21 Effect of temperature on the degradation rate of rhodamine B dye using Ag-doped TiO_2 nanofibers having 2.0 wt% $AgNO_3$ under UV irradiation

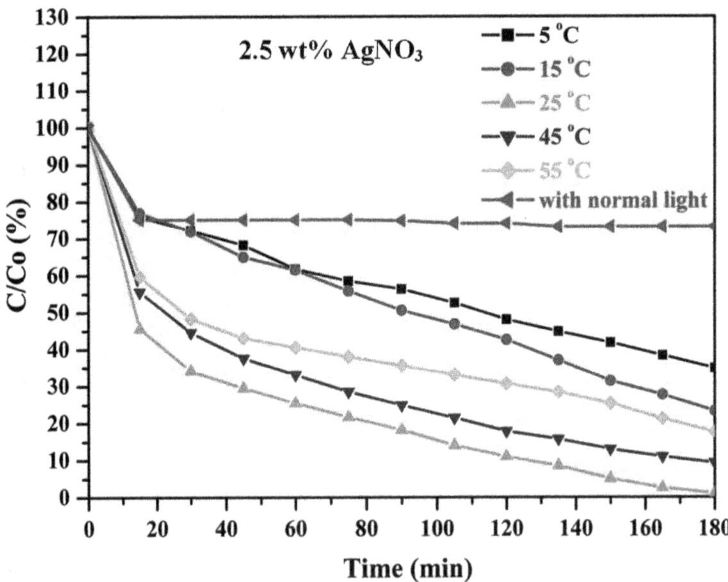

Fig. 22 Effect of temperature on the degradation rate of rhodamine B dye using Ag-doped TiO_2 nanofibers having 2.5 wt% $AgNO_3$ under UV irradiation, and also at room temperature under the normal light

However, more increase in the temperature leads to produce kinetic energy able to move away the dye molecules from the active zones before achieving the degradation process. Therefore, for the nanofibers obtained from electrospun mats having silver nitrate more than 1.5 wt% the optimum temperature is 25 °C followed by 45 °C and then 55 °C which means inverse relationship between the photodegradation and the temperature due to increase the kinetic energy of the molecules. In summary, in case of nanofibers, the temperature affects the photo-degradation process from the kinetic energy point of view. However, in case of nanoparticles, beside the kinetic energy, temperature can enhance the activation energy which improves the photodegradation process.

Based on the introduced hypothesis; the initial rate should show a positive temperature effect over the nanofibers with high silver content. However, some results (Fig. 13) do not match this expected behavior because the experimental plan was established to detect the first concentration after 10 min. Because the surface is very active, it is believed that if the first samples, in case of nanofibers, were drawn after very short time (e.g. 2 min) the initial rate will be found temperature dependent.

Figure 24 displays the photodegradation of the methylene blue dye using Ag-doped nanoparticles; (A) and nanofibers; (B) at the optimum temperatures (i.e. 55 and 25 °C, respectively). As shown in the figure, the nanofibers have distinct performance compared with the nanoparticles at all silver contents. As shown in

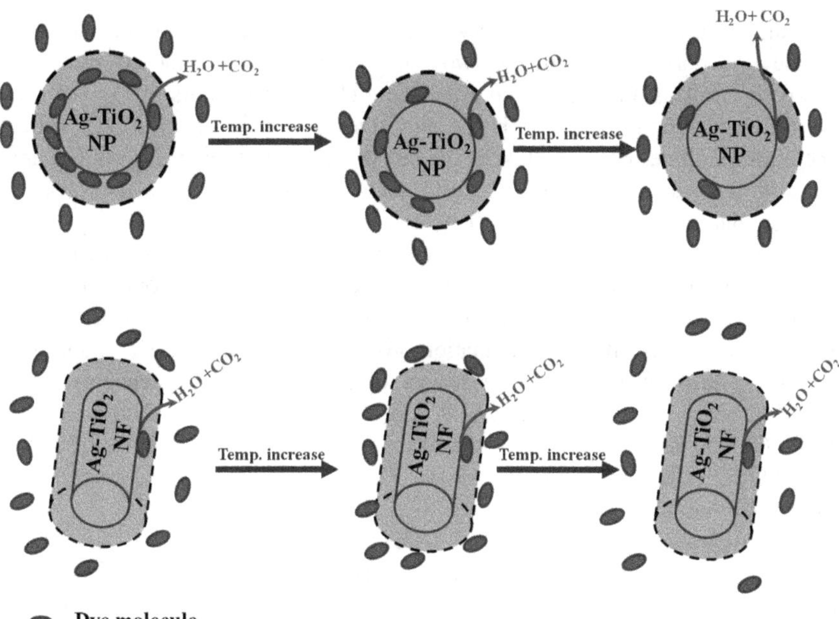

Fig. 23 Schematic diagram illustrates the influence of the temperature on the photodegradation process in case of using the Ag-doped TiO_2 nanoparticles and nanofibers

the figure, 92 % from the dye was degraded in 10 min when Ag-doped TiO_2 nanofibers obtained from electrospun mats containing 2.0 wt% silver nitrate were used as photocatalyst. As aforementioned in the introduction section, silver-doping is a known strategy to enhance the photocatalytic activity of the titanium oxide nanostructure. Therefore, the photocatalytic activity of the both utilized formulations was strongly enhanced upon silver addition as shown in all the obtained results.

The results obtained in Fig. 24 supports the aforementioned hypothesis about the strong activity of the nanofibers surfaces compared to the nanoparticles. Therefore, for the same silver content, distinct difference in the photodegradation performance was obtained as shown in all results. A proposed explanation for this finding can be as follow: as the photodegradation process mainly bases on electrons transfer through the photocatalyst so the structure providing high surface to volume ratio is expected to have better performance because it supplies good electron mobility. Mathematically, it is known that the particulate shape (sphere) has the lowest surface to volume ratio. Therefore, the nanofibrous shape provides more space for the electrons which reflects higher activity compared to the nanoparticles.

4.5 Influence of Silver Doping on the TiO_2 Crystal Structure

Titanium oxide has three popular crystal structures; anatase, rutile and brookite. Anatase is a polymorph with the two other minerals. The minerals rutile and brookite as well as anatase all have the same chemical formula, TiO_2, but they have different structures. Rutile is the more common and the more well known mineral of the three, while anatase is the rarest. Anatase shares many of the same or nearly the same properties as rutile such as luster, hardness and density. However, due to structural differences, anatase and rutile differ slightly in the crystal habit. The phase change from anatase polymorph of titania to rutile one has been the subject of considerable interest and the focus of many groups' activities over the years. Figure 1 shows the effect of silver-doping on the crystal structure of the obtained powder after the calcination process. In case of silver-free solution, the results affirm formation of pure anatase titanium dioxide, existence of strong diffraction peaks at 2θ values of 25.09°, 37.65°, 38.44°, 47.89°, 53.89°, 55.07°, 62.40°, 68.70°, 70.04° and 75.00° corresponding to the crystal planes of (101), (004), (112), (200), (105), (211), (204), (220), (220) and (215), respectively indicates formation of anatase titanium dioxide [JCPDS card no 21-1272]. Addition of small amount of silver nitrate (2 wt%) did not affect the crystal structure of the titania, however pure silver metal was formed due to decomposition of the silver nitrate upon heating. The extra peaks at 2θ values of 38.11°, 44.29°, 64.43° and 77.48°, corresponding to the crystal planes of (111), (200), (220) and (311), respectively affirm presence of silver metal [JCPDS card no 04-0783]. In Fig. 25, the main peaks of anatase and silver were marked as A and S, respectively.

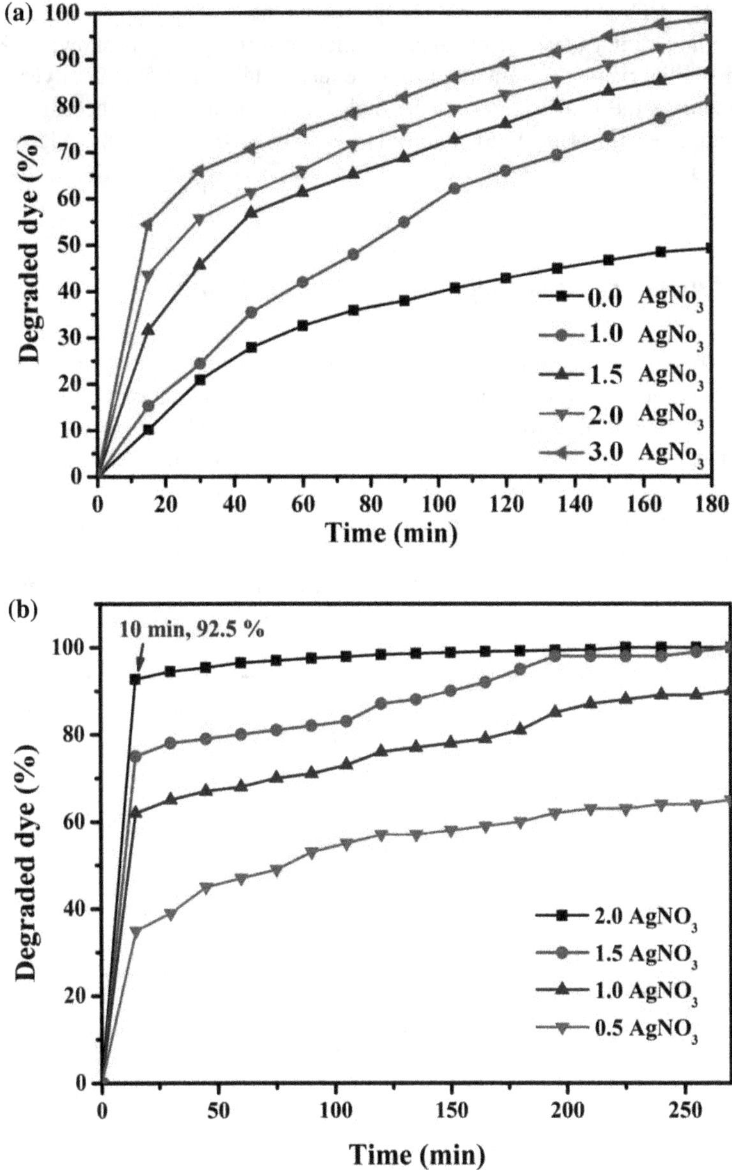

Fig. 24 Photodegradation of methylene blue dye using different Ag-doped TiO_2 in nanoparticulate; **a** and nanofibrous; **b** at 55 and 25 °C, respectively

Increase the silver nitrate content to 3 wt% in the original electrospun solution led to partial phase change in the titania crystal structure. The standard peaks of rutile can be observed at 2θ values of 27.45°, 36.09°, 41.23°, 54.32°, 56.64°, and 69.08° corresponding to the crystal planes of (110), (101), (111), (211), (220), and (301), respectively [JCPDS card no 21-1272]. Interestingly, the results indicate that increase the amount of silver enhances formation rutile phase as shown in Fig. 1 (the main rutile peaks were marked by R letter). Addition of 5 wt% silver nitrate led to increase the rutile content to be higher than the anatase. According to the XRD data, the ratio of rutile to anatase is ~ 75 and 440 % when the added silver nitrate was 3 and 5 wt%, respectively.

4.6 Influence of Silver Doping on the Morphology

The electrospinning technique involves the use of a high voltage to charge the surface of a polymer solution droplet and thus to induce the ejection of a liquid jet through a spinneret. Due to bending instability, the jet is subsequently stretched by many times to form continuous, ultrathin fibers. It is widely used for production of many polymeric nanofibers. Moreover, the electrospinning process has been exploited to produce metal oxides nanofibers by calcination of electrospun mats obtained from completely miscible sol–gel solutions. Electrospinning of a sol–gel composed of Ti(IsO) and PVAc/DMF solution has been carried out in our lab in

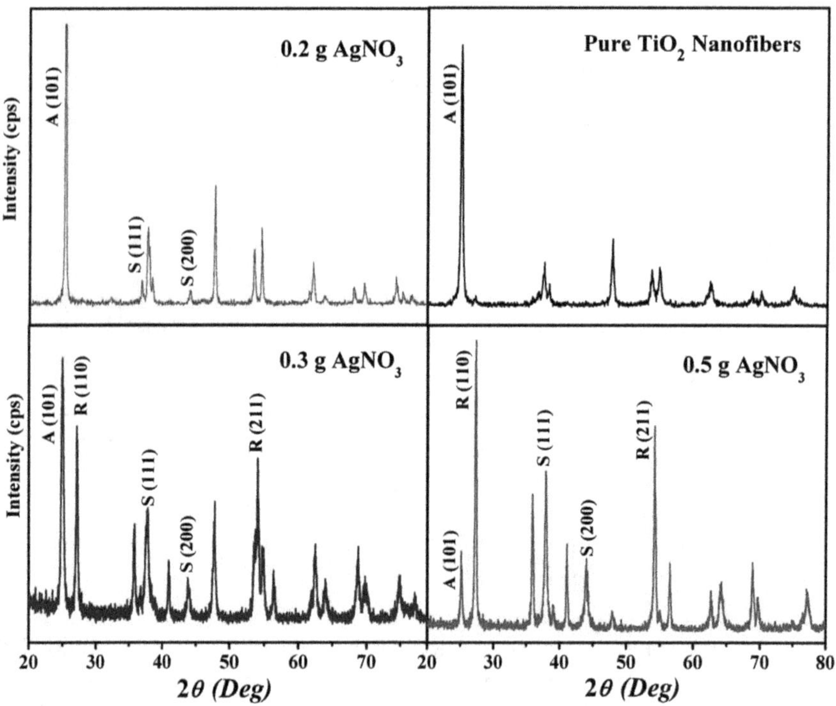

Fig. 25 Effect of silver content on the crystal structure of TiO$_2$

Fig. 26 Effect of silver content on the nanofibrous morphology of TiO$_2$

previous study [62], the resultant electrospun nanofibers have good morphology. Addition of silver nitrate to the sol–gel does not affect the morphology [5]. Accordingly, well morphology electrospun nanofibers mats were obtained from AgNO$_3$/Ti(Iso)/PVAc solutions at all the utilized AgNO$_3$ contents (data are not shown). Figure 26 shows the morphology of the resultant powder after the calcination process. As can be observed in Fig. 26a and b which demonstrate the obtained product from silver-free electrospun solution, well and smooth TiO$_2$ nanofibers were obtained, these nanofibers composed of pure anatase as could be concluded from the XRD results (Fig. 25). Incorporation of small amount of silver does not have considerable impact on the nanofibrous morphology as shown in Fig. 26c and d which represent the FE-SEM images of the powder obtained from calcination of an electrospun solution containing 2 wt% AgNO$_3$. However, the surface became little rough (Fig. 26d).

5 Conclusion

Ag-doped TiO$_2$ nanofibers can be prepared by calcination of electrospun nanofiber mats obtained from electrospinning of titanium isopropoxide, silver nitrate and poly(vinyl acetate) sol-gels. Also, drying, grinding and calcination of the same sol-gels lead to produce Ag-doped TiO$_2$ nanoparticles. The obtained nanostructure is having both of antase and rutile phases if the silver nitrate content is more than 3 wt%. Rutile phase content can be increased by increasing the silver content in the original electrospun solution. However, the nanofibrous morphology is strongly affected by the silver-content, excess silver causes to destroy the 1D structure. Silver-doping of the titanium oxide nanofibers greatly enhances the photocatalytic activity of this interesting material. The temperature has positive impact on the photodegradation when the photocatalyst is exploited in the form of nanoparticles due to modification of the kinetic and the activation energies. However, in case of the nanofibrous morphology, the temperature has negative impact due to the super activity of the surface compared to the nanoparticles. Increase the temperature leads to enhance the kinetic energy of the dyes molecules which leads to escape the molecules from the active zones surrounding the nanofibers. Overall, this study strongly recommends utilizing using the photocatalyst in the form of nanofibers at room temperature as the maximum degradation can be obtained at these conditions.

Acknowledgments This work was financially supported by the Ministry of Education, Science Technology (MEST) and National Research Foundation of Korea (NRF) through the Human Resource Training Project for Regional Innovation and "Leaders in Industry-University Cooperation". We thank Mr. T. S. Bae and J. C. Lim, KBSI, Jeonju branch, and Mr. Jong- Gyun Kang, Centre for University Research Facility, for taking high-quality FESEM and TEM images, respectively.

References

1. Karunakaran, C., Abiramasundari, G., Gomathisankar, P., Manikandan, G., Anandi, V.: Cu-doped TiO2 nanoparticles for photocatalytic disinfection of bacteria under visible light. J. Colloid Interface Sci. **352**, 68–74 (2010)
2. Ou, H.-H., Lo, S.-L.: Effect of Pt/Pd-doped TiO2 on the photocatalytic degradation of trichloroethylene. J. Mol. Catal. A: Chem. **275**, 200–205 (2007)
3. Vorontsov, A., Stoyanova, I., Kozlov, D., Simagina, V., Savinov, E.: Kinetics of the photocatalytic oxidation of gaseous acetone over platinized titanium dioxide. J. Catal. **189**, 360–369 (2000)
4. Li, F., Li, X.: The enhancement of photodegradation efficiency using Pt-TiO2 catalyst. Chemosphere **48**, 1103–1111 (2002)
5. Kanjwal, M.A., Barakat, N.A.M., Sheikh, F.A., Khil, M.S., Kim, H.Y.: Functionalization of electrospun titanium oxide nanofibers with silver nanoparticles: strongly effective photocatalyst. Int. J. Appl. Ceram. Technol. **7**, E54–E63 (2010)
6. Li, X., Li, F.: Study of Au/Au^{3+}−TiO$_2$ photocatalysts toward visible photooxidation for water and wastewater treatment. Environ. Sci. Technol. **35**, 2381–2387 (2001)
7. Tian, Y., Tatsuma, T.: Mechanisms and applications of plasmon-induced charge separation at TiO$_2$ films loaded with gold nanoparticles. JACS **127**, 7632–7637 (2005)
8. Haynes, C.L., Van Duyne, R.P.: Plasmon-sampled surface-enhanced Raman excitation spectroscopy. J. Phys. Chem. B **107**, 7426–7433 (2003)
9. Hodak, J.H., Martini, I., Hartland, G.V.: Spectroscopy and dynamics of nanometer-sized noble metal particles. J. Phys. Chem. B **102**, 6958–6967 (1998)
10. Zhao, G., Kozuka, H., Yoko, T.: Sol—gel preparation and photoelectrochemical properties of TiO$_2$ films containing Au and Ag metal particles. Thin Solid Films **277**, 147–154 (1996)
11. Sung-Suh, H.M., Choi, J.R., Hah, H.J., Koo, S.M., Bae, Y.C.: Comparison of Ag deposition effects on the photocatalytic activity of nanoparticulate TiO$_2$ under visible and UV light irradiation. J. Photochem. Photobiol., A **163**, 37–44 (2004)
12. Wu, T., Liu, G., Zhao, J., Hidaka, H., Serpone, N.: Photoassisted degradation of dye pollutants. V. Self-photosensitized oxidative transformation of rhodamine B under visible light irradiation in aqueous TiO$_2$ dispersions. J. Phys. Chem. B **102**, 5845–5851 (1998)

13. Sobana, N., Muruganadham, M., Swaminathan, M.: Nano-Ag particles doped TiO_2 for efficient photodegradation of Direct azo dyes. J. Mol. Catal. A: Chem. **258**, 124–132 (2006)
14. Arabatzis, I., Stergiopoulos, T., Bernard, M., Labou, D., Neophytides, S., Falaras, P.: Silver-modified titanium dioxide thin films for efficient photodegradation of methyl orange. Appl. Catal., B **42**, 187–201 (2003)
15. Herrmann, J.M., Tahiri, H., Ait-Ichou, Y., Lassaletta, G., Gonzalez-Elipe, A., Fernandez, A.: Characterization and photocatalytic activity in aqueous medium of TiO_2 and Ag-TiO_2 coatings on quartz. Appl. Catal., B **13**, 219–228 (1997)
16. Damm, C., Israel, G.: Photoelectric properties and photocatalytic activity of silver-coated titanium dioxides. Dyes Pigm. **75**, 612–618 (2007)
17. Shie, J.L., Lee, C.H., Chiou, C.S., Chang, C.T., Chang, C.C., Chang, C.Y.: Photodegradation kinetics of formaldehyde using light sources of UVA, UVC and UVLED in the presence of composed silver titanium oxide photocatalyst. J. Hazard. Mater. **155**, 164–172 (2008)
18. He, X., Zhao, X., Liu, B.: The synthesis and kinetic growth of anisotropic silver particles loaded on TiO_2 surface by photoelectrochemical reduction method. Appl. Surf. Sci. **254**, 1705–1709 (2008)
19. Barakat, N.A.M., Abadir, M.F., Nam, K.T., Hamza, A.M., Al-Deyab, S.S., Al-Deyab, S.S., Baek, W.-i., Kim, H.Y.: Synthesis and film formation of iron-cobalt nanofibers encapsulated in graphite shell: magnetic, electric and optical properties study. J. Mater. Chem. **21**, 10957–10964 (2011)
20. Barakat, N.A.M., Khil, M.S., Sheikh, F.A., Kim, H.Y.: Synthesis and optical properties of two cobalt oxides (CoO and Co_3O_4) nanofibers produced by electrospinning process. J. Phys. Chem. C **112**, 12225–12233 (2008)
21. Kanjwal, M., Barakat, N., Sheikh, F., Baek, W.-i., Khil, M., Kim, H.: Effects of silver content and morphology on the catalytic activity of silver-grafted titanium oxide nanostructure. Fibers Polym. **11**, 700–709 (2010)
22. Arbiol, J., Cerda, J., Dezanneau, G., Cirera, A., Peiro, F., Cornet, A., Morante, J.: Effects of Nb doping on the TiO_2 anatase-to-rutile phase transition. J. Appl. Phys. **92**, 853–861 (2002)
23. Li, X., Wang, H., Wu, H.: Phthalocyanines and their analogs applied in dye-sensitized solar cell. Funct Phthalocyanine Mol. Mater. 229–273 (2010)
24. Hegde, R.R., Dahiya, A., Kamath, M.: Nanofiber nonwovens, June (2005)
25. Chowdhury, M.M.R.: Electro spinning process nano fiber and their application
26. Wu, H., Lin, D., Zhang, R., Pan, W.: Facile synthesis and assembly of Ag/NiO nanofibers with high electrical conductivity. Chem. Mater. **19**, 1895–1897 (2007)
27. Doshi, J., Reneker, D.H.: Electrospinning process and applications of electrospun fibers. J. Electrostat. **35**, 151–160 (1995)
28. Shin, Y., Hohman, M., Brenner, M., Rutledge, G.: Experimental characterization of electrospinning: the electrically forced jet and instabilities. Polymer **42**, 09955–09967 (2001)
29. Han, T., Yarin, A.L., Reneker, D.H.: Viscoelastic electrospun jets: Initial stresses and elongational rheometry. Polymer **49**, 1651–1658 (2008)
30. Xia, Y., Yang, P., Sun, Y., Wu, Y., Mayers, B., Gates, B., Yin, Y., Kim, F., Yan, H.: One-Dimensional Nanostructures: synthesis, characterization, and applications. Adv. Mater. **15**, 353–389 (2003)
31. Law, M., Goldberger, J., Yang, P.: Semiconductor nanowires and nanotubes. Ann. Rev. Mater. Res. **34**, 83–122 (2004)
32. Sun, Y., Khang, D.Y., Hua, F., Hurley, K., Nuzzo, R.G., Rogers, J.A.: Photolithographic route to the fabrication of micro/nanowires of III–V semiconductors. Adv. Funct. Mater. **15**, 30–40 (2004)
33. Li, D., Xia, Y.: Electrospinning of nanofibers: reinventing the wheel? Adv. Mater. **16**, 1151–1170 (2004)
34. Reneker, D.H., Chun, I.: Nanometre diameter fibres of polymer, produced by electrospinning. Nanotechnology **7**, 216 (1999)
35. Chronakis, I.S.: Novel nanocomposites and nanoceramics based on polymer nanofibers using electrospinning process—a review. J. Mater. Process. Technol. **167**, 283–293 (2005)

36. Sheikh, F.A., Barakat, N.A.M., Kanjwal, M.A., Park, S.J., Park, D.K., Kim, H.Y.: Synthesis of poly (vinyl alcohol)(PVA) nanofibers incorporating hydroxyapatite nanoparticles as future implant materials. Macromol. Res. **18**, 59–66 (2010)
37. Kc, R.B., Kim, C.K., Khil, M.S., Kim, H.Y., Kim, I.S.: Synthesis of hydroxyapatite crystals using titanium oxide electrospun nanofibers. Mater. Sci. Eng., C **28**, 70–74 (2008)
38. Aoi, K., Aoi, H., Okada, M.: Synthesis of a poly (vinyl alcohol)-based graft copolymer having poly (ε-caprolactone) side chains by solution polymerization. Macromol. Chem. Phys. **203**, 1018–1028 (2002)
39. Kim, C.H., Khil, M.S., Kim, H.Y., Lee, H.U., Jahng, K.Y.: An improved hydrophilicity via electrospinning for enhanced cell attachment and proliferation. J. Biomed. Mater. Res. B Appl. Biomater. **78**, 283–290 (2006)
40. Tang, Z., Wei, J., Yung, L., Ji, B., Ma, H., Qiu, C., Yoon, K., Wan, F., Fang, D., Hsiao, B.S.: UV-cured poly (vinyl alcohol) ultrafiltration nanofibrous membrane based on electrospun nanofiber scaffolds. J. Membr. Sci. **328**, 1–5 (2009)
41. Chuang, W.Y., Young, T.H., Yao, C.H., Chiu, W.Y.: Properties of the poly (vinyl alcohol)/ chitosan blend and its effect on the culture of fibroblast in vitro. Biomaterials **20**, 1479–1487 (1999)
42. Lai, Y., Sun, L., Chen, C., Nie, C., Zuo, J., Lin, C.: Optical and electrical characterization of TiO_2 nanotube arrays on titanium substrate. Appl. Surf. Sci. **252**, 1101–1106 (2005)
43. Lai, Y., Chen, Y., Zhuang, H., Lin, C.: A facile method for synthesis of Ag/TiO_2 nanostructures. Mater. Lett. **62**, 3688–3690 (2008)
44. Xu, M.W., Bao, S.J., Zhang, X.G.: Enhanced photocatalytic activity of magnetic TiO_2 photocatalyst by silver deposition. Mater. Lett. **59**, 2194–2198 (2005)
45. Hufschmidt, D., Bahnemann, D., Testa, J.J., Emilio, C.A., Litter, M.I.: Enhancement of the photocatalytic activity of various TiO_2 materials by platinisation. J. Photochem. Photobiol., A **148**, 223–231 (2002)
46. Nikolajsen, T., Leosson, K., Bozhevolnyi, S.I.: Surface plasmon polariton based modulators and switches operating at telecom wavelengths. Appl. Phys. Lett. **85**, 5833–5835 (2004)
47. Huang, P., Wu, F., Zhu, B., Gao, X., Zhu, H., Yan, T., Huang, W., Wu, S., Song, D.: CeO_2 nanorods and gold nanocrystals supported on CeO_2 nanorods as catalyst. J. Phys. Chem. B **109**, 19169–19174 (2005)
48. Zhang, L., Yu, J.C.: A simple approach to reactivate silver-coated titanium dioxide photocatalyst. Catal. Commun. **6**, 684–687 (2005)
49. Herrmann, J.M., Disdier, J., Pichat, P.: Photoassisted platinum deposition on TiO_2 powder using various platinum complexes. J. Phys. Chem. **90**, 6028–6034 (1986)
50. Mulvaney, P., Giersig, M., Henglein, A.: Electrochemistry of multilayer colloids: preparation and absorption spectrum of gold-coated silver particles. J. Phys. Chem. **97**, 7061–7064 (1993)
51. Herrmann, J.M.: Termodynamic considerations of strong metal-support interaction in a real $PtTiO_2$ catalyst. J. Catal. **118**, 43–52 (1989)
52. Herrmann, J.M., Disdier, J., Pichat, P.: Effect of chromium doping on the electrical and catalytic properties of powder Titania under UV and visible illumination. Chem. Phys. Lett. **108**, 618–622 (1984)
53. Cozzoli, P.D., Comparelli, R., Fanizza, E., Curri, M.L., Agostiano, A., Laub, D.: Photocatalytic synthesis of silver nanoparticles stabilized by TiO_2 nanorods: a semiconductor/metal nanocomposite in homogeneous nonpolar solution. JACS **126**, 3868–3879 (2004)
54. Tian, R., Wang, X., Li, M., Hu, H., Chen, R., Liu, F., Zheng, H., Wan, L.: An efficient route to functionalize singe-walled carbon nanotubes using alcohols. Appl. Surf. Sci. **255**, 3294–3299 (2008)
55. Mai, L., Wang, D., Zhang, S., Xie, Y., Huang, C., Zhang, Z.: Synthesis and bactericidal ability of Ag/TiO_2 composite films deposited on titanium plate. Appl. Surf. Sci. **257**, 974–978 (2010)
56. Marques, H., Canário, A., Moutinho, A., Teodoro, O.: Work function changes in the Ag deposition on TiO_2 (110). Vacuum **82**, 1425–1427 (2008)

57. Barakat, N.A.M., Kanjwal, M.A., Al-Deyab, S.S., Chronakis, I.S., Kim, H.Y.: Influences of silver-doping on the crystal structure, morphology and photocatalytic activity of TiO_2 nanofibers. Mater. Sci. Appl. **2** (2011)
58. Barakat, N.A.M., Kim, B., Park, S.J., Jo, Y., Jung, M.-H., Kim, H.Y.: Cobalt nanofibers encapsulated in a graphite shell by an electrospinning process. J. Mater. Chem. **19**, 7371–7378 (2009)
59. Barakat, N.A.M., Woo, K.-D., Kanjwal, M.A., Choi, K.E., Khil, M.S., Kim, H.Y.: Surface plasmon resonances, optical properties, and electrical conductivity thermal hystersis of silver nanofibers produced by the electrospinning technique. Langmuir **24**, 11982–11987 (2008)
60. Barakat, N.A.M., Hamza, A., Al-Deyab, S.S., Qurashi, A., Kim, H.Y.: Titanium-based polymeric electrospun nanofiber mats as a novel organic semiconductor. Mater. Sci. Eng., B (2011)
61. Barakat, N.A.M., Shaheer Akhtar, M., Yousef, A., El-Newehy, M., Kim, H.Y.: Pd-Co-doped carbon nanofibers with photoactivity as effective counter electrodes for DSSCs. Chem. Eng. J. (2012)
62. Ding, B., Kim, C.K., Kim, H.Y., Seo, M.K., Park, S.J.: Titanium dioxide nanofibers prepared by using electrospinning method. Fibers Polym. **5**, 105–109 (2004)
63. Xiao, Q., Zhang, J., Xiao, C., Tan, X.: Photocatalytic decolorization of methylene blue over $Zn_{1-x}Co_xO$ under visible light irradiation. Mater. Sci. Eng., B **142**, 121–125 (2007)
64. Mascolo, G., Comparelli, R., Curri, M., Lovecchio, G., Lopez, A., Agostiano, A.: Photocatalytic degradation of methyl red by TiO_2: comparison of the efficiency of immobilized nanoparticles versus conventional suspended catalyst. J. Hazard. Mater. **142**, 130–137 (2007)

Foam-Glass-Crystal Materials

O. V. Kazmina and B. S. Semukhin

Abstract This chapter presents a comprehensive study of the structure, physical and mechanical properties of FGCM obtained from natural siliceous materials. The scientific principles of controlled formation of FGCM macrostructure and technological aspects of its production have been formulated in this chapter. At this, the strength of foamed materials exceeds the strength of the foamglass 3–4 times as much due to the effect of nanoscale crystals in a vitrified matrix. It has been shown that the physical and mechanical properties of foamglassceramic material depend on the number and the size of a crystalline phase. The increase of mechanical strength as compared to the foamglass can be provided by the particle sizes of a crystalline phase of less than 1 μm. Maximum strength depending on the volume rating of the residual crystalline phase accounts for 25 % for less than 1 μm size, and 5–7 % at a size decrease down to 300 nm. With the increase of the crystalline phase size up to 10 μm and higher, the compressive strength of material decreases.

1 Introduction

The problems of energy saving are the foremost in the industrial and construction sectors of economics in all developed countries. The analysis of the experience shows that one of the most effective ways to address the energy saving is the use of high-performance heat-insulating materials. Currently, the promising building materials are materials which combine high heat-proofing and structural

O. V. Kazmina (✉)
National Research Tomsk Polytechnic University, 30 Lenin Avenue, Tomsk, Russia634028,
e-mail: kazmina@tpu.ru

B. S. Semukhin
Institute of Strength Physics and Materials Science of the Siberian Branch of RAS, 2/4
Akademichesky Avenue, Tomsk, Russia634021,
e-mail: bss@ispms.tsc.ru

J. Njuguna (ed.), *Structural Nanocomposites*, Engineering Materials,
DOI: 10.1007/978-3-642-40322-4_6, © Springer-Verlag Berlin Heidelberg 2013

properties, fire safety and durability. The significant scientific and practical experience of their production and use was acquired. At present, there is a vast literature on heat- and sound-insulating construction materials. It covers such aspects as technological, environmental, physical and chemical production of these materials.

One of the high-efficient heat-insulating materials that meet the requirements of environmental safety is foam glass.

Industrial waste utilization and reduction of industrial atmospheric emissions is closely connected with technical and economic problems of resource and energy saving as well as environmental protection aspects. Foam glass production is one of the efficient trends of utilization of both industrial and domestic waste glass. At the same time, the problem of cullet recycling is relevant as before, because it is so far one of the most hardest to utilize hard domestic waste.

The amount of cullet required for foam glass production is obtained by the traditional glass technology including glass manufacturing in industrial furnaces. The glass melting process is connected with generation of a considerable amount of gaseous dust air and chimney gas emissions containing end products of fuel combustion (CO_2, N_xO_y and others), and stack solids (B_2O_3, Fe_2O_3, As_2O_3). Therefore, taking into account the number of production and economic features of glass manufacturing, the foremost task is a pre-synthesis of quenched cullet from natural or industrial raw materials using energy-saving technology in order to escape the glass melting process. Low-temperature synthesis of quenched cullet is, on the one hand, economically feasible due to the absence of such energy-intensive glass manufacturing operations as fining and homogenization of the glass mass. On the other hand, this is the way to lower hazardous emissions in terms of furnace end gases.

Foam glass produced on the basis of glass cullet is composed of 100 % silicate glass which is not, practically, influenced by chemical reagents, is not a nutrient medium for fungus, mold, and is an ideal barrier for insects and rodents. Resistance of foam glass to putrefaction and the absence of a nutrient medium for mold and spread of mold and fungus are especially important when the foam glass is used in closed nonaerated space of roofs, walls, and basements. The absence of organic matter ensures the escaping of situations connected with the destruction of heat-insulating materials under the influence of the bioactive environment. Due to the specific structure of the foam glass with closed impermeable pores, at negative or variable temperatures and actuation of steam migration inside the material, a possibility of accumulation of water particles and steam is excluded as well as their condensation on the surface of interpore partitions which can be destructed in the weaker areas as a result of water and glass interaction. Therefore, one of the most important properties of the base glass is its chemical resistance.

Chemical resistance of glass is its capability to withstand the destructive effect of water, moisture, atmospheric gases, salt solutions, and various chemical reagents. Chemical resistance of glass mainly depends on its chemical composition, particularly, on the content of silicon and alkali metal oxides. Introduction of silicon oxide into glass considerably increases the its resistance while alkali metal oxides decrease it. When changing the glass composition, its chemical properties are also changed. As a rule, potash soda-ash glasses increase chemical resistance of glass.

At this stage, the study of ecotoxicological indicators of FGC (foam-glass-crystal materials), which is produced from natural and industrial raw materials based on low-temperature synthesis of quenched cullet, is conducted in comparison with the commercial types of foam glass synthesized from secondary glass cullet, e.g. FOAMGLAS® (Pittsburgh Corning www.foamglas.com). The ecotoxicological assessment procedure should pay a special attention to the study of toxic substances effect on the environment which can emit from the material while in operation.

FOAMGLAS® emits no harmful substances, such as formaldehydes, styrenes, or fire-proofing compounds, fibres, chlorofluorocarbons (CFC/HCFC) etc. that are detrimental to health. Ecological and biosafety of FOAMGLAS® offers optimum construction solutions not only for rooms where a high degree of air purity is required (hospitals, museums, schools, offices, waiting rooms, high-tech production facilities, etc.) but also in locations where there are special bacteriological and hygienic requirements (slaughterhouses, cheese factories, dairies, industrial kitchens, canteens, restaurants, swimming pools, etc.). All approved accessory products for the systems also aim to satisfy low-emission standards.

The BTU institute (Buro fur technischen Umweltschutz, Reiskirchen, Germany www.btu-umwelt.de) has gathered data on different parameters (e.g. length of service-life, primary energy content, price, thermal conductivity, costs of disposal or recycling and technical or application polyvalence) of the principal industrially-manufactured insulation products in the market. By applying a mathematical formula and with the weighting coefficients (per category from 1 to 5, with 1 being "negative" and 5 "very positive"), an examination of the parameters under consideration leads to different results per group of products and is expressed by the profitability index (R). The higher the R-value, the better the product evaluation.

For FOAMGLAS® the profitability index—a characteristic value for ecological and economic evaluation, involving a number of parameters—leads to the value 45.5. This result positions FOAMGLAS® amongst the leading products in overall economic/ecological assessment.

FGC falls within the scope of the Law of the Russian Federation (the Technical Regulations of the Safety of buildings and facilities), which claims the following: Building materials and products must not extent damage effect on humans and the environment; they must not contain and emit harmful substances in such concentrations or combinations that could influence directly or indirectly humans, or animals, or plants; concentration of harmful substances in the air of working area and habitable rooms must not exceed the normative values stated by the RF legislation in the field of sanitary-epidemiological welfare of population.

Physicotechnical properties of foam glass are specified mainly by its production technique, chemical composition and foaming mixture, the kind and the quantity of gasifier, sponging and fritting modes. Changing these factors one can obtain material of different density, strength, heat conductivity, water absorption [1-3]. Depending

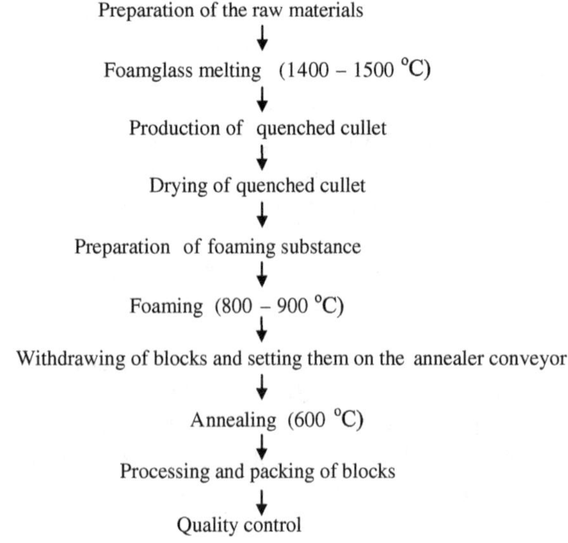

Fig. 1 Flow chart of the powder mode of foam glass production (two-stage process)

on its application, foam glass is classified into heat-insulating, sound-insulating and specific high-temperature. European production widely practises heat-insulating foam glass with density of 200 kg/m^3. Foam glass production technique includes silicate glass grinding, mixing with foaming additive of carbonic or carbonate type and thermal treatment of the mixture at 750–850 °C temperatures.

Foam glass industrial production is mainly based on the powder mode of two-stage process, the flowsheet of which is given in Fig. 1 [4]. The advantages of the two-stage production mode are as follows: minimum sponging time; a complete use of fritting furnace space due to minimum clearance between the glass blocks arranged; the minimum number of forms to be returned to production upon a completion of the sponging process neglecting the long cycle of foam glass fritting. Figure 2 illustrates the plan of the foam glass area designed by the Glamaco Company (Germany).

The global experience shows that it is not always feasible to utilize foam glass in the form of blocks or plates. In a number of cases, the foam glass is used in the form of gravel or crushed stone for building, e.g. as heat-insulating bulk material or concrete filler. A great experience has been accumulated in the field of utilizing foam-glass crushed stone Schaumglas Schotter and gravel Poraver. Production of small granules of heat-insulating material is always more profitable from the viewpoint of heat exchange. In producing block-like foam glass, up to 80 % of the length of the tunnel furnace and, as a consequence, up to 80 % of time is spent on the slow cooling of material. The quenching and productivity improvement result in considerable internal tensions and destruction of blocks. It is the principle of quenching which foam-glass crushed stone Schaumglas Schotter production is being based on: the foam glass is leaves the furnace at a rather high speed, then additionally cooled, and as a result of this crushed stone is obtained.

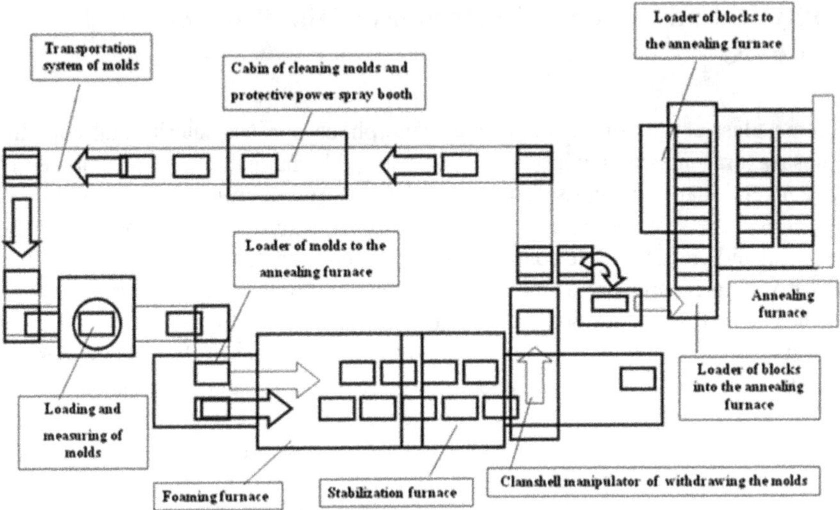

Fig. 2 Plan of the foam glass area in the two-stage mode of production designed by the Glamaco company (Germany)

Raw material used to obtain foam glass according to the classic scheme is broken glass, the composition of which meets the specific requirements. Firstly, they use glasses of a wide melting range, the so-called 'long' glasses. Secondly, the glass must contain a component which at the process temperatures would interact with carbon so as to provide gasification. At the same time, broken glass still remains one of the most difficultly utilized components of hard domestic waste. For example, in 2001 the percentage of its recycling in the developed West European countries ranged from 24 to 92 %. Only profiled broken glass undergoes the effective recycling while the low-grade one is stored at hard domestic waste grounds. It leads to the inevitable interaction between the glass and precipitation which results in scavenging of Na^+ ions from the glass surface and formation of alkaline solutions. Especially, it concerns small fractions which possess the largest specific surface and, therefore, define scavenging of the alkaline component to a great extent that negatively influences the environment.

Thus, profiled broken glass of the certain composition is employed to manufacture the qualitative foam glass; its lack is made up by the traditional glass technology. The glass production requires substantial material and energy expenditures. Therefore, taking into account a number of technological and economic glass manufacturing features, the foremost task is to provide the preliminary synthesis of the quenched cullet from the natural or anthropogenic raw materials using the energy saving technology at temperatures not above 950 °C escaping the power-intensive process of glass melting.

This chapter presents a comprehensive study of the structure, physical and mechanical properties of FGCM obtained from natural siliceous materials.

2 Physical and Chemical Principles of the Production of FGC

A foam-glass-crystalline material is an amorphous porous material that contains nanostructural units and exhibits increased strength characteristics as compared to the conventional foam glass manufactured from a glass cullet. The possibility of controlling the structure of foam-glass-crystalline materials is determined by the chemical and granulometric compositions of the initial batch; the nature of the basic glass forming component of the batch, the technology used for preparation of the batch, which brings into operation of the mechanisms and effects of mechanical activation of the mixture; and the regime of heat treatment of the batch under the conditions providing the low-temperature synthesis of the glass granulate.

The glass granulate is the initial raw material for the subsequent foaming and manufacturing of foam-glass articles with specified characteristics. According to the phase composition, this intermediate product is a glassy material containing the residual quartz crystals. The properties of foam-glass-crystalline materials depend not only on the composition and structure of the phases of the glass granulate, which are formed in this material, but also on the size and mutual arrangement of the structural units at the microscale level.

A two-stage method for producing foam glass via the intermediate product (quenched cullet) synthesized by thermal treatment of the mixture of the certain composition was developed at Tomsk Polytechnic University [5]. This product acts as the raw material for the following sponging and obtaining foam-glass-crystal products with the pre-set characteristics. According to the phase composition the quenched cullet represents a vitrified product with residual crystal inclusions which define the density and strength of the end items. The principal procedure for obtaining foam-glass-crystal materials is given in Fig. 3, which outlines three main technological stages, two of which relate to the quenched cullet. The idea of the technology developed lies in the following principles:

- Special preparation of raw materials allows synthesizing a glass ceramic at temperatures not over 950 °C which is the raw material for the foam glass;

Controlling the formation of nano and microstructure of the interpore partition allows regulating the strength and density of the end item. Relatively low temperatures of the glass phase synthesis carried out according to the designed mode, promotes not only energy consumption decrease, but also that of the carbonic acid release. The working operation of the mixture compaction makes possible to lower the air pollution of the working area as well as the general dust emission to the atmosphere.

Compositions of container glasses or the like are considered to be the optimal for foam glass manufacture. The content of oxides required for these glasses consists of the following elements (mas. %): SiO_2 67–72; Al_2O_3 1–6; CaO 7–11; MgO 1–4; Na_2O 14–15. It has been experimentally stated that the most perspective

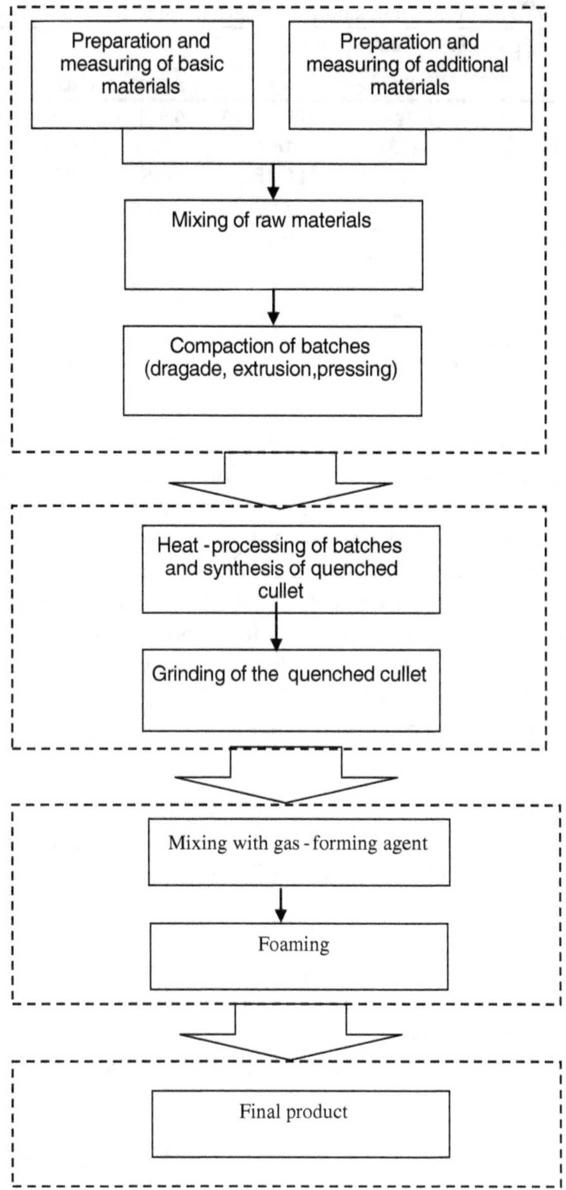

Fig. 3 The principal procedure for obtaining foam-glass-crystal materials

raw material is the breakage of a dark green bottle while less manufacturable is the breakage of float-glass. Table 1 gives typical compositions of silicate glasses used to manufacture foam glass in different countries. Oxides included in the glass composition influence the sponging mode, namely: the temperature, the process duration and also such properties as crystallization, viscosity and surface tension [6].

Table 1 Chemical composition of glasses used for foam glass production

Composition	Content, mass. %						Country
	SiO_2	Na_2O	Al_2O_3	CaO	MgO	B_2O_3	
1	72–73	15.5–16.5	0.5–1	6–6.5	3.5–4	–	Russia
	(73)	(16.1)	(0.8)	(6.3)	(3.8)		
2	55–72	11–18	3–16	9–12		–	Czech
	(64.8)	(14.8)	(9.7)	(10.7)			
3	72–74	13–17	0–2	8–12	–	–	Great Britain
	(73.7)	(15.2)	(1)	(10.1)			
4	70–72.7	14	2–7	4.9–7.6	3	0–3.6	France
	(70.5)	(14)	(4.5)	(6.3)	(2.9)	(1.8)	

In brackets given are average values translated into 100 % glass composition

The use of the raw materials alternative to the broken glass significantly expands the technological possibilities of the foam glass production. The diverse natural and/or anthropogenic feed stock can be utilized as raw materials. The composition and properties of siliceous and silica-alumina raw materials allow synthesizing the quenched cullet on their basis with the following processing until the porous material will be obtained. Controlling the component composition of the mixture and the temperature conditions for thermal treatment, heat-insulating materials can be obtained being rather various in their properties.

The analysis of scientific publications and research findings provided have shown that in order to obtain a low-temperature quenched cullet, the choice of the mixture composition should be carried out in relation to the following conditions:

- The amount of glass formers' oxides and alkali metal oxides in the quenched cullet must be sufficient for a sustainable glass formation, i.e. the content of SiO_2 is in the range of 60–75, and the content of alkali metal oxides is in 13–22 mass % range;
- The number of the glass bath must not be less that 70 % that is necessary to provide plastic state of the mass at the sponging stage;
- The liquid phase must be of the optimal viscosity (103–106 Pa•s) within the sponging range;
- The temperature for the liquid phase formation (glass bath) must not exceed 950 °C.

With a purpose to select the basic glass composition and diminish the amount of preliminarily experiments so as to obtain the quenched cullet, carried out was the analysis of state diagrams of the ternary system $Na_2O–CaO–SiO_2$ which is the basis for a production of the industrial glasses of more oft-used compositions, namely: container glass, sheet glass, etc. In accordance with the state diagram the industrial silicate glass area is close to the equilibrium field of crystals of wollastonite, devitrite, tridymite. It is these phases which are often demonstrated in glass crystallization. Almost all industrial compositions get one elementary phase triangle $Na_2O \times 2SiO_2–Na2O \times 3CaO \times 6SiO_2–SiO_2$ and begin to melt at 725 °C that corresponds to the low-temperature system eutectic.

The analysis of the diagram has shown that the following three groups of compositions correspond to the above stated conditions: (1) compositions with the varying ratio of the basic oxides and the fixed amount of SiO_2 (73 %); (2) compositions with the varying content of CaO and SiO_2 at the fixed amount of the alkaline component (21 %); (3) compositions with the varying content of Na2O and SiO_2 at the fixed amount of alkali-earth metal (5 %) [7].

The research findings have shown that the optimal glass content appropriate to obtain the foam glass under the low-temperature technology, is the composition with rather high content of SiO_2 (72–74 mass %) and the ratio of basic oxides CaO/Na_2O within the range of 0.2 ÷ 0.9 that corresponds to the sufficient viscosity of the glass bath and stable glass formation.

For test experiments the compositions characterized by the ratio of the basic oxides were selected within the following limits: the amount of Na_2O ranges from 14.5 to 21 %; the amount of CaO ranges from 5 to 13 %. The state diagram of $Na_2O–CaO–SiO_2$ system illustrates that the given compositions limit the area of glass formation with the various component ratio.

The change of the composition along the line connecting points 1 and 2 is accompanied by the decrease of Na_2O and the increase of the melting temperature which does not exceed 950 °C.

The first composition is eutectic with melting temperature of 725 °C; in the second composition the share of the liquid phase forming in the temperature range of 800–950 °C changes from 45 to 80 %. For all that, one can expect that real temperatures of mixture treatment will be lower due to the impurity substances presented in raw materials. To obtain the quenched cullet based on siliceous raw materials the following two boundary compositions were chosen (mass %): SiO_2—74; CaO—5; Na_2O—21 and SiO_2—73; CaO—11; Na_2O—16.

To synthesize the glass phase of the chosen compositions the two-component mixture is required which includes the siliceous and alkaline material in the form of soda ash. In the lack of alkali-earth oxides in raw materials, the mixture is corrected by the additional introducing of chalk or dolomite.

The adequacy of siliceous raw material for quenched cullet production is defined by three main indicators: chemical, phase and granulometric compositions. In this case, siliceous materials with the following oxide content (mass %) are the most adequate for quenched cullet production: silicon not less than 83; aluminum up to 7; the total amount of alkali-earth oxides up to 13; ferrum not over 10.

The phase content of siliceous materials is of significant meaning for low-temperature synthesis of the glass phase. The presence of amorphous component SiO_2 creates favorable background for solid-phase reactions and allows lowering temperature of glass synthesis in comparison with crystalline silica. In this case, the most optimal are high-silica amorphous rocks because the silicate glass is an amorphous material which contains over 70 % of SiO_2. However, it should be taken into consideration that silicate glasses which contain coloring oxides tend to be crystallized, for example, ferric oxide at the repeated thermal treatment. The presence of the residual crystal phase in the quenched cullet may promote crystallization which is promoted as well by the high specific surface of initial mixtures. Therefore, it is important to take into account the qualitative ratio between the crystal and amorphous phase components in the siliceous mixture component.

Below given are the experimental findings for siliceous raw materials which meet the stated requirements in terms of such materials as sand, marshallite, diatomite and flask (Table 2). The chemical composition of these materials is represented by rather a high content of SiO_2 and the presence of impurities which are not harmful because of their inclusion in the glass composition and taking into account in the mixture calculating.

The experiment data showed that the residual content of crystals remains at the level of 10–23 %; the largest amount of the glass phase has the quenched cullet based on marshallite and flask. The mixture based on the material with amorphous SiO_2 shows more higher chemical activity at the stage of silicate and glass formation.

It has been stated that with the increase of the mixture treatment temperature the content of the crystal phase decreases for all compositions. In case of using diatomite and flask, the temperature decreases down to 850 °C with the further stabilization of the residual silica amount. Obviously, it is connected with relatively low temperature of the mixture melting and the early approach of high-

Table 2 Properties of siliceous raw materials and quenched cullet produced on their basis

Indicators	Siliceous raw materials			
	Silica sand	Marshallite	Diatomite	Flask
Chemical composition, mass %:				
SiO_2	98.15	95.7	86.44	83.00
$Al_2O_3 + TiO_2$	0.73	2.10	5.30	5.25
$CaO + MgO$	0.09	1.4	1.27	3.52
Fe_xO_y	0.05	0.27	1.60	2.72
Phase composition, volume, %:				
Crystal forms SiO_2	98	95	14	13
Amorphous opal	–	–	70	57
Silica-alumina crystal phases	2	5	16	30
Specific surface of prepared rock, cm^2/g	3,484	5,500	14,510	16,150
Maximum temperature of mixture treatment, °C: Thermal treatment of mixture				
Composition (I)	930	875	830	810
Composition (II)	880	840	850	825
Crystal phase amount for quenched cullet, mass %: Quenched cullet properties				
Composition (I)	16	1	13.7	2.65
Composition (II)	23	10.38	14.92	6.14
Plastic state temperature of quenched cullet, °C:				
Composition (I)	800	748	777	740
Composition (II)	850	823	845	841

viscosity liquid phase which takes disjoining effect and lowers the dissolution of crystal SiO_2. The glass formation process for sand mixtures is not complete (the amount of crystal phase is 23 %). Thus, it is advised to treat the mixture at higher temperatures or provide the additional preparation of the mixture using mechanical activation.

Thus, in order to obtain the quenched cullet at relatively low temperatures one can utilize fine-dispersed siliceous materials in the capacity of the mixture component taking into account the following factors:

- The use of quenched cullet with the glass phase amount of over 80 % allows obtaining foam-glass-crystal material of up to 250 g/cm^3 density; 3 MPa strength and 6 % water absorption;
- The amount of the glass phase of the quenched cullet increases along with the increase of amorphous component in the siliceous mixture component.

3 Foam-Glass-Crystal Materials Structure

In this section, the experimental results of the analysis of FGC structure on different structural levels are presented. To determine the structure of the samples obtained, the optical and electron microscopy was used. The elemental composition of interpore partitions was detected with of high resolution scanning electron microscope JSM-7500FA and equipped with X-ray microanalyzer. Before recording, the samples were covered with a thin platinum layer. Recording was carried out at the following modes: accelerating voltage of the electron beam of 10–15 kV; working distance of 20–50 mm; and magnification up to 100,000 times.

The X-ray structural analysis of the phase composition of quenched cullet and foamed material was carried out using DRON-3 M diffractometer with Cu $K\alpha$ radiation with monochromatization of a diffracted beam by a pyrolitic graphite crystal.

The energy characteristics of a submicron crystalline phase differ from those of the starting material [8]. A maximal enhancement in the strengthening of material might be expected if the crystalline phase microstructure is far from equilibrium. According to Academician N. P. Lyapishev, a material having specific properties would be composed of crystallites or of a mixture of nanometer-sized crystals and amorphous phase. Indeed, the foam glass material manufactured from glass granulate is in conformity with the latter principles [9].

The experimental dependence illustrated in Fig. 4 was obtained for the test samples, which had a predetermined content of residual crystalline phase (5 % on the average) and differed in the size of crystalline phase particles. By assuming that the minimal critical size of crystalline phase particles is 10 nm, the maximal theoretical strength was calculated for foam glass material; the value obtained is 5.0 MPa. In the case of foam glass whose amorphous matrix contains no crystalline phase this value is not greater than 1.5 MPa [10].

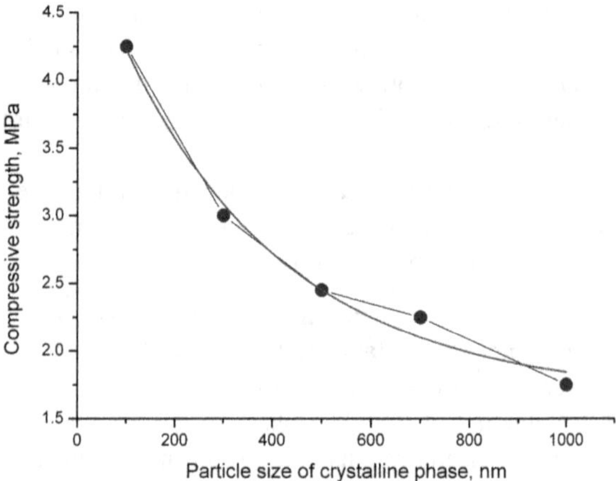

Fig. 4 The strength of foam glass material versus particle size of crystalline phase

The particle size of residual crystalline phase can be no further reduced; therefore, one can employ an alternative method for strengthening of amorphous matrix by changing its internal structure. Compressive strength values were obtained experimentally for all the test samples of foam glass material. This data suggests that an enhancement in the strength of materials (up to 4.3 MPa) was achieved due to the occurrence in their interpore partitions of spherical formations (spheroids).

High resolution electron microscope images show spheroids having sizes of 60–160 nm, which occur in the amorphous matrix of these materials. No formations of this kind were found to occur in the interpore partitions of common foam glass and glass ceramics whose matrix contains residual quartz particles having size >200 nm (Fig. 5). Figure 5 clearly illustrates a complex hierarchy of single spheroids and groups of spheroids. An X-ray dispersion analysis suggests the occurrence of nonuniformly distributed silicon in the amorphous matrix of the interpore partition: the maximal silicon concentration is observed in the vicinity of interpore partition and the minimal, within the interpore partition (Fig. 6). The formation of spheroids would evidently cause redistribution of silicon in the interpore partition. It is thus concluded that the main silicon-containing structural elements are nanospheroids which gather mostly in the vicinity of interpore partition.

It is thus found that the structural changes of a strengthened foam glass material are due to a change in the structure of its amorphous matrix rather than to the occurrence and structure of residual crystalline phase. This conclusion seems inconsistent with the conventional scheme discussed in the literature [11-13]. However, the experimentally observed behavior of the glass ceramic is in accord with a structure, which enables minimization of the energy of the entire material volume at the expense of the main (amorphous) component.

In the course of foaming process, the structure of amorphous matrix would change spontaneously; in so doing, spheroids are formed. Here a remark is in

Fig. 5 The electron microscope image of spheroids in the interpore partition of foam glass material

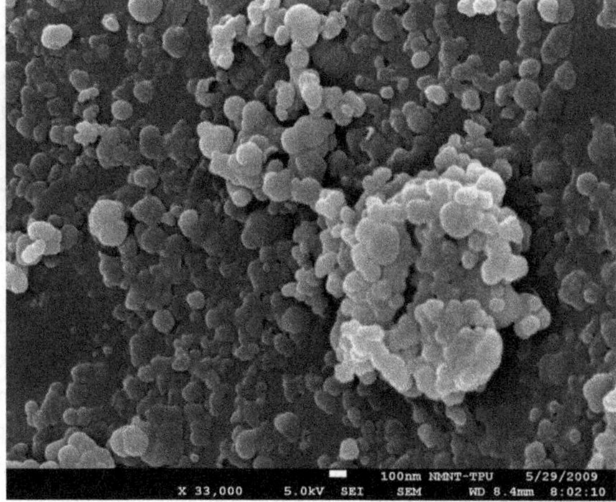

Fig. 6 Silicon content distribution in the interpore partition

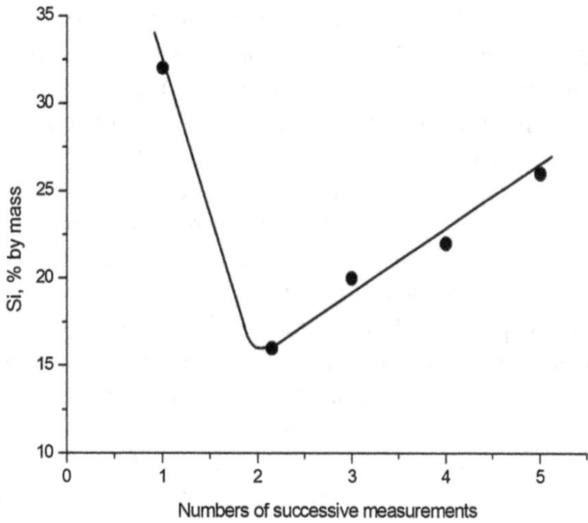

order: when dealing with minerals, spheroids are called 'globules'. The experimental evidence for the occurrence of globules in silica and the mechanisms involved in their formation are discussed elsewhere [14]. It is assumed that a spherical globule has a loosely packed body, while its core has a close-packed arrangement of SiO_4 tetrahedrons. The general aspect of a globule is that of a 3D particle having a quartz- or crystobalite- type structure; on its surface are arranged unidimensional SiO chains, i.e. mono-, di- and trimers of silicic acid or sodium silicate. Such a globule is taken to be an intermediate state between crystalline and amorphous matter. Abundant experimental evidence for silica globule formation is available [15, 16]. X-ray diffraction analysis and thermography data suggests that the globule structure is a set of randomly distributed nanocrystallites and amorphous areas.

Using IR spectroscopy technique, investigations were carried on for the test samples of silica gel and synthetic amorphous aluminosilicate. The spectra obtained show a band 1,200 cm^{-1}, which corresponds to globule formation [14]. This suggests that on the crystal surface have appeared fragments of SiO_4 tetrahedrons, which make an angle of 180° with one another; some of these fragments are composed of two or three tetrahedrons joined together to make a kind of chain.

The IR spectra obtained for glass ceramic samples are found to show a new line 1,249.6 cm^{-1}, which is absent from the IR spectra of foam glass. On the base of IR spectra and micrographs obtained one can evidently assume that the amorphous component of the interpore partition has a globular structure. It is for the first time that globules are discovered in the amorphous matrix of foam glass material. It is these globules that are responsible for a sudden enhancement in the strength of foam glass material. It is emphasized that the microstructure of interpore partition is composed of amorphous matrix, which contains residual quartz particles (≤ 5 %) having size ~ 200 nm and globules having size ≤ 100 nm.

The strength of amorphous materials is generally attributed to the action of several mechanisms, with the main mechanisms being shearing and local jumps. In the former case, the deformation is assumed to involve shearing of one amorphous cluster (globule) against another to cause dissipation of energy [17]. In the latter case, it is assumed that a jump-wise change of atoms positions would occur within a cluster (globule). It was proposed recently that, in addition to the above mechanisms, one should taken into account another factor, which might also be responsible for material strengthening, i.e. nanostructure formation within the deformation bands [18-22]. In our experiments the strengthening of foam glass materials is largely due to two processes: (1) a decrease in the particle size of residual crystalline phase and (2) formation of nanospheroids (globules) in the amorphous matrix. In the former case, an enhancement in the strength of material is ≤ 3 MPa and in the latter case, one can achieve theoretical strength ≤ 5 MPa. In the course of deformation, the dissipation of energy would mainly occur in material areas containing close-packed globules, with the volume fraction of globules being ≤ 95 %.

A new type of deformation strengthening has been observed for the first time. This involves spontaneous formation of nanometer-sized spherical globules within the amorphous matrix of the deforming material. The model of strengthening proposed for amorphous materials in is validated by the experimental results discussed herein.

4 Mechanical Properties

Foamglass is one of the universal high-performance heat and sound-insulating materials whose scope of use as a structural material is impartially limited by the compressive strength indicator (not over 1.5 MPa) that depends on the nature and structure of material, i.e. 'glass foam' with the pore partition thickness up to 50 mkm.

The foam glass-crystal material high-strength characteristics as compared to the traditional foam glass technology based on the use of cullet. The increase of mechanical strength of foam glass allows widening its scope of use as a structural material.

The research findings of the fracture process of foam glass-crystal samples obtained *in vitro* as well as the relation between the structure and mechanical strength are discussed in this Report.

There are two extreme cases of a body behavior at the external load application. In the first case, which corresponds to an ideal solid, deformation is proportional to the stress imposed. The second extreme case is viscous fluid for which the deformation rate equals to the applied load divided by the coefficient of viscosity. Amorphous solids are neither perfectly elastic nor perfectly viscous ones and combine both elasticity and viscosity properties. It means that complete deformation of the amorphous solid is likely to consist of two parts—elastic and viscous. In case of foam glass, this phenomenon appears to the specific degree because the 'classical' glass is fractured in perfectly brittle way according to the Griffiths' model, while the foam glass and the foam glass-crystal material display both elastic and viscous components.

To analyze the mechanical properties, and specifically to define the tensile ultimate strength of foam glass, some tests were conducted by means of the universal machine 'Instron 1185' with the load range between 0–100 N and 0–100 kN. The measurement accuracy was ±0.5 % of the display value, or ±0.25 % of the load scale used. In compliance with the recommendations given by the Russian State Standards (ГОСТs), P EH 826–200 "Heat-insulating products used in construction. Method of compression characteristics' detection", tests were carried out at rate of 2 mm/min. Foam glass samples in the form of rectangles or cubes underwent the test for compression down to their complete fracture with registering strain diagram in the automatic mode. For the comparative analysis three kinds of samples were chosen, namely: foam glass-crystal materials, industrial foam glass and laboratory foam glass obtained from cullet.

Figure 7 presents the generalized view of dependencies σ–$f(\varepsilon)$ which shows that the dependence characters for foam glass samples obtained from cullet are similar, while those ones of the foam glass-crystal material are different. The value of the tensile ultimate strength of the foam glass-crystal samples is considerably higher (2–3 times) as compared to the foam glass. This significant difference can be explained by the material structure including the size of pores and interpore partitions, the composition and structure of the amorphous component.

To describe the deformation curve of the foam glass-crystal material the synergetic approach was applied, which was described by Prof. Olemsky in his work on fluid vitrification [23]. It was shown in this work that within the synergetic equations the modified Maxwell equation is in compliance with the experiment for the viscoelastic medium:

$$\sigma = -\sigma/\tau_\sigma + g\,\sigma\,\varepsilon\,T \tag{1}$$

Fig. 7 Compression test
curves of foam glass
materials

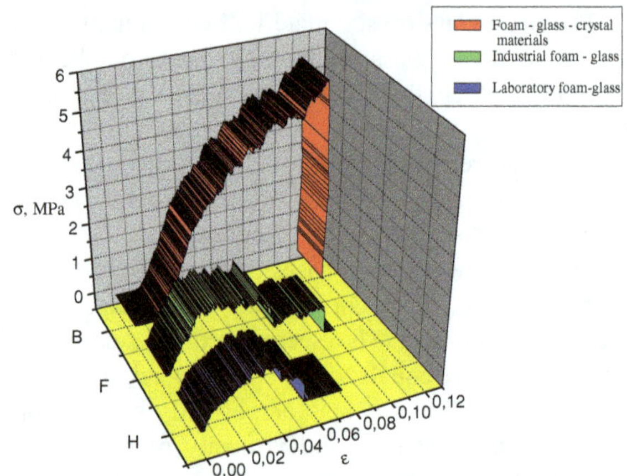

The first summand is responsible for the dissipative process of stress relaxation to the equilibrium value, and for the microscopic Debye time of 10^{-12} s at that. This time appears in the well-known Zhurkov formula which describes the interrupting time and rupture stress σ_p relation [24]:

$$\tau = \tau_0 \exp[(U_0 - \gamma\,\sigma_p) = \mathrm{k}\,\mathrm{T} \tag{2}$$

The second summand of Eq. (1) defines the self-organization process.

Thus, the elementary deformation curve $\sigma\,(\varepsilon)$ includes not one but two sections. The first section is elastic, Hookian, and has a wider slope angle; the second section is more gradient and responsible for plastic deformation processes. Taking into account the deformation module defect at transition from Hookian stage to the plateau of the plastic flow adequately reflects the vitrification process in terms of synergistic concepts. In the case described, the deformation curve of the foam glass-crystal material actually consists of two segments specified by the suggested synergetic model and Zhurkov's model (Fig. 8).

It should be noted that in compression testing of foam glass, the abrupt fraction of samples typical for brittle materials has almost never been observed. With the increase of the load the sample becomes deformed, and at the same time, thin partitions of cells are sequentially fractured on the fixed thrust face. The received glass powder is pressed into the anew fractured cells.

The elemental composition of interpore partitions was detected by the electronic microscope JSM-7500FA equipped with the power-dispersion microanalyzer *Edax*. It is shown in the figure that nearby the interpore partition the silicon concentration is the highest while in the centre it is minimal. This high content of silicon must result in formation of crystal inclusions of SiO_4 in the boundary layer of the interpore partition. The presence of crystal phase in amorphous matrix increases the strength of the interpore partitions. In applying the load to the sample during the compression test process, the partitions collapse.

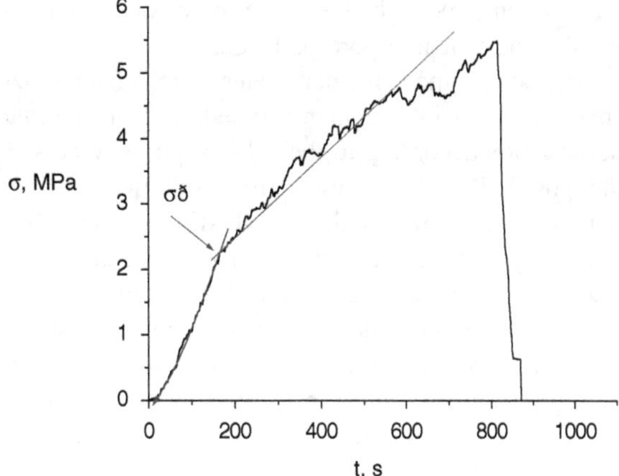

Fig. 8 Deformation *curve* (by time) of the foam glass-crystal material

The relation of grain size and strength for non-deformed materials is similar to Hall–Petch relation for the yield strength of ductile metals: if average grain size could be decreased even further to the nanometer length scale the yield strength would increase as well. However, in case of porous materials, this correlation should take place rather between the sizes of interpore partitions than the pores themselves because all mechanical properties of the material depend on the properties of partitions themselves.

Figure 9 illustrates the curve of average size of the interpore partition in the foam glass-crystal material depending on stress. One can see that this dependence is similar to Hall–Petch relation.

The important feature of cellular material affecting the mechanical strength is the porosity character that includes spatial arrangement of pores (packing),

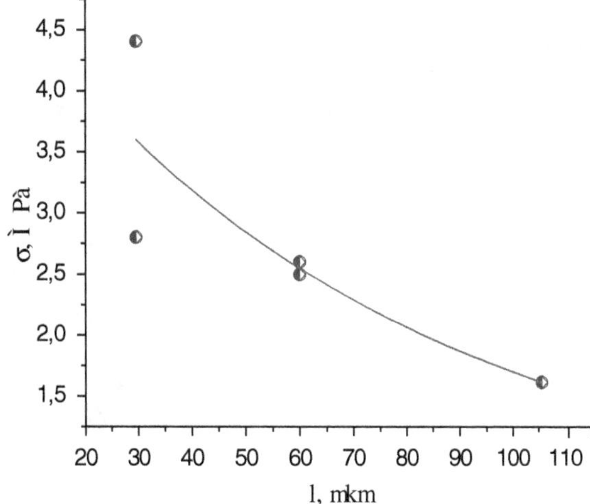

Fig. 9 Dependence of the average size of the interpore partition upon the ultimate strength

distribution of pores by their size (pore combination of various sizes), the form and the thickness of interpore partitions.

The pore shape is the parameter which characterizes a degree of deformation of spherical pores into regular polyhedrons. One can judge about the degree of pore deformation according to the cellular porosity bulk: the 75–80 % excess indicates the possibility of transition from spherical pores to polyhedrons. The higher porosity the more regular polyhedrons should be. The system tends to such porosity parameters that provide formation of pores with a dense smooth surface. The porosity increase is achieved at various sizes and aspheric shapes of pores. Products with more homogenous honeycomb structure of pores possess higher strength. A polydisperse character of pore distribution by their sizes provides the high probability of uniform location of smaller pores between the larger ones.

It is worth noting, that the pore distribution by sizes in the foam glass-crystal material is similar to that one of the structural elements presented in nanocrystalline materials. This distribution of hierarchically subordinate microvolumes of materials should have an impact on strength properties (Fig. 10).

At the same time, the size and the shape of pores are connected with mechanical properties. Figure 11 presents the diagram of the pore size (average value according to the image) depending on the definite fracture stress. This linear dependence has the high correlation coefficient.

Thus, the defect structure of pore types in cellular materials can characterize the performance of foam glass products. At the load applied to cellular materials the pressure transmits along the partitions of pores which form arches. Due to this, the structure of linearly loaded material always has the bulk stress state. It is known that at big arch sizes high stresses occur. Arch sizes in the structure of cellular articles are in direct relation to the pore sizes. Therefore, cellular materials possess higher strength characteristics when they have pores of smaller sizes, all other conditions being equal.

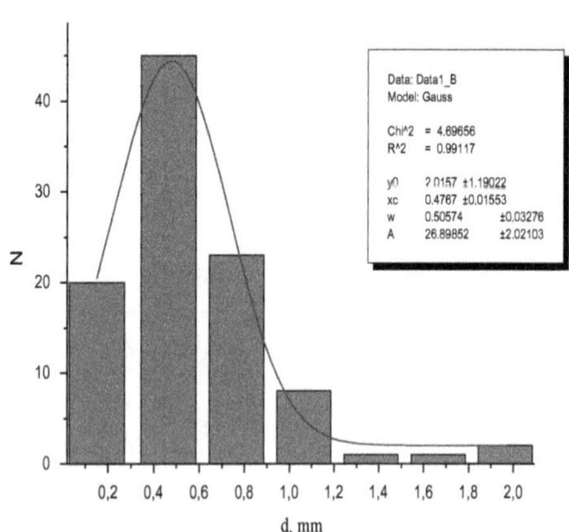

Fig. 10 Pore distribution by the sizes in the foam glass-crystal material

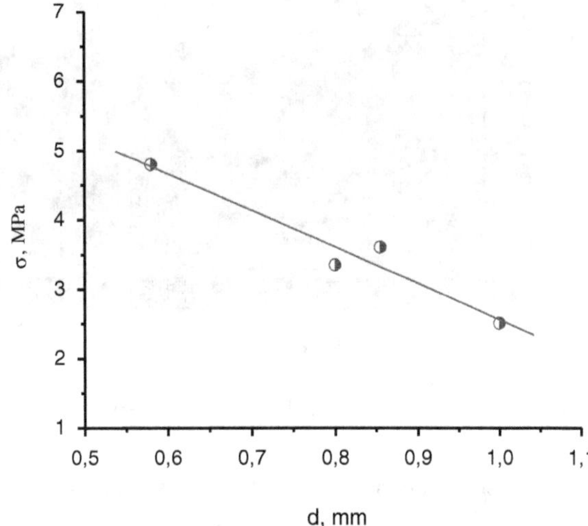

Fig. 11 Dependence of the average pore size on the ultimate strength of the foam glass-crystal material

However, the analysis of micrographs illustrating the foam glass-crystal materials obtained from mixtures of diverse compositions makes obvious the advantage of location of spherical pores of two different diameters, the so-called two-modal filling manner. This layout implies filling of the free space between the spherical pores of larger sizes with those ones of smaller sizes. It can be seen well in Fig. 12 which represents the microstructure of various foam glass-crystal materials.

Moreover, modification of pore shapes, i.e. the tendency to a honeycomb distribution of hexagonal pores can considerably increase the mechanical properties of cellular material. Figure 13 illustrates two pictures of materials with different pores and different ultimate strengths.

According to the research findings in relation to the fracture process of foam glass-crystal samples obtained in vitro it has been stated:

- foamglass-crystal samples have higher strength as compared to the foamglass obtained from cullet;

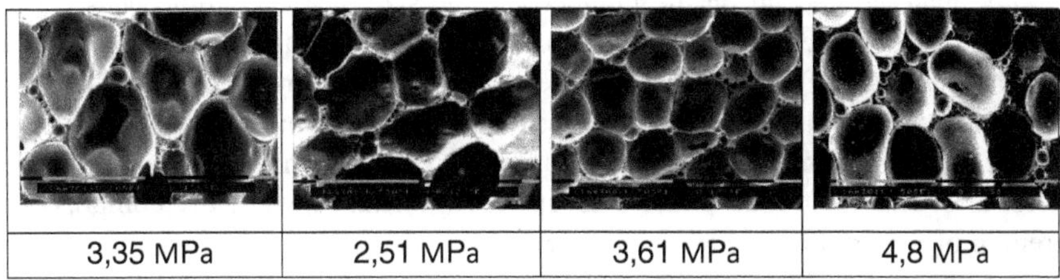

| 3,35 MPa | 2,51 MPa | 3,61 MPa | 4,8 MPa |

Fig. 12 Sizes and shapes of pores belong to various foam glass-crystal materials and the corresponding strength

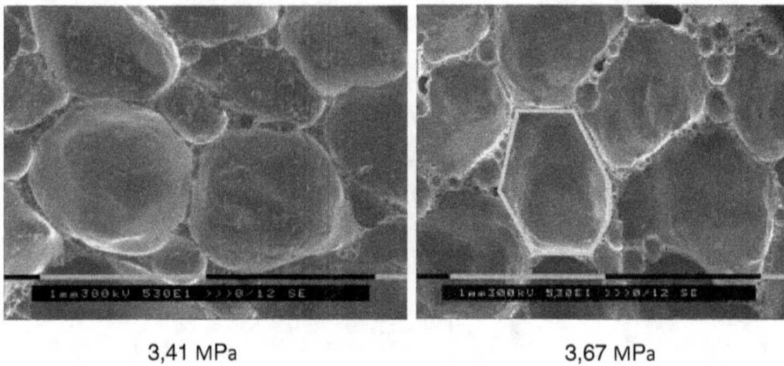

Fig. 13 Hexagonal shape of pores of the foam glass-crystal materials and their strength

- the fracture mechanism for foamglass-crystal samples is well described by the synergetic models of deformation of a quasi-viscous (amorphous) solid;
- the relation between the interpore partitions of foamglass-crystal material and the ultimate strength is similar to Hall–Petch relation;
- the fracture stress is in direct proportion to the pore size whose diminishing raises the strength of cellular materials.

5 Conclusion

The scientific principles of controlled formation of FGCM macrostructure and technological aspects of its production have been formulated in this chapter. A novelty of research findings is consisted in the development of physic-chemical principles of obtaining materials by a formation of nano and microcrystalline phase in a vitrified matrix of the silicate system. This method allows using a wide range of natural and industrial raw materials without glass melting unlike the traditional foam glass production. At this, the strength of foamed materials exceeds the strength of the foam glass 3–4 times as much due to the effect of nanoscale crystals in a vitrified matrix.

It has been shown that the physical and mechanical properties of foam glass ceramic material depend on the number and the size of a crystalline phase. The increase of mechanical strength as compared to the foam glass can be provided by the particle sizes of a crystalline phase of less than 1 μm. Maximum strength depending on the volume rating of the residual crystalline phase accounts for 25 % for less than 1 μm size, and 5–7 % at a size decrease down to 300 nm. With the increase of the crystalline phase size up to 10 μm and higher, the compressive strength of material decreases.

The FGC technology can be recommended as a safe technology for a large-scale industrial introduction. The research of this technology justifies both the safety of methods and high quality of products. At the same time, the problem of energy efficiency of technology have been solved (lowering energy costs by 1.3 times) and decrease of hazardous emissions to the atmosphere at the production stage (by 2 times on average).

Production of materials using this technology is supposed to be carried out at temperatures 850–900 °C what is significantly lower as compared to keramzit technology (1,150–1,200 °C) and traditional cullet-based foam glass (1,400–1,500 °C).

The development of compositions and technology of FGC which combines heat-insulating and constructive properties widens the nomenclature of construction products and facilitates the problem solving of heat-insulating material production which meet the requirements of fire and environmental safety.

References

1. Sokolova, S.N., Vereshchagin, V.I.: Die Verwendung von Zenosphiren in der Produktion von Demmstoffen. Ibausil, 20–23 September 2006, Weimar, Deutschland, Band **2**, S. 0861–0866 (2006)
2. Mueller, A., Vereshchagin, V.I., Sokolova, S.N.: Characteristics of lightweight aggregates from primary and recycled raw materials. Constr. Build. Mater. **22**, 703–712 (2008)
3. Lebullenger, R., Chenu, S., Rocherullé, J., Merdrignac-Conanec, O., Cheviré, F., Tessier, F., Bouzaza, A., Brosillon, S.: Glass foams for environmental applications. J. Non-Cryst. Solids **356**, 2562–2568 (2010)
4. Kazmina, O.V., Vereshshagin, V.I., Semukhin, B.S., Abijaka, A.N.: Low-temperature synthesis of granular glass from mixes based on silica-alumina containing components for obtaining foam materials. Glass Ceramic. **66**(9–10), 314–344 (2009)
5. Kazmina, O.V., Vereshshagin, V.I., Abijaka, A.N.: Expansion of the raw-materials source base for foam-glass-ceramic materials production. Build. Mater. **7**, 54–56 (2009)
6. Kazmina, O.V..: Principles of low-temperature technology of producing foam glass-ceramics from silica raw materials. Methods Technol. Silica. **17**, 7–17 (2010)
7. Kazmina, O.V., Semukhin B.S.: Safety processes for foamglass-crystal materials producing in a new two-stage technology. Vestn. TSU Architec. Build. **1**, 123–132 (2012)
8. Sarkisov, P.D.: Production of multifunctional crystalline glass materials by directional crystallization of glass. Mendeleyev PKhTU, p. 218 (1997)
9. Lyakishev, N.P.: Nanocrystalline structures—a novel perspective of construction material development. Vestn. Russ. Acad. Sci. **73**(5), 422 (2003)
10. Kazmina O.V., Vereshshagin V.I., Semukhin B.S.: Structure and strength of foam-glass-crystalline materials produced from a glass granulate. Glass Phys. Chem. **37**(4), 29–36 (2011)
11. Wu, J.P., Boccaccini, A.R., Lee, P.D., Rawlings, R.D.: Thermal and mechanical properties of a foamed glass-ceramic material produced from silicate wastes. Eur. J. Glass Sci. Technol. Part A Glass Technol. **48**(3), 133–141 (2007)
12. Glezer, A.M.: Amorphous and nanocrystal structures: similarities, differences, intertransfers. Russ. Chem. J. **46**(5), 57–63 (2002)
13. Gusev, A.I., Rempel, A.A.: Nanocrystaline Materials, p. 224. Fizmatlit, Moscow, ISBN 5-9221-0039-4 (2001)
14. Chukin, G.D.: Surface Chemistry and Disperse Silica Structure, p. 172. Paladin Publishing House, UK, « Printa » (2008)

15. Golubev, E.A: Hypomolecular Structures of Natural-Ray Amorphous Materials, pp. 155. Publishing House of RAS (Urals Branch), Ekaterinburg (2006)
16. Golubev, E.A.: Scanning probe microscopy data on the globular structure of higher anthraxolites. Doklady Akademii Nauk. **425**(4), 519–521 (2009)
17. Golubev Ye. A.: Scanning probe microscopy in researches of micro- and nanostructure in noncrystalline geomaterials. Microsc. Microanal. **9**(S3), 304–305 (2003)
18. Beryozkin, V.I.: Fullerenes as nuclei of soot particles. FTT. **42**(43) 567–572 (2000)
19. Argon, A.S.: Inelastic deformation and fracture of glassy solids. In: Mughrabi, H (ed.) Material Science and Technology, p. 63. VCH, Weinheim (1993)
20. Jiang, W.H., Atzmon, M.: Acta Mater. **51**, 4095 (2003)
21. Jiang, W.H., Atzmon, M.: Scripta Mater. **54**, 333 (2006)
22. Jiang, W.H., Pinkerton, F.E., Atzmon, M.: Appl Phys **93**, 9287 (2003)
23. Olemsky, A.I., Khomenko, A.V.: Phenomenological equations of fluid vitrification. J. Eng. Phys. **70**(6), 6–9 (2000)
24. Zhurkov S.N., Betekhtin V.I., Bakhitbaev A.N.: Solid State Phys. **11**(3), 690 (1969)

High Temperature Polymer Nanocomposites

K. Balasubramanian and Manoj Tirumalai

Abstract The chapter on High Temperature Polymer Nanocomposites mainly covers the advancements made in the research on varied composite materials and their novel innovations for high temperature applications. The chapter begins with a prelude on existing polymers of thermoplastic or thermosets and also a combination of the two to exploit the dual advantage of both high temperature thermoplastics and thermosets. The polymer composites are developed and implemented for varied temperature regimes in the range of 120–250°C, 250–350°C specially suited for aerospace applications. Further research work is under progress to explore the material suitable for temperatures above 350°C. The initial section explains on the applicability of polymer composites having matrices such as either liquid crystalline polymer, cyanates, polyimides or bismaleimides which are reinforced with either the carbon, glass or aramid fibres especially continuous fibre which provides improved material and physical properties. The polymer-composites are assessed for performance of the matrices and the composites not only for their strength or stiffness but also for their resistance to cracking, minimum loss of weight, brittleness due to cross-linking and other properties that may degrade the performance of the composites over the long period of high temperature applications. This section mentions on the literature reviews having come up with research in progress on other types of resins for high temperature applications namely the oligomers such as the acetylene terminated polyimides or the norbornene terminated polyimides. The second section is concerned with the polymer-nanocomposites for high temperature applications. In this section, the role of nanofillers in the enhancement of composite properties is discussed. The various forms of nanofillers are the nanoclay, nanofibres, carbon

K. Balasubramanian (✉)
Head: Materials Engineering, DIAT(DU), Girinagar, Pune, Maharashtra 411025, India
e-mail: balask@diat.ac.in

M. Tirumalai
Department of Airforce, MILIT, Girinagar, Pune, Maharashtra 411025, India
e-mail: meetkbs@gmail.com

J. Njuguna (ed.), *Structural Nanocomposites*, Engineering Materials,
DOI: 10.1007/978-3-642-40322-4_7, © Springer-Verlag Berlin Heidelberg 2013

nanotubes, Polyhedral OligoSilsesquioxanes (POSS) and nano-oxides where they reinforce polymer chain at molecular scale as against the carbon fibres of macroscopic scale seen in polymer-composites. In a nanocomposite the thermal properties depends on the filler's nature, rate and dispersion in a matrices. The applications have been as fire retardant materials or for re-entry applications in aerospace vehicles. Research on nanofillers is still at its nascent stage that ample scope is available for exploring the potential of various nanofillers for high temperature applications and as fire retardant materials. Some polymer thermosets such as PMR-15 (Polymerization from Monomeric Reactants) are the most extensively used resin for applications where long term thermal stability is required at 300°C. This section dwells on other progress under way in the field of nanocomposites as high temperature nanocomposites. The last section presents a review on advanced materials for high temperature applications, especially applicable for the next generation aerospace vehicles. The issues concerning adhesives for joining of surfaces exposed to high temperatures, the effective role of nano-fillers for improving fracture toughness at high temperature etc. It is seen that the most promising resin for high temperatures has been the polyimides (315°C). This resin has high Tg (400°C) and other good characteristics such as good micro cracking resistance, low moisture absorption and low toxicity. There are various grades of polyimides under development. Overall, the chapter provides a glimpse on the effective developments in the field of polymer-composites and nanocomposites for high temperature applications especially for aerospace vehicle and automotives.

1 Prelude on Polymer Composites for High Temperature Applications

1.1 Introduction

1. The appearance of polymers in 19th century and the popularity of composites since last decade has revolutionised the scope of materials with substitution of metals and non-ferrous alloys by polymer-composites. The area of application has been from the least significant parts of aerospace vehicles, medical equipments, household equipments, technology equipments for automobiles, and other science and engineering products to the most significant parts [1]. The application has been progressively being made for high temperature requirements [2–7].
2. The application of fibre reinforced composites has been mainly seen on aerospace vehicles where the parts are subjected to a temperature range of 150–400 °C. The parts affected with such temperature range are the airframe skin, the aero-engine parts and the surrounding areas, and the parts exposed to hyper velocity or supersonic speeds. The commonly used matrices are the liquid crystalline polymers, cyanates, polyimides and bismaleimides [2, 8, 5, 6, 9]. The

matrices are reinforced with carbon, glass or aramid fibres especially continuous fibre which provides improved material and physical properties [2, 8, 9]. The epoxy based carbon or the glass or the aramid composites have been extensively used for load bearing structural applications on the aircraft. The future generation aircraft or any other high temperature applications of the range of 120–400 °C [5] would require the matrices to withstand the brunt of the high temperatures based on the area of application. In case of aircraft the high temperature zones have been the aircraft skin of supersonic/hypersonic aircraft, the leading edges of the wing and control surfaces and around the aero-engine regions even at the exhaust end where high exhaust plumes are exposed on the exhaust plates which are subjected to temperature stresses and strain.

1.2 High Temperature Composite Specialty

3. The matrices which are most popular will be the polyimides. The polymer matrices-composite behaviour depends on the interfacial bondage. The composite needs to retain properties at high temperatures, maintain glass transition temperature (~ 300 °C) without being affected by the presence of the moisture i.e. low moisture absorption of nearly 4 %, retain thermal stability, overcome ageing affects and do not get affected by oxidation. The polymer matrices are developed in the high temperature ranges viz, 120–250 °C, 250–350 °C and 350 °C and above [5].
4. The polymer matrices such as the cyanate esters (CE), BMI, phenolics, polyimides have been used for temperature application between the range 120–250 °C, for the range 250–500 °C the polyimides are found suitable and for temperatures beyond 350 °C the research is in progress to find the required material [5].
5. The epoxies were found to mostly suit for a temperature application of 135 °C limits. The BMI has been found to be thermally stable and having thermo-oxidative stability [6]. The various mechanical properties of a Polymer Composite Matrices as applicable to mainly aircraft structures have been the tension providing the longitudinal modulus (GPa)/strength (MPa), transverse modulus (GPa)/strength (MPa), the compression also providing the longitudinal modulus (GPa)/Strength (MPa) and transverse modulus (GPa)/strength (MPa). The carbon fibre has the Young's Modulus given in the range 220–230 GPa, Ultimate Tensile Strength 3 GPa. Over the period the advancement in polymer composite research has been in improving these properties by performing structural modifications in the polymers like the bismaleimides by reducing the cross-link density and having cross-linking through chain extension. The properties are further refined for damage tolerance by adding fine particles of polyimides which are high temperature materials. The primary structures which are subjected too high temperatures due to kinetic heating during flight have such composites of carbon/bismaleimide that can sustain a temperature of 180 °C.

1.3 Research Review on High Temperature Composites

6. Promising work is in progress at various countries to improve the thermal stability and curing behaviour of high temperature polymer-composites. In one such example, is the research going on with the addition of amines with bismaleimides/carbon. The other areas of application of high temperature composites have been especially on aerospace vehicles for the radomes, stealth aircraft composites, antennae covers where the surface temperature are high and need to be contained. The Cyanate esters though less superior than the bismaleimides have better dielectric properties and low absorption of moisture. The field has been further expanding from the usage of epoxies as matrice material to thermoplastics. Of the two types of thermoplastics namely the amorphous or the crystalline, it is seen that the crystalline thermoplastics are finding large applications. The crystalline thermoplastic show slow decrease in modulus above T_g (Glass Transition Temperature). The thermoplastics show high fracture toughness (1–2.5 kJ/m^2) and impact strength. Thermoplastics have shown lower compressive strength than the thermosets, however the resistance to environmental effects and better inter laminar shear stress show better applications especially for area of impact and compression. The dual properties of both amorphous and crystalline thermoplastics have been extracted when combined; examples are PEEK, PEI etc [8, 5].

7. The process of improving the high temperature polymer performance is by polymerising more of aromatic heterocyclic units and minimise the aliphatic contents. In the temperature range of 250–350 °C, the groups of polyimides have been used be it the condensation or the addition types. Most of these are commercially available in United States. The condensation type has certain drawback due to volatiles that skill is required to produce low void composites.

8. The polyimides resins obtained through the polymerisation of monomer reactants have been taken for composite applications exposed to high temperatures in the range of 260–288 °C. Therefore in polyimides of condensation types, the reduction in aliphatic content has improved the thermo-oxidative stability and improved toughness due to reduced crosslinking.

9. The thermoplastic matrices under high temperature applications yield to cracking and loss of weight and are subjected to repeat thermal cycling. The compressive and shear properties of the matrices get adversely affected with high temperature exposures.

10. Long term applications of polymer-composites for high temperature applications need to assess the performance of the matrices and the composites not only for their strength or stiffness but also for their resistance to cracking, minimum loss of weight, brittleness due to cross-linking and other properties that may degrade the performance of the composites over the long period of high temperature applications. The high temperature polymer-composites have been extensively researched for applications on aero-space vehicles especially for the skin, aero-engine parts and engine areas (nozzle, liners etc.,)

and the re-entry requirements for the space vehicles (ablative materials) and armament ammunitions (anti ballistic missiles). The research is mainly seen to improve composite composition by improved matrices synthesisation, selection of matrix material with compatible fibre reinforcement, catalysts, additives and solvents to achieve higher glass transition temperature and other mechanical properties useful for prolonged high temperature applications. The degradation due to ageing needs to be considered for improving the materials properties for longer application endurance with time.

11. The other mostly used thermoset-composite will be the phenolics which though are not suitable for load bearing structure are found applicable for high temperature applications especially as ablative and erosion resistance materials in composite form with carbon fibre/silica reinforcements. The other improvement in thermal property will be the char yield which normally is assessed for 70 % at a high temperature depending on the resin composition. Several works are in progress toward this to improve the temperature limits for the char yield of 70 %. The experiments of phenolic-carbon or silica have shown this property at 700 °C with the resin cured by functional groups like maleimides propogylether and phenyl ethynyl while the resin has shown further improvements with cyanates with the temperature limits raising up to 1,000 °C.

12. Literature reviews have come up with research in progress on other types of resins for high temperature applications namely the oligomers such as the acetylene terminated polyimides or the norbornene terminated polyimides which are experimented for suitable replacement as matrices for the phenolics. The process limitations and cost economy has held back the actual applications. As on date, the research still continues on these other oligomers to arrive at cost effective and better polymer-composites for high temperature applications.

13. Therefore polymer-composites have found its usage for various high temperature applications especially connected to aerospace vehicles have been the aircraft skin, engine nacelles, panels, firewalls, engine ducts, nozzle liners, ablative materials, fins for ammunition stores (missiles), cases for helicopter gear box etc. The temperature applications cover a range of 150–250 °C i.e. the aircraft skin areas and 250–350 °C for the engine areas. The thermoplastics such as BMI, CE, and Thermoplastic Polyimides are for lower temperature applications (150–250 °C) and the toughened polymerisation of monomer reactants and the oligomers are for higher temperature ranges (250–350 °C). The latest research on liquid crystalline polymers and phthalonitrile are found to be the latest resins of interest for high temperature applications which are cost effective and less complex in processing with possible adaptation for fine tuning with the autoclave moulding followed for epoxy composites.

14. The applications may be extended to any other field of science and engineering such as the automotives, the ships, thermal and nuclear engineering [10, 11] etc.

2 Polymer-Nanocomposites for High Temperature Applications

2.1 Introduction

1. The polymer composites gave an insight to the various mixtures of polymers with primary reinforcements such as the carbon, glass or the aramid fibres and the silicates for the high temperature applications. The search for improved material properties for varied science and engineering applications gave rise to the birth of nanofillers as substitutes for carbon fibres and other forms of composite reinforcements [12]. The nanofillers are of the size less than 100 nm which are functionalised or surface modified for better interfacial adhesion with the bulk matrices which may be either thermoplastic or thermosets.
2. The nanofillers due to their superior multi-functional properties are being researched as an effective replacement for the exiting carbon fibres and other primary structural reinforcement materials under application for high temperature applications [13, 14]. The drawbacks in polymer–composite mixture of process or synthesisation complexity and cost have been taken care in the use of polymer-nanocomposites. There are several literatures available that describes the nanofiller synthesisation, characterisation and properties [15, 16] along with the [17, 18], various applications of these nanocomposites (polymer-nanofiller) in medical, bio-engineering [17, 18], automotive, civil engineering, nuclear and thermal plants and maximum in aerospace vehicles.
3. The polymer-nanocomposites have been even researched as repair agent for high temperature polymer-composites. One such example is the Bisphenol E Cyanate Ester with alumina nano-particles ($\sim$40 nm) [19]. The repair of laminate composites are undertaken by use of such polymer-nanocomposites by injecting low viscosity polymer-nanocomposites which provides desired toughness and better mechanical properties. The filler will have high T_g (>270 °C) and decomposition temperature which permits the repair of high glass transition temperature composites.

2.2 High Temperature Nanocomposite Salients

4. The various forms of nanofillers are the nanoclay, nanofibres, carbon nanotubes, Polyhedral OligoSilsesquioxanes (POSS) and nano-oxides. The nanofillers reinforce polymer chain at molecular scale as against the carbon fibres of macroscopic scale seen in polymer-composites. One of the significant applications of polymer-nanofiller composites have been as an effective fire retardant material. Nano-composites therefore are a group of nano materials wherein a nanoparticle is dispersed in a matrices resin or multi-phases. The nanoparticle

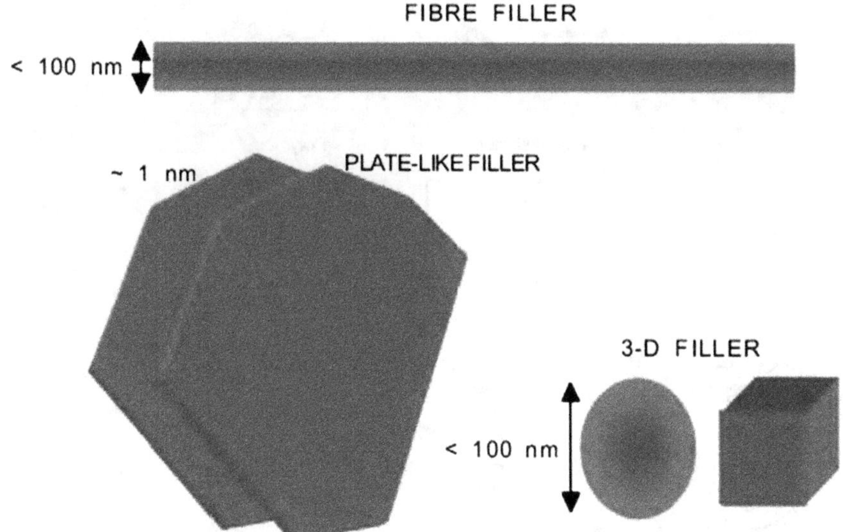

Fig. 1 Nanofiller scale [20]

could be of 1D (a plate) or a 2D (tubes and fibres) or a 3D (nano-metric silica beads) sized nano particle of dimension <100 nm (Fig. 1) [20].

5. The effect in significant properties is achieved by addition of nanofillers in the range of 1–10 wt%. These nanofillers may be mixed in addition to the conventional additives and composite fillers such as carbon black, carbon fibres, silicates etc. The nanoalumina as a filler for repair of composites and the nano antimony tin oxides as good fire retardant material have been found suitable for high temperature applications.

6. The blend of polymers-nanofillers is seen to be better whenever the nanofillers are surface modified or functionalised for good interfacial adhesion and cross-linking. The nanofillers when added either forms regions in the resin matrix forming multi-phases or the polymers intercalates between the nanofillers or exfoliates and disperses throughout the polymer resin (Fig. 2) [21, 20].

2.3 Research Review on Nanocomposites for High Temperature Applications

7. It is seen from the literatures that functionalisation of carbon nanotubes with Poly (acryloyl chloride), PACL reacts well with epoxy monomers that the epoxy grafted CNT adheres well to the epoxy resin matrices that the nanocompsoite formed has better thermal–mechanical properties. This is achieved with low wt% addition of CNTs (i.e. 0.1 to 1 wt%). Experiments have shown

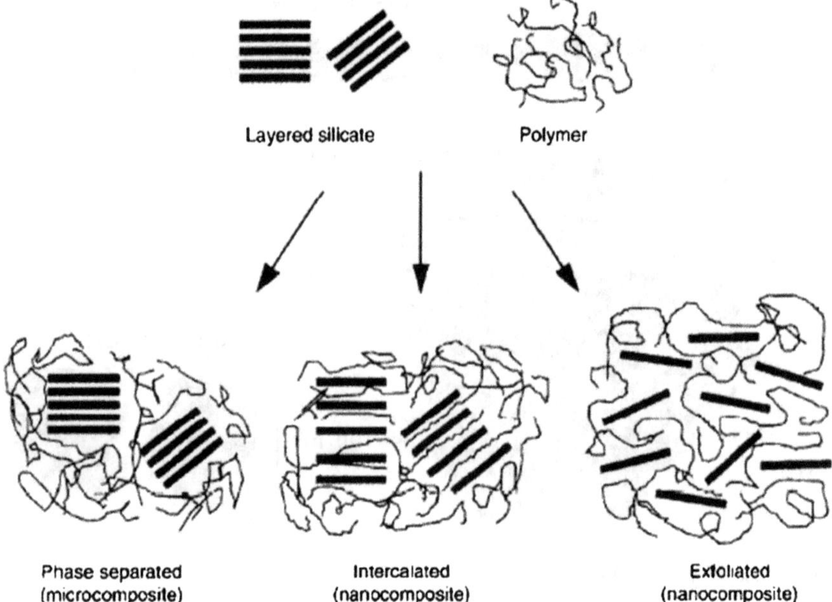

Fig. 2 Three main types of layered nanofiller (say silicates) in polymer matrix [21]

that the addition of Multi Walled CNTs (0.5 %) in epoxy resins have improved the T_g from 167 to 189 °C.

8. In a nanocomposite the thermal properties depends on the filler's nature, rate, and dispersion in a matrices. The combination of traditional composite fibres with nanocomposites has proved to be more beneficial with capabilities to sustain larger temperature range than the traditional composites. One such example is the addition of organophile montmorillonite.

9. The high temperature polymers are usually composed of aromatic or heterocyclic rings which are blended by flexible linking groups. The literature studies have brought out that high temperature polymers have high T_g and good mechanical properties. The chemical structures of commonly used high temperature polymers as brought out in the previous sections are shown in Fig. 3a–d [5, 19, 22].

10. PMR-15 (Polymerization from Monomeric Reactants) thermoset polymer is the most extensively used resin for applications where long term thermal stability is required at 300 °C. This polymer has its own disadvantages such as complex processing with high cure temperature needs and toxic starting materials. Group of Cyanate esters is the other popular polymer which is finding applications in the high temperature applications especially for aerospace and microelectronics sectors.

11. The various research objectives on the high temperature nanocomposites have been the following [19]:

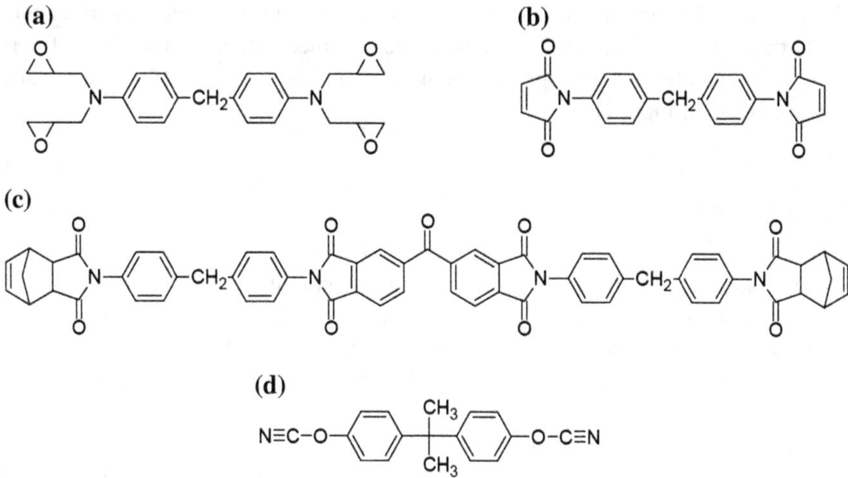

Fig. 3 Chemical structure for high temperature thermosetting polymers. **a** Epoxy resin: MY270. **b** Bismaleimide: 4,4′-bismalimidodiphenyl methane. **c** PMR polyimide: PMR-15 resin. **d** Cynate ester: 2,2′-bis(4-cyanatophenyl)propane [5, 19, 22]

11.1. **Curing Kinetics of Polymer**. This is investigated by Differential Scanning Calorimetry. In this many reaction models are taken to model isothermal curing. The model is analysed for activation energy and reaction orders.

11.2. **The Effect of Nanofillers on the Curing Kinetics of Polymer**. In this the cure kinetics of polymer with varying loadings of nanofiller is tested. The catalyst is not added to the mix. The isothermal curing of the polymer/nanocomposite is determined.

11.3. **Rheology and Dynamic Mechanical Analysis of polymer/nanocomposites**. Nanofillers are functionalised which are characterised by using Fourier Transform IR and Thermogravimetric Analysis (TGA). The dispersion of functionalised nanofillers in the polymer matrices is found to be good. Here, the rheology and rheokinetics of the polymer suspension is evaluated. The Dynamic Mechanical Analysis determines the thermo-mechanical properties for nanocomposites with bare nanofiller and functionalised nanofiller. The change in glass transition temperature is found and the dispersion is seen in a transmission electron microscope.

11.4. **Creep Behaviour of Nanocomposite**. The creep property of nanocomposite is studied through tensile creep testing using DMA. The tensile creep interval is measured at isothermal conditions. The temperature time superposition principle predicts the long term creep behaviour in a high temperature environment.

12. The nanofiller having increased number of hydroxyl groups on the surface will have greater catalytic effects on the cured polymer. The addition of nanofillers will bring about a change in chemistry, curing, crystalline structure, and matrix chain mobility.

2.4 Applications

13. One such application of high temperature conditions will be as ablative materials in nozzles of high speed space vehicles. The selection of filler irrespective of micron or nano size is of importance. The selection of composites/nanocomposite should be such that there is proper bonding of the ablative material with the metal substrate of the nozzle and also serve the purpose of correct transfer of the hot gases of the propellant through the nozzle without damage to the metal structure. The process of manufacture of this ablative material will govern the adequacy of the density, percent resin content, compressive strength, inter laminar shear strength, thermal conductivity, coefficient of thermal expansion, and tensile strength. The example of such a material showing adequate properties for high temperature ablative material is the Carbon-phenolics [23]. This carbon fibre of micron size when replaced by nanofiller will enhance the composite properties many folds due to the better interfacial adhesion between the matrix and the filler with the generation of large surface area and high aspect ratio. Also the thermal conductivity improves due to low percolation thresholds which could be further improved by functionalising the nanofillers. The other ablative materials used depending on the location in the Solid Rocket Motor (SRM) nozzle are the Glass cloth-phenolic and the Silica cloth–phenolic. The ablative properties of C-Ph material can also be improved by adding zirconium diboride particles as additives (Fig. 4) [24].

14. In common a high temperature material especially of nanocomposites show improved thermal stability and high Hot Distortion Temperature (HDT). The thermal stability is improved by commonly incorporating nanoclay into the polymer matrix as it performs as a superior insulator and mass transport barrier to the volatile products generated during decomposition.

15. The research literatures have brought out that the elastic and thermal properties of CNTs vis-a-vis carbon reinforcing fibres have been found to be nearly four times high. The CNTs are uniquely stiff fibres with tensile strength ranging above 50 GPa with the strength to weight ratio being nearly four times larger than the carbon reinforcing fibres. The SWCNT has a greater strain as compared to any other structural material unlike the negligible strain seen in polymer-composite material where the fracture is sudden without prior sign of failure. The thermal conductivity is also high in the range of 1,750–5,800 W/mK.

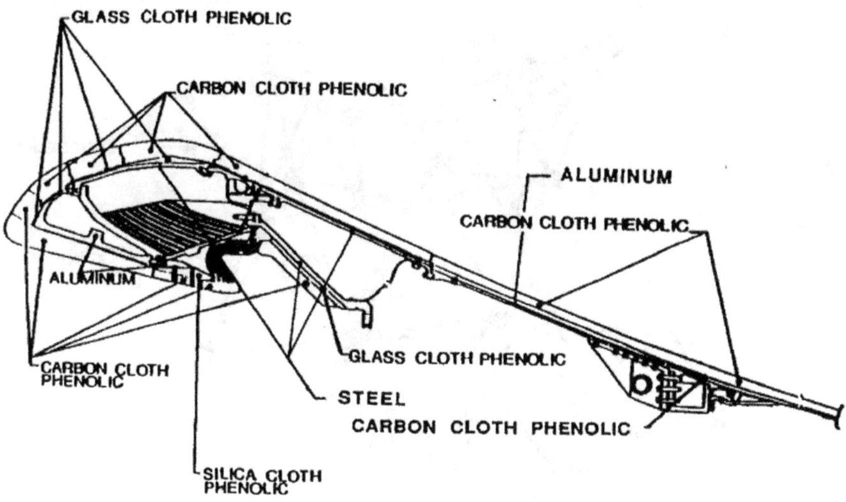

Fig. 4 Ablative materials in a solid rocket motor of a nozzle [24]

16. *Layered silicates as fillers.* Nanocomposites are layered with silicates of a thickness of around 1 nm. The aspect ratio (length to thickness) is high over 1,000 due to the lateral dimension ranging wide. The layer forms gap between them leading to interlayer. The interlayer distances are increased by the exchange of cationic surfactants thereby reducing the action of these layers form stacks with a gap in between them called the interlayer or the gallery. Isomorphic substitution within the layers (Mg^{2+} replaces Al^{3+}) generates negative charges that are counterbalanced by alkali or alkaline earth cations situated in the interlayer [20] (see Fig. 5). The inorganic cations within the interlayers can be substituted by other cations. The exchanges with cationic surfactants, such as bulky alkylammonium-ions, increase the spacing between the layers and reduce the surface energy of the filler. Therefore these modified fillers (called organoclays) are more compatible with polymers and form polymer-layered silicate nanocomposites. Montmorillonite, hectorite and saponite are the most commonly used layered silicates [21, 20, 25,26, 17, 18].
17. Nanoclay has been widely researched as potential nanofiller for flame retardancy [14]. The nanocomposite materials such as the organoclay plus aluminium oxide and series of polypropylene plus organoclay have found applications as flame retardant materials. These materials have been commercialised. Even the nanofillers like the CNTs or any other alternatives as the Graphite Oxides or the Graphenes have been experimented to enhance the mechanical, electrical and flammability properties of the otherwise traditional polymer-composites. The graphene nanofillers in range of ceramics have improved their toughness making them suitable for high performance structural applications.

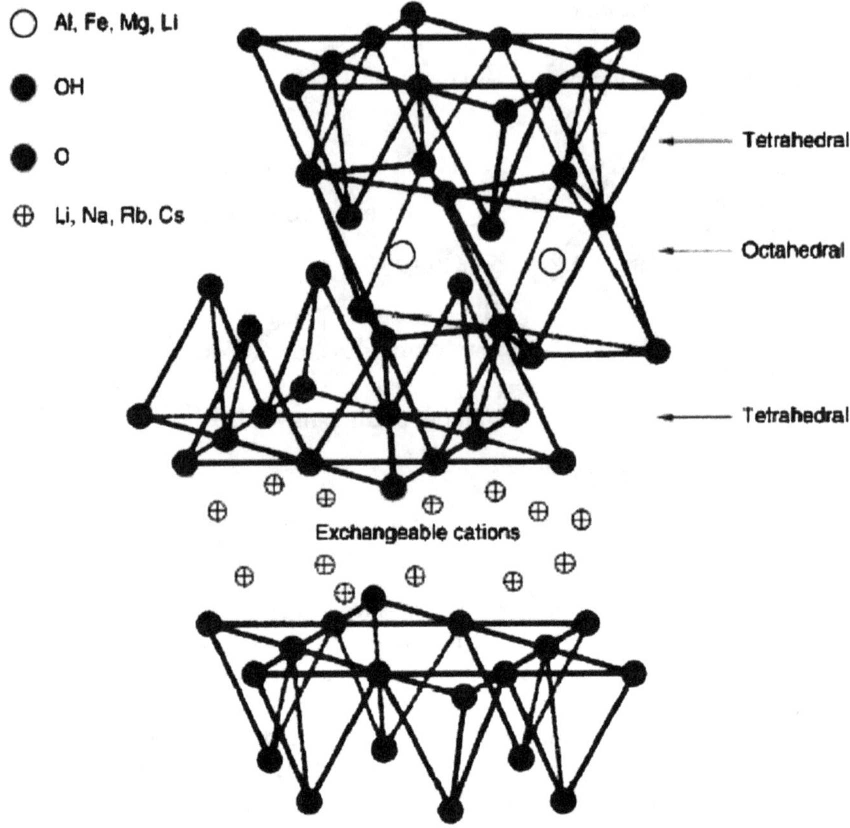

Fig. 5 Layered silicates [21, 20]

18. The nanoceramics are finding its foothold as candidates for high temperature applications. The reasons of having nanocomposites over traditional polymer-composites are for their following benefits: -
  (a) Improved strength, modulus and dimensional stability.
  (b) Higher thermal stability and HDT (Heat Distortion Temperature).
  (c) Better flame retardant properties and low smoke emissions
19. Nanoceramics have shown to have the abovementioned properties and a suitable material for engine applications subjected to temperatures of around 1,400 °C with strength maintained at 300 MPa. The material also exhibits negative creep rate a desirable property for high temperature applications. The other salients exhibited by nanoceramics over the traditional polymer-ceramic composites are the high resistance to oxidation, sub-critical crack growth and thermal shocks.
20. Nanocomposites such as SiC (Silicon Carbide)/Si_3N_4 (Silicon Nitride) composites, are another material exhibiting improved strength, creep and fracture toughness [10, 11, 27–32]. The material retains strength at higher temperatures

and has good resistance to creep. The Fig. 6 shows the microstructure of polycrystalline Silicon Carbide (SiC)-Silicon Nitride (Si3N4) nanocomposites. The structure contains filler Si_3N_4 of 0.8–1.5 µm and the polymer SIC of 200–300 nm. The grain boundary thickness is seen to be around 50 nm. The other similar types of nanocomposites suitable for high temperature applications or under research will be the Graphene or Graphite Oxide/SiC, carbon nanofibres/SiC-Aluminium Oxide (Al_2O_3)/and $TiN-Si_3N_4$.

21. The ongoing research on nanoceramics at Multiphysics Laboratory at Purdue mainly focuses in understanding the performance of high temperature based nanoceramics especially of nitrides and carbides. The materials so developed will be helpful for power generation applications, nuclear applications, and aerospace applications.

22. The research work is focusing on following areas.

 (a) To develop materials of low conductivity by studying effects of material morphology changes on thermal conductivity using simulation techniques such as Molecular Dynamics (MD).

 (b) To develop ceramic nanocomposites (nitrides and carbide based) capable of withstanding high temperature environments even meant for nuclear applications. The advanced ceramic nanocomposites materials are researched to have the capability to withstand temperatures of the order of approx 1,800 K and

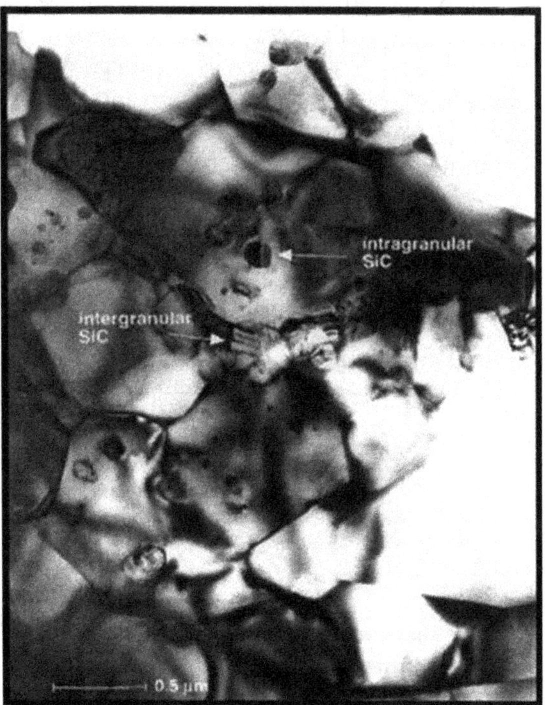

Fig. 6 Actual microstructure of a SiC-Si3N4 nanocomposite [31]

above [33]. The usage of such materials is experimented for various functions like the Multi-functional material or as structures for extreme temperature applications (Nuclear) [10, 11] or as sensors in high temperature zones.

23. The other area of research in nanodomain for high temperature applications have been the nanocomposites of amorphous (Si, O and C) with nano SiC or C fillers [30]. The research has shown that these materials have required properties and characteristics for a wide range of high temperature usage. The nanocomposite exhibits ceramic like properties of negative creep rate and anti-oxidant. The Fig. 7 shows one of the applications of PyroSic. The pyroSic have been found much better than the conventional CFRPs.

24. The pyrolysis of pre-ceramics has been found to be simple as compared to the conventional processing of CFRP. The tooling and process are seen to be less complex as compared to Ceramic Matrix Composites.

25. The other processes evolved for improvement of ablative resistance of composites such as the C-Phenolics has been by introduction of Zirconium diboride particles [34, 35]. Sufficient research is in progress on this subject with various combinations of nanofiller added to C-Ph composites to improve ablation resistive property and also provide reduced smoke emission [23, 35]. The resistance is due to the formation of oxides of Zirconium and Barium. The research in this field has shown that the ablative rate has been higher as compared to Carbon fibres in C-Ph composites. The ablative resistance properties have been improved by adding modified phenolics in the C-Ph matrix [34–39]. The modification is obtained by treating with boric acid or POSS (Poly Oligomeric Silsequioxanes), or H_3PO_4. This treatment gives a reduced erosion rate thereby retarding ablation. Research has seen further improvements in linear ablation rate by having composites of SiC Ceramic, carbon fibres and boron modified phenolics. The Nanosilica powder modified rayon-based composites show better ablation resistance, reduced thermal conductivity and higher ILSS (inter-laminar shear strength) at a controlled quantity. In this, process of improvement involves the modification of either carbon fibres or the phenolics. Researchers find that the effects of ZrB2 on the ablation performance of C–Ph composites are still unclear. The research

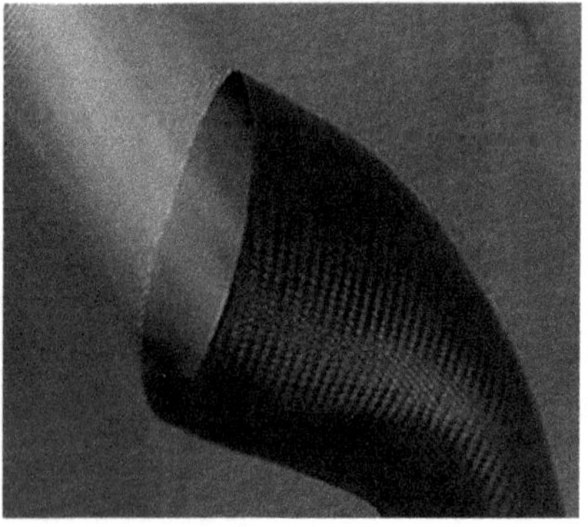

Fig. 7 Exhaust duct made out of PyroSic

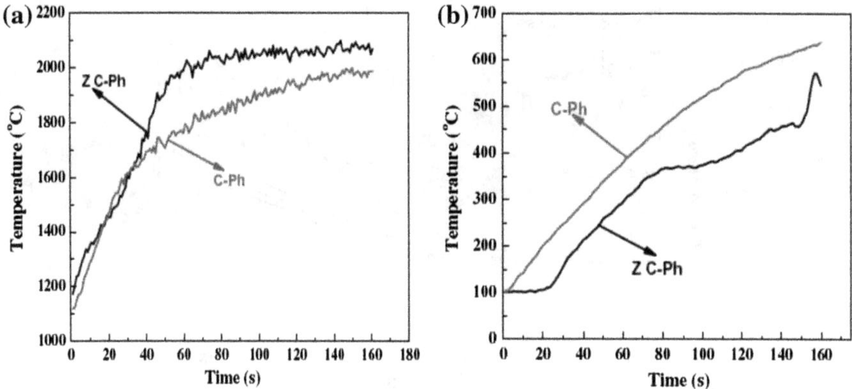

Fig. 8 Surface temperatures of Z C-Ph composites and C-Ph composites (**a**) the ablated surface (**b**) the back surface [35]

papers have shown that the linear ablation rate of Z C-Ph composite has reduced by 79 % compared with the un modified C-Ph composite [35]. The ablation surface though at higher temperature during ablation process the back surface temperature is found to be lower by 100 °C as compared to unmodified composites. This is shown in the Fig. 8. Therefore this material may become the main material for high temperature structural applications.

26. The other promising research for better ablative material or as heat shields for re-entry applications has been the resol type phenolic resin with kaolinite and asbestos cloth nanocomposite [40]. The material has been tested for an external heat flux of 8×10^9 W/m^2 and 3,000 K. The material was tested for its ablation, charr formation and thermal degradation performance for its usage as heat shields. The temperature is withstood by the formation of a strong refractory char on the insulator surface. Insulator later degrades to a char layer at higher temperatures due to pyrolysis. Recent advances in polymer layered silicate nanocomposites are seen as potential ablative materials [41, 42]. The reason being on pyrolysis, the organo–inorgano nanostructure reinforcing the polymer are transformed into a uniform ceramic char, leading to increased resistance to oxidation and mechanical erosion in comparison with composite materials.

27. *Ablation performance.* Experiments on ablation performance of composites (asbestos-Ph) and nanocomposites (Asbestos-Ph-Kaolinite) [40] have shown the following:-

   (a) The back surface temperature of a nanocomposite is found approx 43 % lower than composites (Fig. 9) [40].
   (b) The Heat Release Rate (HRR) and mass loss @ 80 KW/m^2 (Fig. 10) [40] show that to some stage i.e. up to 100 s of the Oxy-acetylene flame test both composites and nanocomposites behave thermally same. After 100 s, with the surface temperature increasing to 1,000 °C the rate of mass loss and HRR reduces for a nanocomposite. The nanocomposites show a reduction of nearly 35 % HRR and 22 % mass loss.

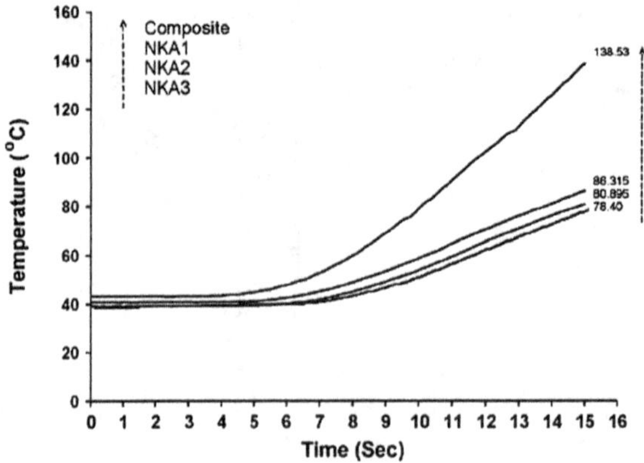

Fig. 9 Temperature distribution of back surface (composite and nanocomposite) [40]

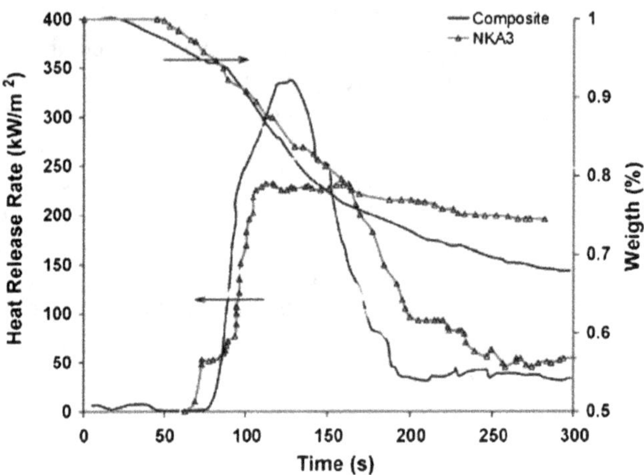

Fig. 10 Comparison of HRR and mass loss @ 80 KW/m^2 [40]

3 Advanced Thermal Protection Materials for Next Generation Aerospace Vehicle Designs

3.1 Introduction

1. The research is under progress at NASA to establish high temperature materials for various aerospace applications such as the structures, adhesives that are used for bonding of composites and the thermal protection system namely fire retardants (coatings). The work is progressing to reduce the thickness of the thermal protection system by having high temperature resistant structures and adhesives.

3.2 Review on Advanced Materials for High Temperature Applications (HTA)

2. The high temperature adhesives are those that have the working temperature ranging from 315–400 °C (600–700 F). The adhesive must also satisfy the bond-line temperature requirement between the TPS-substrate of being above the traditional limit of 250 °C. The adhesive's bond-line (Fig. 11) integrity needs to be maintained for high temperatures and high heating rates [43].

3. The structure which is of honeycomb structure has graphite or aluminium-epoxy face sheet and aluminium honeycomb. This material may be improvised by using nanofillers with low wt% loadings and deriving many fold benefits of thermo-mechanical properties. The thickness of the thermal protection system depends on the allowable temperature for the bond line. The research will be with the aim to increase the bond line temperature (>250 °C) so as to reduce the TPS and hence increase the payload with overall reduction in structural weight. The need for high temperature adhesives will be to withstand high temperatures for short period of time and for one or more loading cycles. The test on adhesive bonding with substrate is carried out to check whether the failure will be due to pure adhesive peeling off from substrate surface which is the adhesive failure or the failure of the adhesive which demonstrates the cohesive failure. The desirable test is to check the cohesive failure which speaks more on the bond line integrity. Some clippings of the NASA tests are shown below at Figs. 12 and 13.

4. The primary requirement of the adhesive is to retain some if not all of its strength at high temperatures. Polymer-composites such as Polyimides group, Cyanate ester group, BMI have shown good strength at high temperatures and the properties profoundly improve when nanofillers such as nanoclay, nanoalumina, nanotubes etc. are added.

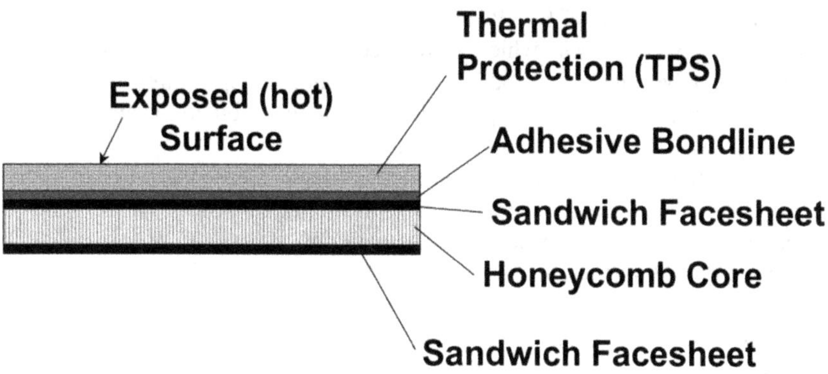

Fig. 11 Aerovehicle structure [43]

Fig. 12 Adhesive failure [43]

Fig. 13 Cohesive failure [43]

5. The high temperature structures mainly of carbon reinforced fibres with resin for polymer-composites may be replaced with nanofillers or additionally added to the existing phases for strength improvement. The limiting factor at high temperatures is the performance of matrices and their interfacial binding with the nanofillers or the composite fibres. The most promising resin for high temperatures have been the polyimides (315 °C). This resin has high T_g (400 °C) and other good characteristics such as good micro cracking resistance, low moisture absorption and low toxicity. There are various grades of polyimides under development.

6. The applicability of a material depends on its performance during the coupon tests which are carried out for determining the tensile strength/modulus, shear strength/modulus (Fig. 14), for laminates the de-lamination at high temperatures which is seen in case of CFRP composites [43],

7. The other type of resin which is less complex for processing will be the modified polycyanates. The ablators for advanced temperature systems have been the silicon based or the phenolic based with nanofllers or high grade of fibres to increase the stiffness and strength.

8. The potential impact of advancing composites will replace the usage of metals to composites for high temperature applications especially in the field of aerospace vehicles. A typical impact of materials is given in Fig. 15 [44].

9. The use of high temperature composites/nanocomposites will definitely reduce the weight of the structure and increase the payload capacity for useful requirements of space research. The high temperature capabilities will make it possible to have entire aerospace vehicles to be of composites. An example of a

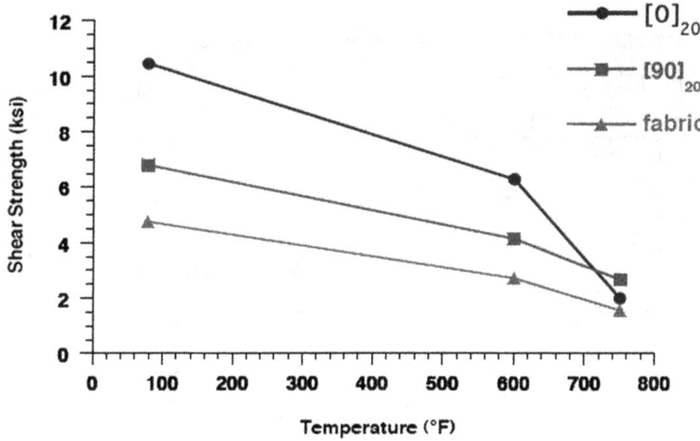

Fig. 14 Effect of temperature on shear strength on three different laminates [43]

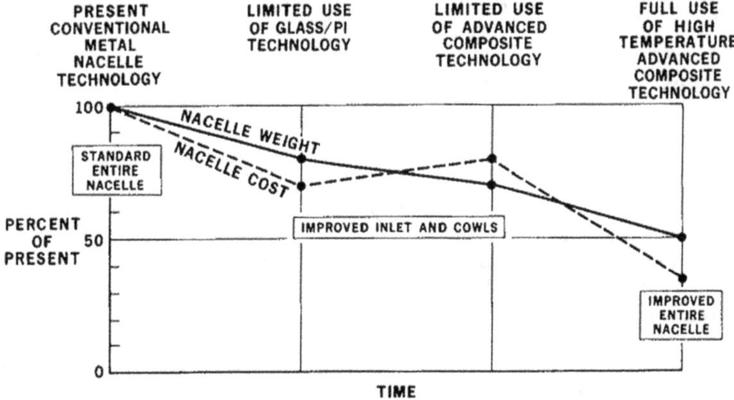

Fig. 15 Impact of change in materials with time [44]

engine nacelle of an aero-vehicle of becoming fully composite is shown in Fig. 16 [44].

10. The challenge ahead is to select the right material for the required high temperature applications. The nanocomposites/composites/hybrid materials etc. are the various many options available which will replace the position of metals in a structure and systems in all fields of science and engineering.

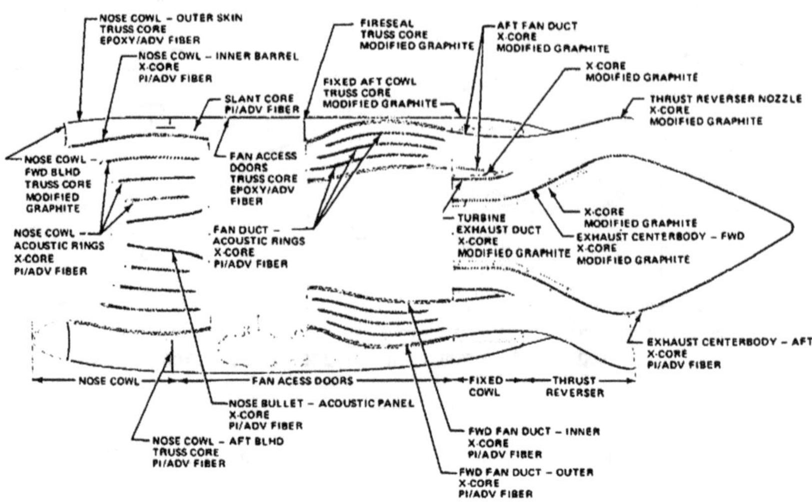

Fig. 16 Impact of advanced composite materials [44]

References

1. Cuppoletti, J.: Metal, ceramic and polymeric composites for various uses, ISBN 978-953-307-353-8, InTech, 20 July (2011)
2. Gan, M., et al.: Degradation of high temperature composites under thermal spike loads, Sampe Europe Conference & Exhibition
3. Dimitrienko, Y.I.: Thermomechanics of Composites under High Temperatures. Kluwer Academic Publishers, Dordrecht (1999)
4. Hoebergen, A., Holmberg, J.A.: Vacuum Infusion, ASM Handbook, vol. 21, pp. 501–515. Composite, Materials Park (OH) (2001)
5. Mangalgiri, P.D.: Polymer-matrix composites for high-temperature applications. Defence Sci. J. **55**(2), 175–193 (2005)
6. Boyd, J.D., Chang, G.E.C.: Bismaleimide composites for advanced high-temperature applications. Int. SAMPE Symp. **38**, 357–369 (1993)
7. Manocha, L.M.: High performance carbon–carbon composites. Sadhana **28**(1–2), 349–358 (2003)
8. Potter, K., Foster, S., Batho, Y.: High temperature high performance composites. www.bris.ac.uk/composites (2012). 17 April 2012
9. Composite materials for space and high temperature applications. www.umeco.com
10. Simos, N.: Composite materials under extreme radiation and temperature environments of the next generation nuclear reactors. www.intechopen.com (2011)
11. Bonal, J.-P., Kohyama, K., van der Laan, J., Snead, L.: Graphite, ceramics and ceramic composites for high-temperature nuclear power systems. MRS Bull. **34**, 28–34 (2009)
12. Tomar, V.: Nanotechnology thought leaders series, school of aeronautics and astronautics, Purdue University (2011)
13. Marquis, D.M.: Properties of nanofillers in polymer. www.intechopen.com (2011)
14. Morgan, A.B.: Polymer-clay nanocomposites: Design and application of multi-functional materials. Mater. Matters **2**, 20–25 (2007)

15. Frank, K.: Book review: Polymer nanocomposites—processing, characterization, and applications. J. Eng. Fibers Fabr. 3(3) (2008)
16. Tzeng, S.S.: Evolution of microstructure and properties of phenolic resin-based carbon/carbon composites during pyrolysis Mater. Chem. Phys. 73, 162–169 (2002)
17. Sambarkar, P.P., et al.: Polymer nanocomposites: An overview. Int. J. Pharm. Pharm. Sci 4(2), 60–65 (2012)
18. Alexandre, M., Dubois, P.: Polymer-layered silicate nanocomposites: Preparation, properties and uses of a new class of materials. Mater Sci. Eng. R 28, 1–63 (2000)
19. Sheng, X.: Polymer nanocomposites for high-temperature composite repair, Iowa State University Digital Repository @ Iowa State University (2008)
20. Kurahatti, R.V., et al.: Defence applications of polymer nanocomposites. Defence Sci. J. 60(5), 551–563 (2010)
21. Hussain, F., et al.: Review article: Polymer-matrix nanocomposites, processing, manufacturing, and application: An overview. J. Compo. Mater. 40(17), 1511–1575 (2006)
22. Takeichi, T., et al.: High-performance polymer alloys of polybenzoxazine and bismaleimide. Polymer 49, 1173–1179 (2008)
23. Srikanth, I., Daniel, A., Kumar, S., Padmavathi, N., Singh, V., Ghosal, P., et al.: Nano silica modified carbon—phenolic composites for enhanced ablation resistance. Scr. Mater. 63, 200–203 (2010)
24. Marshall SpaceFlight Cen, Application of Ablative Composites to Nozzles for Reusable Solid Rocket Motors, practice no. Pd-ed-1218 page 1 of 8 (1989)
25. Lao, S., Ho, W., Ngyuen, K., Koo, J.H.: Flammabiltiy, mechanical, and thermal properties of polyamide nanocomposites. In Proceedings of SAMPE 2005, Seattle, WA (2005)
26. Zhang, Z., Yang, J.L.: Creep resistant polymeric nanocomposites. Polymer 45, 3481–3485 (2004)
27. Walker, L.S., et al.: Toughening in graphene ceramic composites. ACS Nano. Org. 5(4), 3182–3190 (2011)
28. Poyato, R., et al.: Aqueous colloidal processing of single-wall carbon nanotubes and their composites with ceramics. Nanotechnology 17, 1770–1777 (2006)
29. Stankovich, S., Dikin, D.A., Dommett, G.H.B., Kohlhaas, K.M., Zimney, E.J., Stach, E.A., Piner, R.D., Nguyen, S.B.T., Ruoff, R.S.: Graphene-based composite materials. Nature 442, 282–286 (2006)
30. Tomar, V., Gan, M.: Temperature dependent nanomechanics of si-c-n nanocomposites with an account of particle clustering and grain boundaries. Submitted to. Int. J. Hydrogen Energy (2009)
31. Niihara, K., Izaki, K., Kawakami, T.: Hot-pressed Si3N4-32%SiC nanocomposites from amorphous Si-C-N powder with improved strength above 1,200 °C. J. Mater Sci. Lett. 10, 112–114 (1990)
32. Wan, J., Duan, R.-G., Gasch, M.J., Mukherjee, A.K.: Highly creep-resistant silicon nitride/silicon carbide nano–nano composites. J. Am. Ceram. Soc. 89(1), 274–280 (2006)
33. Niihara, K.: New design concept for structural ceramics-ceramic nanocomposites. J. Ceram. Soc. Jpn: Centennial Memorial Issue 99(10), 974–982 (1991)
34. Trick, K.A., Saliba, T.E.: Mechanisms of the pyrolysis of phenolic resin in a carbon/phenolic composite. Carbon 33, 1509–1515 (1995)
35. Chen, Y., Chen, P., Hong, C., Zhang, B., Hui, D.: Improved ablation resistance of carbon–phenolic composites by introducing zirconium diboride particles. Compos. B 47, 320–325 (2013)
36. Vaia, R.A., Price, G., Ruth, P.N., Nguyen, H.T., Lichtenhan, J.: Polymer/layered silicate nanocomposites as high performance ablative materials. Appl. Clay Sci. 15, 67–92 (1999)
37. Abdalla, M.O., Ludwick, A., Mitchell, T.: Boron-modified phenolic resin for high performance application. Polymer 44, 7353–7359 (2003)
38. Corral, E.L., Walker, L.S.: Improved ablation resistance of C–C composites using zirconium diboride and boron carbide. J. Eur. Ceram. Soc. 30(11), 2357–2364 (2010)

39. Pulci, G., Tirillò, J., Fossati, F., Bartuli, C., Valente, T.: Carbon–phenolic ablative materials for re-entry space vehicle: Manufacturing and properties. Compos. Part A: Appl. Sci. Manuf. **41**(10), 1483–1490 (2010)
40. Bahramian, A.R., et al.: High temperature ablation of kaolinite layered silicate/phenolic resin/asbestos cloth nanocomposite. J. Hazard. Mater. **150**, 136–145 (2008)
41. Kashiwagi, T., Harris, R.H., Zhang, X., Briber, R.M., Cipriano, B.H., Raghavan, S.R., Awad, W.H., Shields, J.R.: Flame retardant mechanism of polyamide 6-clay nanocomposites. Polymer **45**, 881–891 (2004)
42. Vaia, R.A., Price, G., Ruth, P.N., Nguyen, H.T., Lichtenhan, J.: Polymer/layered silicate nanocomposites as high performance ablative materials. Appl. Clay Sci. **15**, 67–92 (1999)
43. Collins, TJ, et al.: High-temperature structures, adhesives, and advanced thermal protection materials for next-generation Aeroshell design
44. Golub, M.A., Parker, J.A.: Polymeric materials for unusual service conditions. J. App. Poly. Sci., Appl. Poly. Symp. No. 22 Board (1973)
45. Srikanth, I., et al.: Mechanical, thermal and ablative properties of zirconia, CNT modified carbon/phenolic composites. Compos. Sci. Technol. **80**, 1–7 (2013)
46. Zhang, X., Xu, L., Du, S., Han, W., Han, J.: Crack-healing behavior of zirconium diboride composite reinforced with silicon carbide whiskers. Scripta Mater. **59**, 1222–1225 (2008)

Energy Absorption and Low-Velocity Impact Performance of Nanocomposites: Cones and Sandwich Structures

James Njuguna, Sophia Sachse and Francesco Silva

Abstract The increasing need for high performance structures, in the energy and transport industry, demands a continuous development of new engineering materials. Unique mechanical properties together with low specific weight can be achieved by the combination of various constituent materials into one macroscopic composite material. Coupling of the high strength reinforcement with supporting matrix creates a novel material with the improved characteristics, which could never be obtained using either of the constituents separately. These types of materials are particularly desirable in structures where a high strength to weight ratio is of great importance. In this chapter, two case studies are provided one on nanophased sandwich composites with polyurethane/layered silicate foam cores and the other on thermoplastic glass-fibre and nano-silica reinforced nanocomposites.

1 Introduction

Over the last decade increased amount of research in the field of composite materials proved that addition of nano-sized fillers, rather than micro-sized fillers, can significantly enhance mechanical properties of the polymeric materials. Composite material is usually defined as a 'nano', if one of the constituents possess at least one dimension in the range of 1–100 nm. The unique properties, of the material reinforced with nano-particles, come from the large number of interfacial effects, existing due to the high surface-area-to-volume ratio of the

J. Njuguna (✉)
Institute for Innovation, Design & Sustainability, Robert Gordon University,
Aberdeen AB10 7GJ, UK
e-mail: j.njuguna@rgu.ac.uk

S. Sachse · F. Silva
Cranfield University, MK43 0AL Cranfield, Bedfordshire, UK

J. Njuguna (ed.), *Structural Nanocomposites*, Engineering Materials,
DOI: 10.1007/978-3-642-40322-4_8, © Springer-Verlag Berlin Heidelberg 2013

reinforcement. For the spherical nano-particles and nano-fibres this ratio is irreversibly proportional to their radius, and its value can be even up to 1,000 m²/g. In case of light weight structures, the most widely used nano-reinforcements are silica based particles, due to their good mechanical properties and high thermal stability [1].

Up to date, research works have been conducted to study the influence of the nano-particles on the mechanical behaviour of polymer composites. Several factors influencing the enhancing capabilities of the nano-reinforcements were studied in the literature [2–4]. This includes key parameters such as shape and size of nano-fillers, matrix and reinforcement material, interfacial strength and interphase characteristics, as well as volume fraction and quality of dispersion within the matrix. Mechanical properties and energy absorption characteristics of nano-composites have been mainly characterized by means of tensile, flexural or Izod impact testing.

2 Energy Absorption of Composite Materials

Modern vehicle structures must be able to withstand severe impact loads, at the same time providing safety of the occupants. That is why structural materials used for crashworthy applications must be characterized by the energy absorption capability. In order to ensure survivability of the accident, structure has to dissipate energy in a controlled manner. This is limited by the two factors: induced decelerations and maintaining of a survival space for occupants during a crash [5]. Energy absorbed throughout a collision is defined by the area under the load–displacement curve as shown in Fig. 1 [6].

Analyzing the above graph we can notice that the energy absorbed can be controlled by the value of the force and the deflection. The maximum peak load is limited by the occupants' tolerance and the maximum deflection is limited by the geometry of the structure. In an idealized energy absorbing system induced load should be constant and just below the human tolerance limit. In reality design of

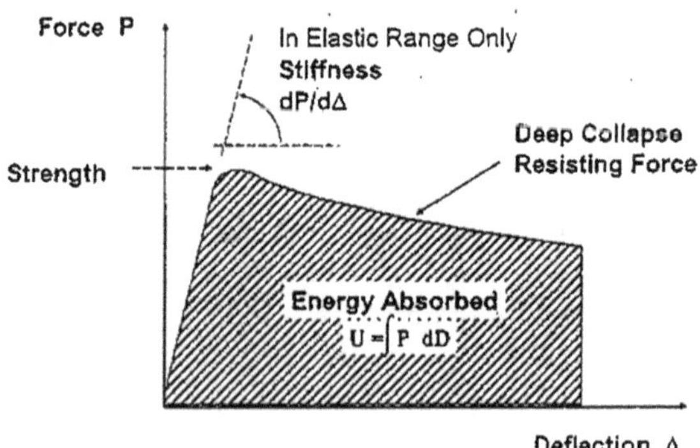

Fig. 1 General load–displacement curve [6]

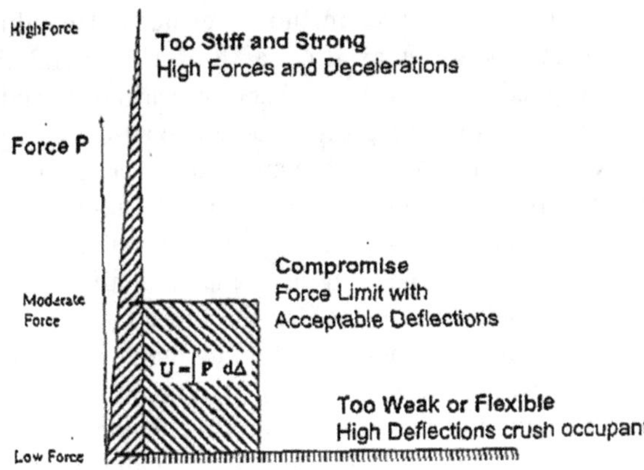

Fig. 2 Various structural responses [6]

crashworthiness structure is always a compromise and trade off. As an example three different structural responses are compared in Fig. 2. In the scenario with very stiff structure induced peak loads can highly exceed allowable limit. In such case occupants will suffer high decelerations while deformations of the structure are small. In case of the weak structure induced peak loads are strongly reduced but large deformations can crash the occupants causing serious injuries or even death. The third scenario is a compromise with the moderate value of the force and acceptable deflections not affecting occupants' space. In this scenario energy absorbed by the structure is maximized but within the limits of allowable load and deformation.

Traditionally metallic materials have been applied for the crashworthy structures due to their ability to sustain plastic deformations. In contrast, composite materials do not exhibit plastic deformations as they are usually brittle. However, if they are properly designed, they can absorb high amounts of impact energy by the progressive failure and delamination.

In order to facilitate a comparison of various crashworthiness materials, several measuring parameters are commonly used, and the most important ones are the specific energy absorption (SEA) parameter and the energy absorption efficiency. The specific energy absorption (SEA) parameter is defined by the amount of energy absorbed per unit mass of crushed material.

$$SEA = \frac{energy\_absorbed}{crashed\_mass} = \frac{\int F dx}{m_c}$$

where F—load, x displacement.

The energy absorption efficiency (EAE) is defined by the ratio of the area under the load–displacement (true energy absorbed) curve to the rectangular area formed by the maximum load and maximum displacement (perfect energy absorbed).

$$EAE = \frac{energy\_absorbed}{max\_load * max\_displacement} \cdot 100$$

The most widely used method to evaluate the ability of a composite material to absorb the energy, is axial collapse of a structural elements. This technique has been applied by many researchers, on various composite materials.

The ability of a composite structure to absorb energy is highly dependent on the mode of fracture. Materials which fail in a progressive manner, with extensive delamination and fragmentation, are able to absorb much higher energies than those materials which tend to fail in a brittle manner. Farley [7–9] found that thermoset composites reinforced with glass and carbon fibres fail progressively in fragmentation and splaying modes. On the other hand, thermoset tubes reinforced with Kevlar, failed in a progressive folding mode. Mamalis et al. [5] who studied polyester cones, cylinders and tubes, reinforced with random orientated glass fibres, divided failure of the samples into four modes: progressive crushing with microfragmentation (Mode I), brittle fracture with catastrophic failure (Mode II and III, depending on the crack form), Progressive folding and hinging, similar to the metallic tubes (Mode IV). The authors observed that significant influence on fracture mode has geometry of the sample. Conical and square tubes with small semi-apical angles (5°–15°) tend to fail in a stable Mode I, whereas samples with large semi-apical angles (20°–30°) were found to fail in a brittle Mode II. They also found that wall thickness, related to number of composite layers, has direct influence on the mode of failure. The collapse mode for large semi-apical angel samples has changed from stable to unstable, with increasing wall thickness. In case of small angel samples, the collapse mode remained the same with increased thickness of the wall.

3 Energy Absorption in Nanocomposites

Modern vehicle structures must be able to withstand severe impact loads, at the same time providing safety of the occupants. For this reason, structural materials used for crashworthy applications must be characterized by the energy absorption capability. Energy absorbed throughout a collision is controlled by the value of the force and the deflection. The maximum peak load is limited by the occupants' tolerance and the maximum deflection is limited by the geometry of the structure. In the scenario with very stiff structure induced peak loads can highly exceed allowable limit and occupants will suffer high decelerations while deformations of the structure are small. In case of the weak structure induced peak loads are strongly reduced but large deformations can crash the occupants causing serious injuries or even death. The third scenario is a compromise with the moderate value of the force and acceptable deflections not affecting occupants' space. In this scenario energy absorbed by the structure is maximized but within the limits of allowable load and deformation.

The most widely used method to evaluate the ability of a composite material to absorb the energy, is axial collapse of a structural elements. The experiments presented in the literature vary in geometry and material of the specimen, as well as parameters of the impact such as: velocity and energy. The most often used geometries of crash-samples are: cylinders, cones and square tubes [5, 7–11]. The materials which have been investigated extensively include: carbon, Kevlar and glass as fibre materials; and epoxy [7–9], PEEK [12], polyester [5] and vinylester [10] as matrix materials.

Low cost thermoplastic polymers, such as polypropylene (PP) and polyamide (PA), are widely used in the aerospace industry because of their good mechanical performances, processing properties and low cost. However, their application as structural materials is limited due to their low impact resistance [13]. Incorporation of various nano-sized filers like; nano-particles (SiO_2, TiO_2, WS_2, $CaSiO_3$, Al_2O_3), carbon nanotubes, and clay nano-plates; can be an appropriate solution to this problem [14]. Injection moulded, short fibre-reinforced thermoplastic composites are the most prevalent composite materials as thermoplastic nano-reinforced structures. Several important factors influencing energy absorption capability of nanocomposites are summarized in the followings:

- *Particles stiffness:* The particles stiffness influences the mechanical properties of polymer matrix nanocomposites. For instance, the impact response of the high density polyethylene with addition of elastic rubber and rigid calcium carbonate ($CaCO_3$) particles was investigated by means of notched Izod impact testing [15, 16]. The results showed that addition of 22 vol. % of elastic rubber causes an increase in notch toughness more than 16 times. However, a decrease by 50–60 % in the Young's modulus and yield stress by 40–50 %, was observed in relation to the net polymer.
- *Particles geometry:* Addition of Al_2O_3 nano-whiskers, glass fibres and wallastonite into polymer matrix improves the fracture toughness significantly, while incorporation of the plate shaped particles of nanoclay, into the same matrix material, was found to decrease it [17]. Favourable effect on the impact toughness was also observed after the addition of amino-functionalized multi wall carbon nano tubes [18] or small amounts of single wall carbon nano tubes [19]. Moreover, it was observed by Kireitseu [20] that the composite impact toughness and stiffness are highly dependent on the modulus of nano-tubes.
- *Volume fraction and inter-particle distance:* Important influence on the impact toughness of nano-composites has the inter-particle distance τ, independently of the reinforcement geometry. Its value is closely related to the concentration φ and average size of the particles d [21]. Zhang et al. [21] found, that if inter-particle distance is smaller than average particle size d, then composite toughness increases significantly, as it are presented in Fig. 3. This phenomenon can be explained by the fact, that distance between particles is small enough to build around them, a three dimensional network of interphase region.
- *Effects of particles size*: Effect of particles size on the mechanical properties of the polyurethane foams was studied by Javni et al. [22]. Incorporation of nano-sized filler was found to increase the compression strength of the foam, and to decrease its rebound resilience. On the other hand the addition of micro-sized fillers was found to lower the hardness and compression strength, at the same time leading to an increase in rebound resilience. This indicates that foams

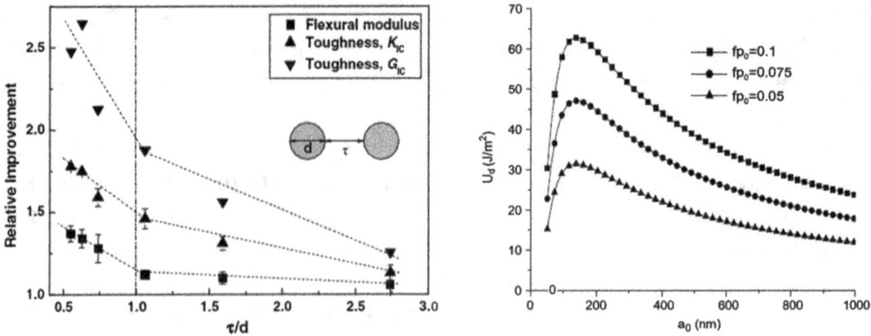

Fig. 3 Improvements in mechanical properties due to the inter-particle distance [21] and energy dissipation against average radius of the filler [23]

reinforced with nano-particles are able to absorb higher amounts of impact energy. According to the analytical studies carried out by Chen et al. [23], energy dissipation due to the interfacial debonding is highly dependent on the size of particles. The material ability to dissipate energy increases significantly with increasing size of the particles up to 140 nm, whereas particles bigger than that indicate gradual decrease in the material performances.

4 Sandwich Composites

Sandwich composites are widely used in wind turbines, automotive bumpers, aircraft engine nacelle, on wings for fuel tanks protection, tail plane panels for protection of stones and pebbles on take-off and landing, naval ships, human vests and helmets for ballistic protection, automotive for collision and heat protection. Unfortunately, any composite materials are susceptible to impact and the damage can be big and tend to increase in time- Research has shown that damage initiation thresholds and damage size on sandwich composites, primarily depend on the properties of the core materials, face sheets, and the relationship between the properties of the cores and those of the facings. Much of the earlier research on sandwich composites under impact loading focused on the honeycomb core (Nomex, glass thermoplastic or glass-phenolic) sandwich constructions. A key problem in the honeycomb sandwich constructions is the low core surface area for bonding. Consequently, expanded foams, (often thermoset) are nowadays preferred to achieve reasonably high thermal tolerance, though thermoplastic foams are also used. In turn, the response of foam core sandwich constructions to impact loading has been studied by many researchers [24–26]. Accordingly, it is now well understood that the response of the foam core sandwich composites strongly depend on the density and the modulus of the foam.

Foams are defined as materials containing gaseous voids surrounded by a denser matrix, which is usually a liquid or solid. Polymer foams can also be defined as either closed cell or open cell foams. In closed cell foams, the foam cells are isolated from each other and cavities are surrounded by complete cell walls. In open cell foams, cell walls are broken and the structure consists of mainly ribs and struts. Generally, closed cell foams have lower permeability, leading to better insulation properties. Open cell foams, on the other hand, provide better absorptive capability. One-step reactive foaming is typical for thermoset polymers. A good example is PU/clay nanocomposite foams, where a physical blowing agent such as pentane is used with monomers and clay nanoparticles. Reaction exotherm leads to a temperature jump and foaming. Most nanocomposite foams to date are synthesized via a two-step process: the nanocomposite is synthesized first and followed by foaming. Readers interested in the synthesis of PU/clay foams are referred to available literature such as Refs. [27, 28] (literature related to the reactive extrusion foaming of various nanocomposites is also covered). Polymeric foams have been widely used as packing materials because they are lightweight, have a high strength/weight ratio, have superior insulating properties and high energy absorbing performance.

A possible way of improving the properties of foam materials is through the inclusion of small amounts of nanoparticles (like carbon nanotubes and nanofibres, TiO_2, nanoclay, etc.) to improve the foam density and modulus properties. Up to now, montmorillonite nanoclays are the best candidate for foam reinforcement due to ease in processing, major thermal–mechanical properties enhancement, wide availability and are relatively cheap [2, 3]. Likewise, polyurethanes (PU) are core materials of choice due to their tailorable and versatile physical properties, ease of manufacture and their low costs. Unfortunately, the use of thermoplastic resins filled with nanoparticles to construct either laminates or foams is relatively new. Moreover, the use of nanoparticles in such laminates or foams in sandwich composite construction is in its infancy but realistic and beneficial. For instance, by using less than 5 % by weight nanoclay loadings, significant improvement on the foam properties (failure strength and energy absorption) can be realised with over 50 % increase in the impact load carrying capacity over the neat foam sandwich [29, 30]. However, since most current research concentrates on the processing and characterization of nanophased foams and evaluation of static properties only, materials data on impact behaviour, failure mechanisms due to impact and impact-structure–property relations is missing. For wide usage of nanophased foams in the sandwich constructions and reduction of weight (while maintaining the same level of protection), proper understanding of their impact behaviour at both high- and low-velocity impact is required.

4.1 Fabrication of Sandwich Panels with Nanophased Cores

Polyurethane foam with various different weight percentages (up to 10 %) of nanoclay have been prepared. The low weight percentages are targeted for infusion to avoid the agglomeration of nanoparticles common in high concentrations.

Preparation of the PU systems modified with MMT consists of three steps—first, polyol blend (polyether RF-551) and polyester (T-425R) mixture from Alfasystems, Brzeg Dolny, Poland, was stirred with powdered MMT (Optibent 987, Süd Chemie AG, Moosburg, Germany). Then catalyst (N,N-dimethyl cyclohexylamine), water and surfactant (SR-321, Union Carbide, Marietta, GA) were added in order to prepare the polyol premix (component A). In the next step n-pentane as a physical blowing agent was added to component A. Component B was polymeric 4, 4'-diphenylmethane diisocyanate (PM 200). It was added to component A and the mixture was stirred for 10 s with an overhead stirrer. Finally, the prepared mixtures were dropped into a mould. All the experiments were performed at ambient temperature of ca. 20 °C.

The sandwich beam samples are fabricated from aluminium face sheets (aluminium grade 24,139, Young's modulus of 79 GPa, 2 mm thick) and the above manufactured 20 mm thick nanophased polyurethane rigid foams. Firstly, the aluminium faceplates were pre-treated and polished. Later they are degreased using acetone for 2 min before applying the adhesive (DP-100 supplied by 3 M$^@$). The surfaces are then dried and the adhesive applied evenly on the foam surface using a glue gun, and the metal skin is laid on top for each side at a time. This is repeated for the other side of the foam after allowing for 24 h of curing time. Once the adhesive is applied the sandwich samples are cured for a further period of 24 h in a press. Basically, the samples are laid on the base of the press between two thick metal plates to ensure pressure is distributed evenly all through the structure. The finished products are 5 specimens of sandwich beams of length 140 mm and width 100 mm for each.

4.2 Low Velocity Impact Tests

All the low impact tests were conducted using an instrumented falling weight impact tester, type 5. The device is equipped with impulse data acquisition system that can acquire data points. Using this machine, impact energy and velocity can be varied by changing the mass and height of the dropping weight. The velocity of the tup is measured just before it strikes the specimen. It is also fitted with pneumatic rebound brake, which prevents multiple impacts on the specimen. During the testing, the specimen is held in the fixture placed at the bottom of the drop tower which provided a clamped circular support span (Fig. 1). The weight of cross-head is maintained at a specific value and it is guided through two smooth guide columns. The impactor end of the drop mass is fitted with an instrumented tup with hemispherical end having a capacity to record the transient response of the specimens. To carry out the impact tests, sandwich composite samples (140 × 100 mm) are placed between the clamps and heights adjusted depending on the desired energy level. The projectile had a 20 mm diameter hemispherical tip. The impact force history obtained during the test was measured using a piezo-electric load cell located above the impactor tip. The amplified signals from the load cell were recorded by the computer. The height is the distance between the tip of the indenter and the top surface of the sample held between the clamps. Once the height required to attain a particular energy level is known, the indenter is moved accordingly to that height before it is dropped on the specimen for the test.

4.3 Results and Discussions

The impact response of sandwich structures with and without nanoclay core and nanophased face sheets were evaluated. Several samples of each set were tested at different energy levels. Transient data including time, load, energy, velocity and deflection were collected for each sample as functions of time. Figure 4 compares typical load and deflection versus time curves for the four specimen types at 45 J impact energy level.

The peak loads were 3,560 N for neat cored sandwich and for 2.5 % MMT loaded core sandwich composites. A slight improvement of ~300 N was observed in peak load for both 5 and 7.5 % MMT loaded samples, indicating that higher MMT loading samples performed better than the lower MMT loaded ones.

The energy absorption at failure point was recorded as 32.5, 32.5, 43.4 and 44 J for 0, 2.5, 5 and 7.5 % MMT loading respectfully. The energy absorption in any material under impact loading is mainly through the elastic deformation in the initial stage with some energy absorbed through friction. Once, the energy level is

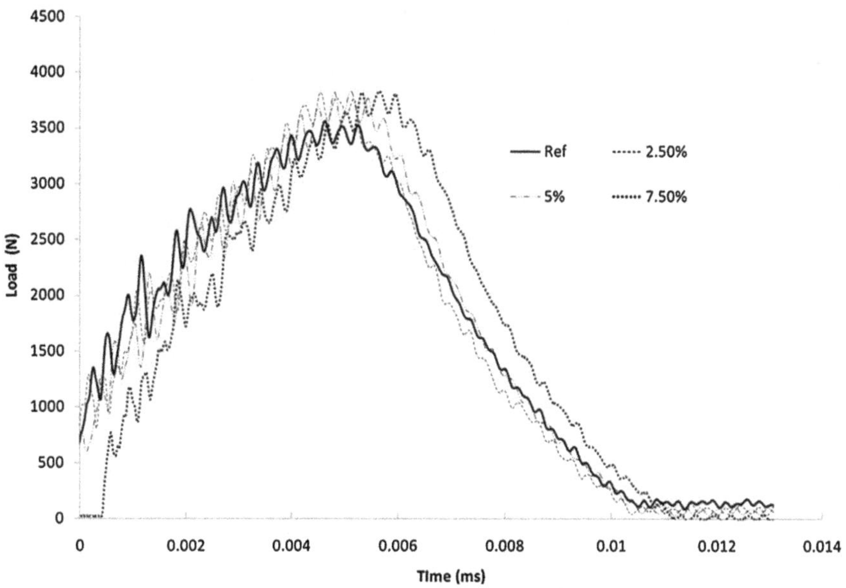

Fig. 4 Comparison of load versus time graph obtained from first impact

beyond the level required for maximum elastic deformation, the structure then dissipates the excess energy in the form of either plastic deformation in case of ductile materials or through various damage mechanisms in the case of brittle materials. As can be seen on Fig. 5, it is evident that the absorbed energy at peak load was within a close range for all samples.

The deflection at peak load is a qualitative indication of the stiffness of the material. The same applies to their maximum deflection recorded, averaging around 2.1 mm but minor diversity between the specimens was found. As expected, the peak loads and deflections were similar at low impact energy. These current results were in contrast to related published work in the literature which reported that at 45 J, the neat core sandwich structures had higher deflection at peak load on nanophased core and fibre-reinforced sandwich composites [30]. The literature further reported that the total time and the time to peak load were also higher for the neat core sandwich samples [30]. However, it should be noted that the work was based on fibre-reinforced composites as face sheets whereas aluminium face sheets have been used in the current work. In the latter case, there was no perforation on the face sheets by the indent striker. It is also apparent that we will need to undertake future tests at higher energy levels above the 45 J thresholds the current tests were conducted at.

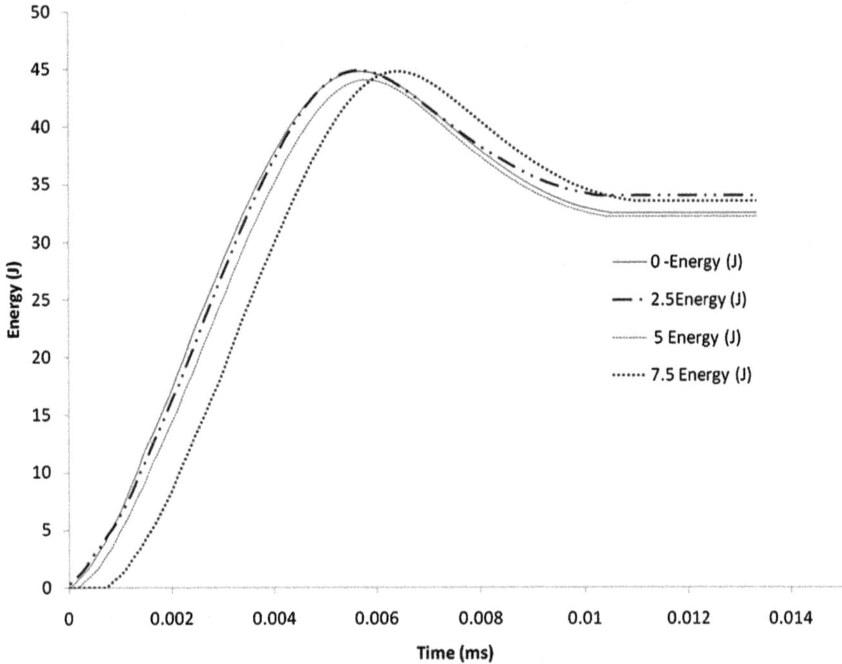

Fig. 5 Energy versus time graph obtained from first impact

A detailed examination revealed no debonding of the skin and core or crack development into the foam cross-section at the mid-plane of the structures over the range of energies considered. This demonstrated the importance of a proper selection of adhesive for particular foams. Typically, core fracture takes place when defects exist in the adhesive layer and at the skin/adhesive/core interfaces or where poor adhesion occurs between the skin and the adhesive. In our case, DP-100$^{@}$ is a thin film and highly viscous. When a one-step process is used to cure the facings to the core, localized cell wall collapse and cell coalescence occur, leading to non-uniform thickness and weaker cores near the skin/core bondline. In such cases, the problem is caused by the high pressure required to cure the facings. In the current work, therefore, a thin layer application was employed followed by compression during bonding hence minimising potential defects in the materials at fabrication stage. Nevertheless, a microscopic inspection revealed some small debonding for both 5 and 7.5 % samples along the skin/foam interface edges in the structure width after the first impact tests. This failure may be attributed parameters that are direct indications of the stiffness of the samples rather than just the adhesive failure. It follows that at impact, as the dented aluminium face sheet yielded inwards, the back face remained intact and no deformations were observed along the cores length. However, debonding occurred along the sandwich composite edges (face plate, width-wise) to make up for the bending displacements thereby forcing the edge lines to split along the skin/core interface. The core at the midsection experienced compression forces but there were no signs of cracks formation for all samples even under microscopy. This may be related to reactive foams and their microcellur cells collapsing to absorb load instead of expected cracks and fracture characteristics, as widely reported in the literature.

Since sandwich structures are often used as energy absorbing structures in damage susceptible areas, we decided to conduct a second drop weight impact recurrence on the previously impacted samples in order to closely replicate repetitive occurrences. Such scenarios are common between structural inspection periods or in incidences where the damage was missed during normal inspection routines. Interestingly, we found out that the nanophased samples recorded significantly higher peak loads than the neat PU cored ones. As can be seen from Figs. 6 and 7, a difference of over 1,000 N in peak load was observed for the 7.5 % samples as compared to that of neat PU foam samples.

It was further noted that the neat foam samples had the highest energy intake at failure point, in agreement with previous observations by Hosur et al. [24]. Still, there was no visible cracking deformation on the surface of the reactive foam cores along the length of the structures. The only visible debonding and crack formation was along the width of the structures which, as eluded to earlier on, may be associated to bending deformation on the face plate due to impact loads. Further investigations by microscopy are currently underway to investigate the level of damage on the microcellular cores' cross-sections in conjunction with in-depth morphology studies.

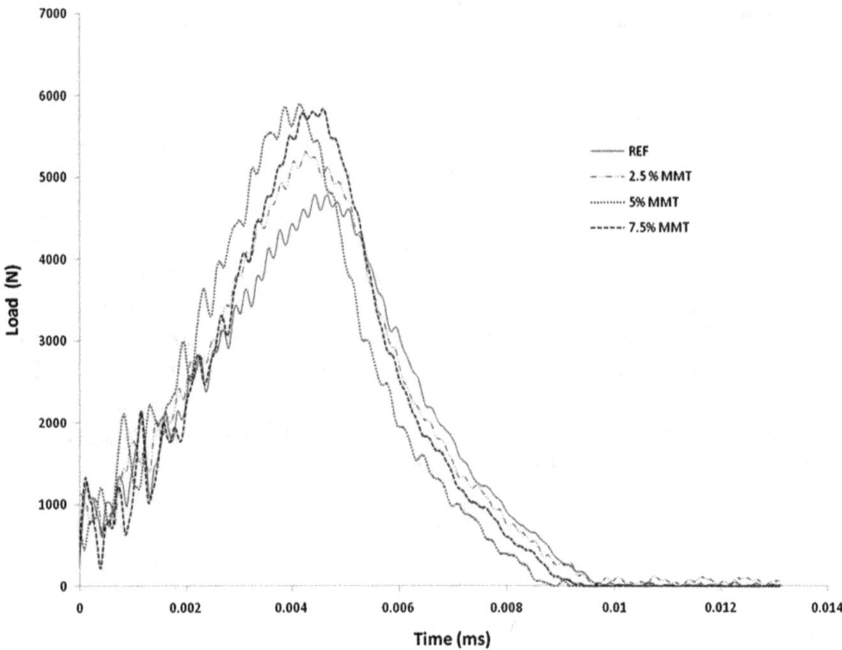

Fig. 6 Load versus time graph obtained from second impact test

5 Thermoplastic Nanocomposites

Numerous researches have been conducted to study the influence of nano-particles on the mechanical behaviour of polymer composites. Several factors influencing the enhancing capabilities of the nano-reinforcements were studied in the literature. This includes key parameters such as: shape [31] and size [32] of nano-fillers, matrix and reinforcement material [33, 34], interfacial strength and interphase characteristics [35], as well as volume fraction [21] and quality of dispersion within the matrix [36].

For the purpose of measuring the energy absorption in composite structures, tube crashing experiments are the most prevailing. The ability of a composite structure to absorb the energy was found to be highly dependent on the mode of fracture. Materials which fail in a progressive manner, with extensive delamination and fragmentation, are able to absorb much higher energies than those materials which tend to fail in a brittle manner. Farley [7–9] found that thermoset composites reinforced with glass and carbon fibres fail progressively in fragmentation and splaying modes. On the other hand, thermoset tubes reinforced with Kevlar, failed in a progressive folding mode. Mamalis et al. [5] who studied polyester cones, cylinders and tubes, reinforced with random orientated glass fibres, divided failure of the samples into four modes: progressive crashing with micro-fragmentation (Mode I), brittle fracture with catastrophic failure (Mode II and III, depending on

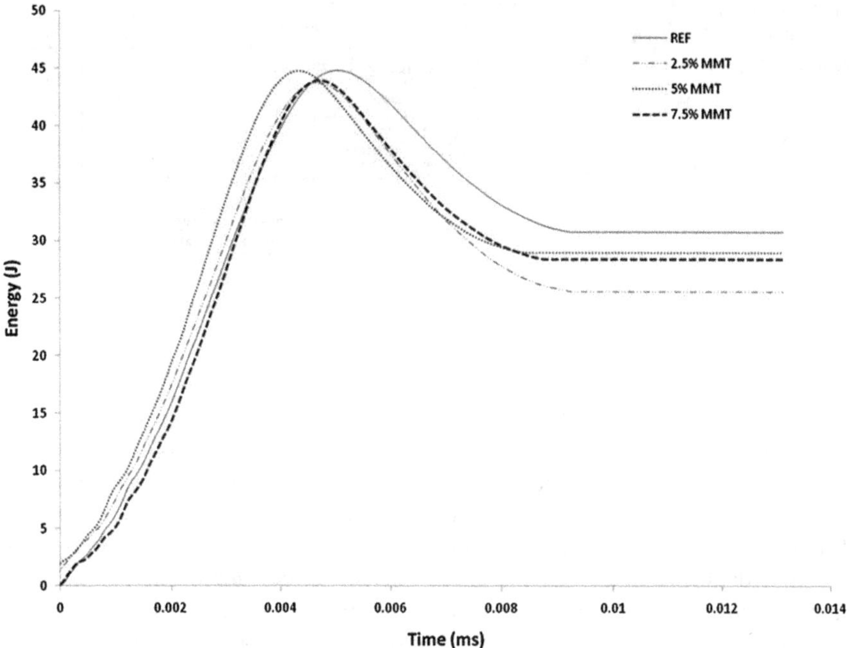

Fig. 7 Energy-time graphs after second impact test

the crack form), progressive folding and hinging, similar to the metallic tubes (Mode IV).

Regarding nanocomposite structures there is a lack of crash experiments conducted on these materials presented in the literature [14]. Energy absorption characteristics of these materials have been mainly characterized by means of compression [37], flexural [38] and Charpy or Izod impact testing [17]. The relationship between mechanical properties of nanocomposite material and energy absorption characteristics of nanocomposite structure is not fully understood. This work therefore aims to correlate changes in the mechanical properties of the material with induced fracture modes and ability of the structure to crash progressively, after the addition of secondary reinforcement.

5.1 Fabrication of Cone-Shaped Nanocomposites Structures

Preparation of the nano and glass reinforced polymer composites was conducted in three main steps: preparation of the nano-composites granulate, mixing and extrusion of the nano and glass reinforced composite granulate and injection moulding of the macro-sample. The flow chart showing the preparation process is shown in Fig. 8.

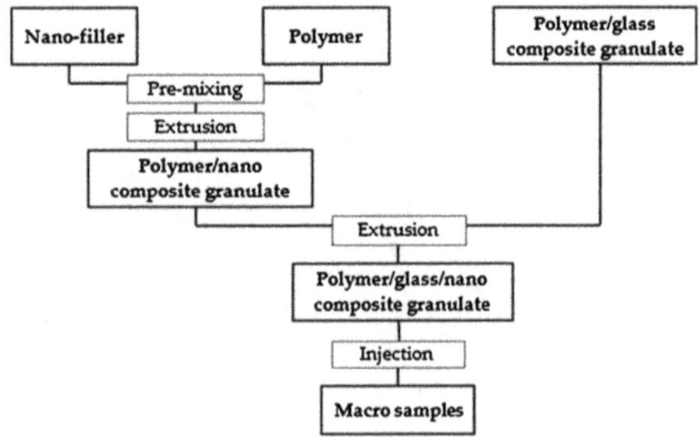

Fig. 8 Material compounding and test sample manufacturing

In order to warrant the highest homogeneity of the composition, nano-reinforcement and polymeric matrix, all in solid (powder) form, were premixed before extrusion. This activity was performed by the use of a turbomixer with rotatory blades. The pre-mixing phase consisted of two steps, the first one at lower speed (1,500 rpm) and the second one at higher speed (3,000 rpm). This choice was made in order to ensure the maximum homogeneity of the premix and, on the other hand, to subject the polymer to a small temperature stress, to improve binding between polymeric matrices and added reinforcements. Subsequently the premixed materials were fed into a twin-screw extruder at a constant predefined rate.

5.2 Crashing Behaviour

Crashing behaviour and energy absorption characteristic of the polymer composites were studied by means of quasi-static compression and dynamic impact tests. Figure 9 show the variation of load with increasing crashed length of the conical structure. Comparing the load–displacement curves, several important comments can be made. In all conducted experiments the initial slope of the load curve is approximately linear. This is associated with the elastic deformation of the structure [39–41]. The first extremism of the curve indicates maximum load supported by the structure, which depends on the material strength. After that point, a sudden drop in load is observed due to the formation of the cracks. Subsequently, a progressive crashing occurs, what is visible as following sharp load peaks. Analyzing the obtained results we can note, that magnitude of the load peaks depends on the crashing characteristic of the material and testing speed. It could be seen that all PA composites and neat PPGF composite, tested under the quasi-static load, had bigger secondary peaks than the initial one. On the other hand, the same materials, tested under the dynamic conditions, induced significantly lower secondary peaks. This difference is associated with the mechanism of cracks propagation. If the crack propagates along the height of the cone, as it can be seen in Figs. 9 and 10, then the load required to crash the sample can be smaller

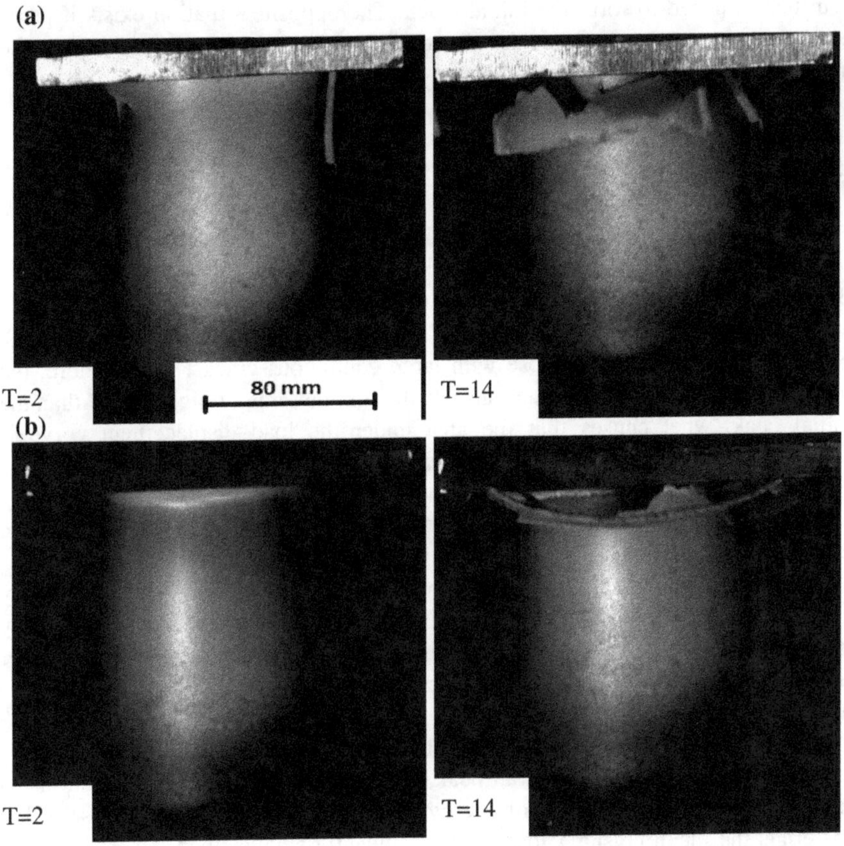

Fig. 9 Crushing characteristics of nanofilled cones. High speed camera records (a) PA/GF/MMT (b) PA/GF/SiO$_2$

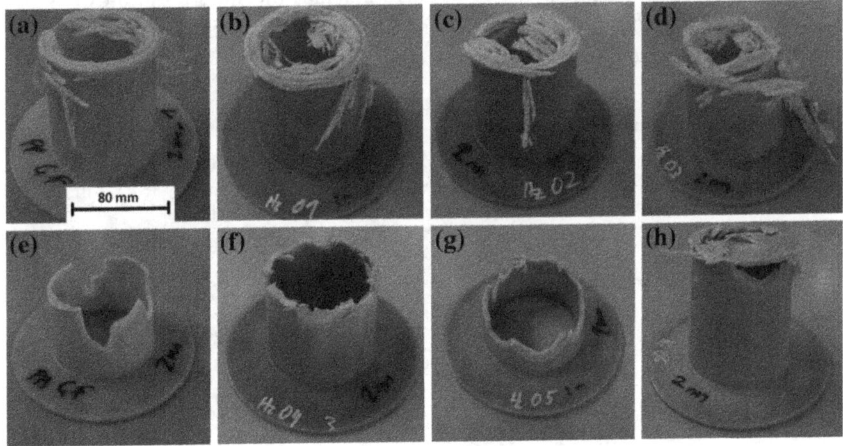

Fig. 10 Dynamic collapse mode in: (a) PP/GF (b) PP/GF/SiO$_2$ (c) PP/GF/MMT (d) PP/GF/GS (e) PA/GF (f) PA/GF/SiO$_2$ (g) PA/GF/MMT (h) PA/GF/GS

than that required to form the initial crack. The opposite situation exists if propagation of the crack stops quickly after the formation, and new cracks have to be initiated in order to crash the sample.

5.3 Energy Absorption Characteristics of Nanocomposite Structures

By relating the energy absorption characteristic with the propagation of the crack, it can be seen that materials that fail in a progressive manner, are able to absorb much higher energies than those with large continuous cracks. Furthermore, the mean crashing load in these materials is either on the same or higher level than the initial peak, what causes that the area under the load–displacement curve is significantly bigger.

Analyzing the results listed in Tables 1 and 2, we can note that crashing characteristic under dynamic load are different from those subjected to quasi-static compressive load.

All materials absorbed similar amount of impact energy, whereas energy absorbed during the quasi-static test is different in each material. This discrepancy is caused by the fact that each material failed with a different crashing length under the impact loading, while under the static loading the crashing length is constant in all experiments

Regarding the loads induced during the impact, we can note that mean crashing load was much closer to the initial peak, in case of all PP composites, what has a direct influence on the amount of energy absorbed by the structure. In case of PA materials the mean crashing load was significantly smaller than the initial peak, indicating weaker impact-energy absorption capabilities. Conclusions

Nanophased reactive polyurethane cores were manufactured and then used to fabricate sandwich structures. It has been found that the incorporation of MMT resulted in higher number of PU cells with smaller dimensions and higher anisotropy index (cross-sections RI and RII). The obtained materials exhibited

Table 1 Quasi-static crashing characteristics

Material	Crash length (mm)	Collapse mode	Initial peak (kN)	Mean crashing load (kN)	Energy absorbed (kJ)	Specific energy (SE) [kJ/kg]	Change in SE (%)
PPGF	86	III	29.74	34.75	2.993	49.4	
PPGF + SiO$_2$	86	III	26.59	17.86	1.489	24.4	−50.7
PPGF + MMT	86	III	24.75	15.39	1.294	21.2	−57.0
PPGF + GS	86	I	22.06	17.66	1.652	26.3	−46.9
PAGF	86	III	47.66	50.44	4.339	58.1	
PAGF + SiO$_2$	86	III	44.61	45.66	4.156	54.5	−6.1
PAGF + MMT	86	III	54.59	40.65	3.232	42.9	−26.2
PAGF + GS	86	III	55.10	45.74	4.117	51.7	−11.0

Table 2 Dynamic crashing characteristics

Material	Crash length (mm)	Collapse mode	Initial peak (kN)	Mean crashing load (kN)	Energy absorbed (kJ)	Specific energy (kJ/kg)	Specific energy increase (%)
PPGF	29.79	I	22.99	14.19	0.365	23.6	–
PPGF + SiO$_2$	31.4	I	25.72	15.41	0.376	22.6	−4.2
PPGF + MMT	36.02	I	20.02	12.86	0.401	20.5	−13.0
PPGF + GS	35.03	I	26.28	13.52	0.403	20.7	−12.3
PAGF	60.5	II	19.99	5.64	0.356	7.7	–
PAGF + SiO$_2$	57.56	III	26.51	8.98	0.432	9.8	27.0
PAGF + MMT	62.61	II	38.82	4.48	0.376	7.7	0.1
PAGF + GS	22.03	III	40.42	15.58	0.320	22.3	188.5

improved parameters in terms of thermal insulation properties. The investigation revealed that nanophased sandwich structures were capable of taking higher peak loads than those made of neat polyurethane cores when subject to low-velocity impact. This was especially true for multi-impact recurrences within the threshold loads and energies studied. It is proposed to investigate the threshold load that initiates delamination damage in the sandwich laminate which is a particularly important property of polymer composite components exposed to water and/or moisture and subjected to low energy impact.

Thermoplastic polymer glass-fibre and nano-silica reinforced composites were investigated as an alternative to polymer glass-fibre composites. The effect of matrix and reinforcement material on the energy absorption capabilities of composite structures was studied in details. The axial dynamic and quasi-static collapse of conical structures was conducted using high energy drop tower, as well as Instron electro-mechanical testing machine. The impact event was recorded using high-speed camera and the fracture surface was investigated with scanning electron microscopy (SEM). Attention is directed towards the relation between micro-fracture process and crack propagation mechanism, and energy absorbed by the structure.

The obtained results indicate an important influence of filler and matrix material on the energy absorption capabilities of the polymer composites. A significant increase in specific energy absorption is observed in polyamide 6 reinforced with nano-silica particles and glass-spheres, whereas addition of montmorillonite caused a decrease in that property. On the other hand, very little influence of the secondary reinforcement on the energy absorption capabilities of polypropylene composites was found.

Acknowledgments The authors would like to acknowledge the European Commission financial support through the FP7 Project- CP-FP, Project Reference No.: 228536– 2. Support from Research Councils Unite Kingdom (RCUK) and the Engineering and Physical Sciences Research Council (EPSRC) equipment pool is also acknowledged.

References

1. Njuguna, J., Pielichowski, K., Desai, S.: Nanofiller-reinforced polymer nanocomposites. Polym. Adv. Technol. **19**(8), 947–959 (2008)
2. Leszczyńska, A., Njuguna, J., Pielichowski, K., Banerjee, J.R.: Polymer/montmorillonite nanocomposites with improved thermal properties: Part II. Thermal stability of montmorillonite nanocomposites based on different polymeric matrixes. Thermochimica Acta. **454**(1), 1–22 (2007/2/15)
3. Leszczyńska, A., Njuguna, J., Pielichowski, K., Banerjee, J.R.: Polymer/montmorillonite nanocomposites with improved thermal properties: Part I. Factors influencing thermal stability and mechanisms of thermal stability improvement. Thermochimica Acta. **453**(2), 75–96 (2007/2/1)
4. Njuguna, J., Pielichowski, K., Desai, D.: Nanofiller-reinforced polymer nanocomposites. Polym. Adv. Technol. **19**(8), 947–959 (2008)
5. Mamalis, A.G.: Crashworthiness of composite thin-walled structural components. Technomic Pub. Co., Lancaster, Pa (1998)
6. Njuguna, J.: Crashworthiness of composite materials, Cranfield University, School of Applied Sciences, lecture notes (2012)
7. Farley, G.L.: Effect of fiber and matrix maximum strain on the energy absorption of composite materials. J. Compos. Mater. **20**(4), pp. 322 (1986)
8. Meressi, T., Paden, B.: Buckling control of a flexible beam using piezoelectric actuators. J. Guidance Control and Dyn. **16**(5), 977–980 (1993)
9. Farley, G. L.: The effects of crushing speed on the energy-absorption capability of composite tubes. J. Compos. Mater. **25**(10), pp. 1314 (1991)
10. Thornton, P.: The crush behavior of pultruded tubes at high strain rates. J. Compos. Mater. **24**(6), pp. 594 (1990)
11. Thornton, P.: Energy absorption in composite structures. J. Compos. Mater. **13**(3), pp. 247 (1979)
12. Hamada, H., Coppola, J., Hull, D., Maekawa, Z., Sato, H.: Comparison of energy absorption of carbon/epoxy and carbon/PEEK composite tubes. Compos. **23**(4), 245–252 (1992)
13. Yuan, Q., Misra, R.: Impact fracture behavior of clay-reinforced polypropylene nanocomposites. Polym. **47**(12), 4421–4433 (2006)
14. Sun, L., Gibson, R.F., Gordaninejad, F., Suhr, J.: Energy absorption capability of nanocomposites: A review. Compos. Sci. Technol. **69**(14), 2392–2409 (2009)
15. Bartczak, Z., Argon, A.S., Cohen, R.E., Weinberg, M.: Toughness mechanism in semi-crystalline polymer blends: I. High-density polyethylene toughened with rubbers. Polym. **40**(9), 2331–2346 (1999)
16. Bartczak, Z., Argon, A.S., Cohen, R.E., Weinberg, M.: Toughness mechanism in semi-crystalline polymer blends: II. High-density polyethylene toughened with calcium carbonate filler particles. Polym. **40**(9), 2347–2365 (1999)
17. Viana, J.: Polymeric materials for impact and energy dissipation. Plast., Rubber Compos. **35**(6–7), 260–267 (2006)
18. Wang, J., Fang, Z., Gu, A., Xu, L., Liu, F.: Effect of amino-functionalization of multi-walled carbon nanotubes on the dispersion with epoxy resin matrix. J. Appl. Polym. Sci. **100**(1), 97–104 (2006)
19. Cooper, C.A., Ravich, D., Lips, D., Mayer, J., Wagner, H.D.: Distribution and alignment of carbon nanotubes and nanofibrils in a polymer matrix. Compos. Sci. Technol. **62**(7–8), 1105–1112 (2002)
20. Kireitseu, M., Hui, D., Tomlinson, G.: Advanced shock-resistant and vibration damping of nanoparticle-reinforced composite material. Compos. B Eng. **39**(1), 128–138 (2008)
21. Zhang, H., Zhang, Z., Friedrich, K., Eger, C.: Property improvements of in situ epoxy nanocomposites with reduced interparticle distance at high nanosilica content. Acta Mater. **54**(7), 1833–1842 (2006)

22. Javni, I., Zhang, W., Karajkov, V., Petrovic, Z., Divjakovic, V.: Effect of nano-and micro-silica fillers on polyurethane foam properties. J. Cell. Plast. **38**(3), pp. 229 (2002)
23. Chen, J., Huang, Z., Zhu, J.: Size effect of particles on the damage dissipation in nanocomposites. Compos. Sci. Technol. **67**(14), 2990–2996 (2007)
24. Caprino, G., Teti, R.: Impact and post-impact behavior of foam core sandwich structures. Compos. Struct. **29**(1), 47–55 (1994)
25. Anderson, T., Madenci, E.: Experimental investigation of low-velocity impact characteristics of sandwich composites, Composite Structures. **50**(3), 239–247 (2000/11)
26. Mines, R.A.W., Worrall, C.M., Gibson, A.G.: Low velocity perforation behaviour of polymer composite sandwich panels. Int. J. Impact Eng. **21**(10), 855–879 (1998)
27. Lee, L.J., Zeng, C., Cao, X., Han, X., Shen, J., Xu, G.: Polymer nanocomposite foams. Compos. Sci. Technol. **65**(15–16), 2344–2363 (2005)
28. Cao, X., James Lee, L., Widya, T., Macosko, C.: Polyurethane/clay nanocomposites foams: Processing, structure and properties. Polymer **46**(3), 775–783 (2005)
29. Mahfuz, H., Islam, M.S., Rangari, V.K., Saha, M.C., Jeelani, S.: Response of sandwich composites with nanophased cores under flexural loading, Composites Part B: Engineering. **35**(6–8), 543–550 (2004/0)
30. Hosur, M.V., Mohammed, A.A., Zainuddin, S., Jeelani, S.: Processing of nanoclay filled sandwich composites and their response to low-velocity impact loading. Compos. Struct. **82**(1), 101–116 (2008)
31. Wetzel, B., Rosso, P., Haupert, F., Friedrich, K.: Epoxy nanocomposites-fracture and toughening mechanisms. Eng. Fract. Mech. **73**(16), 2375–2398 (2006)
32. Subramaniyan, A.K., Sun, C.: Toughening polymeric composites using nanoclay: Crack tip scale effects on fracture toughness. Compos. A Appl. Sci. Manuf. **38**(1), 34–43 (2007)
33. Njuguna, J., Michalowski, S., Pielichowski, K., Kayvantash, K., Walton, A.: Improved car crash protection. SPE Plastic Research Online (2010). 10.1002/spepro.003061
34. Hussain, F., Hojjati, M., Okamoto, M., Gorga, R.E.: Review article: Polymer-matrix nanocomposites, processing, manufacturing, and application: An overview. J. Compos. Mater. **40**(17), pp. 1511 (2006)
35. Ng, C., Schadler, L., Siegel, R.: Synthesis and mechanical properties of TiO2-epoxy nanocomposites. Nanostruct. Mater. **12**(1–4), 507–510 (1999)
36. Ma, J., Mo, M.S., Du, X.S., Rosso, P., Friedrich, K., Kuan, H.C.: Effect of inorganic nanoparticles on mechanical property, fracture toughness and toughening mechanism of two epoxy systems. Polym. **49**(16), 3510–3523 (2008)
37. Guo, Y., Li, Y.: Quasi-static/dynamic response of SiO2-epoxy nanocomposites. Mater. Sci. Eng. A. **458**(1–2), 330–335 (2007)
38. Han, J., Cho, K.: Nanoparticle-induced enhancement in fracture toughness of highly loaded epoxy composites over a wide temperature range. J. Mater. Sci. **41**(13), 4239–4245 (2006)
39. Njuguna, J., Silva, F., Sachse, S.: Nanocomposites for vehicle structural applications" in nanofibers—production, properties and functional applications, Lin, T. (ed.) In Tech Press, (ISBN: 978-953 307 420 7) (2011)
40. Silva, F., Sachse, S., Njuguna, J.: Energy absorption characteristics of nano-composite conical structures, IOP Conf. Ser.: Materials Science and Engineering **40** 012010 doi:10.1088/1757-899X/40/1/012010 (2012)
41. Silva, F., Njuguna, J., Sachse, S., Pielichowski, K., Leszczynska, A., Gianocolli, M.: The influence of multiscale fillers reinforcement into impact resistance and energy absorption properties of polyamide 6 and polypropylene nanocomposite structures. Mater. Des. **50**, 244–252 (2013)

Predictions of Energy Absorption of Aligned Carbon Nanotube/Epoxy Composites

D. Weidt and Ł. Figiel

Abstract The potential of aligned CNT/epoxy nanocomposites towards energy absorption applications was demonstrated using finite element modeling. For that, two cases studies were carried out: (1) prediction of crack resistance characteristics of epoxy reinforced with aligned double-walled CNTs (DWCNTs), and (2) prediction of rate-dependent compressive response of epoxy filled with aligned single-walled CNTs (SWCNTs). It was found that reinforcing epoxy with CNTs can significantly reduce the crack driving force in epoxy and increase strains to failure as a result of the damage propagation at the CNT–epoxy interphase. Particularly, enhancements of shear stiffness, shear strength and mode II fracture energy of CNT–epoxy interphases via CNT functionalization and minor increases of low sp^3-bond densities in the interwall phase of DWCNTs were shown to increase the crack resistance of the nanocomposite. Furthermore, it was found that the linear and nonlinear compressive deformations and thus the energy absorption characteristics of epoxies can be significantly affected by the presence of CNTs. Specifically, the initial stiffness was increased and the post-yield behaviour of the nanocomposite showed enhanced strain stiffening, both with increasing CNT loading and increasing CNT aspect ratio. Additionally, a combined effect of aspect ratio and volume fraction on energy absorption characteristics was found. This suggests that the average aspect ratio of CNTs should be carefully selected in order to maximise the energy absorption for the given CNT volume fraction.

D. Weidt · Ł. Figiel
Department of Mechanical, Aeronautical, and Biomedical Engineering/Materials and Surface Science Institute, University of Limerick, Limerick, Ireland

Ł. Figiel (✉)
Centre of Molecular and Macromolecular Studies, Polish Academy of Sciences, Sienkiewicza 112 90-363 Łódź, Poland
e-mail: lfigiel@cbmm.lodz.pl

J. Njuguna (ed.), *Structural Nanocomposites*, Engineering Materials,
DOI: 10.1007/978-3-642-40322-4_9, © Springer-Verlag Berlin Heidelberg 2013

1 Introduction

Carbon fibre reinforced epoxy laminates are commonly used to build lightweight structures due to their excellent strength/stiffness to weight ratios, and good fatigue and corrosion resistance (when compared to metals). As a result, they have a wide range of applications in transport (aerospace, automotive) and energy-generation (wind and tidal turbines) sectors. Unfortunately, they frequently behave as brittle materials under various types of impact loading as they cannot absorb/dissipate enough energy. This is especially important for laminates used in aircraft industry, which experience a considerable number of impact events during their service life. As a result, matrix cracks can develop due to the brittle nature of epoxy. In combination with a weak interlaminar strength, further opening of these matrix cracks can cause delaminations, and lead to catastrophic failure of the laminates and their structures.

Over the last two decades, attention in composites research has been shifted to nanofillers such as carbon nanotubes (CNTs). The reason for this is that CNTs exhibit exceptional mechanical properties (e.g. stiffness, resilience) [7, 18]. Furthermore, nanoreinforcements possess a specific surface area several orders of magnitude larger than the conventional microfillers [16]. Hence, an enormous interface area is available to enhance the interaction between the matrix and reinforcement at low CNT concentrations. Convinced by the potential of advanced nanoscale reinforcements several researchers used CNTs to improve the damage and failure resistance of CFRP laminates. Various approaches have been pursued, and they include enhancement of the mechanical performance of composite matrices [24]. Others have used CNTs to strengthen the interface between the CFRP pre-preg lamina [13], or have grown them on the microfibre surfaces to produce a stitching effect [23]. We believe that the incorporation of a CNT/epoxy nanocomposite coating onto the laminate surface can help to enhance overall energy absorption characteristics of the laminates. This is expected to have two different origins: (1) enhanced ductile deformation of epoxy in compression due to presence of CNTs and (2) CNT crack bridging mechanisms [12, 22, 25] accompanied by CNT pull-out and interface debonding.

Therefore, in order to demonstrate the potential of aligned CNT/epoxy nanocomposites towards energy absorption applications, computational studies were performed in this contribution. In particular, two case studies were investigated: (1) crack resistance characteristics of the CNT/epoxy nanocomposites with aligned double-walled CNTs (DWCNTs) in the vicinity of a matrix crack, and (2) rate-dependent compressive response of CNT/epoxy systems with aligned single-walled CNTs (SWCNTs). In particular, the influence of different properties of CNT–CNT and CNT–epoxy interphases on the normalized total energy release rate at the matrix crack tip and damage propagation along interphases were investigated in case (1). In case (2), the nanocomposite behaviour at various strain rates ranging from quasi-static to impact rates was predicted and the effects of CNT volume fraction and CNT aspect ratio on the macroscopic compressive stress–strain response were investigated. Those predictions were compared with the response of unfilled epoxy, and the normalized energy absorption characteristics of the nanocomposites in the absence of interphases and interphase damage were evaluated.

2 Energy Absorption of Aligned CNT/Epoxy Nanocomposites in Tension

2.1 Modelling Approach

An increased energy absorption in impact areas dominated by tensile stresses is expected to be caused by CNT crack bridging mechanisms accompanied by CNT pull-out and interface debonding. Hence, a problem of a stationary matrix crack in the vicinity of a DWCNT was studied to investigate the effect of interphase (CNT–epoxy and CNT–CNT) properties on the fracture parameter.

Carbon nanotubes in matrix systems provide load transfers similar to conventional short fibre composites. Lourie and Wagner [15] observed the formation of damage doublets in adjacent straight carbon nanotubes embedded in a polymer matrix. Such damage clusters usually occur in traditional non-continuous fibres reinforced composites due to a redistribution of stress from a failed fibre to its intact neighbour. The authors followed that the fundamentals of continuum mechanics of fibre composites hold to some degree in carbon nanotubes embedded in polymer matrices. The same was assumed in our computational study where continuum finite element (FE) modelling was applied. The continuum approach adopted here offers a compromise between accuracy and computational time when compared to molecular simulations. The FE software ABAQUS/Standard was used.

An axisymmetric unit cell with a circumferential matrix crack of length a was used in this work, as shown in Fig. 1. This represents a plausible scenario, where a matrix crack approaches a 'forest' of aligned (locally or globally) carbon nanotubes without yet touching of their surfaces. A CNT volume fraction, defined as volume ratio between a solid representation of the CNT and RVE, of ~ 1.28 % was assumed. A CNT aspect ratio of $L_{CNT}/R_{CNT} = 50$ was considered here, where CNT length $L_{CNT} = 500$ nm, and CNT radius $R_{CNT} = 5$ nm. This aspect ratio is relatively small compared to typical aspect ratios of CNTs. However, it corresponds to a lower reinforcing limit (worst case scenario), where a debonding length and CNT-bridging effect is small. The Young's modulus and Poisson's ratio of the epoxy were assumed to be 2.5 GPa and 0.4 respectively. The Young's modulus of the CNT was assumed to be 3.36 TPa [19]. The model symmetry resulted in modelling of only one half of the axisymmetric cell. The matrix crack was modelled as a sharp circumferential crack perpendicular to the axis of the CNT. Interphases between adjacent CNT walls and CNT and epoxy matrix were incorporated into the model to account for different CNT–CNT and CNT–epoxy

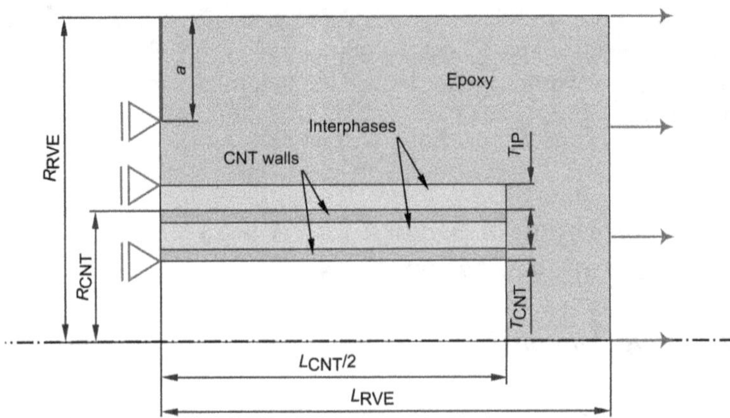

Fig. 1 Schematic view on the 2D axisymmetric RVE. $L_{RVE} = 291$ nm, $R_{RVE} = 41$ nm, $L_{CNT} = 500$ nm, $R_{CNT} = 5$ nm, $T_{CNT} = 0.1$ nm [19], $T_{IP} = 0.24$ nm, $a = 18$ nm

interactions. CAX4 (4-node bilinear axisymmetric quadrilateral elements) and COHAX4 elements (4-node axisymmetric cohesive elements) were used. The stress transfer from the matrix to the top edges of CNT walls was assumed to be negligible, therefore no bonding at the interface between top CNT edges and the matrix was modelled.

CNT–CNT and CNT–epoxy interactions were modelled using the cohesive zone concept through bilinear traction–separation laws (see Fig. 2). Those laws include damage initiation, damage propagation and failure. The damage initiation criterion used here is based on the maximum nominal stress value in shear or normal direction. Damage propagation is represented in terms of the stiffness degradation, and described using a linearly decaying function. The complete set of interphase parameters is shown in Table 1. It was assumed that modes I and II can

Fig. 2 Traction separation law; Energy release rate G_c; Strength S; Stiffness K; Displacement at damage initiation D_i; Displacement at failure D_f

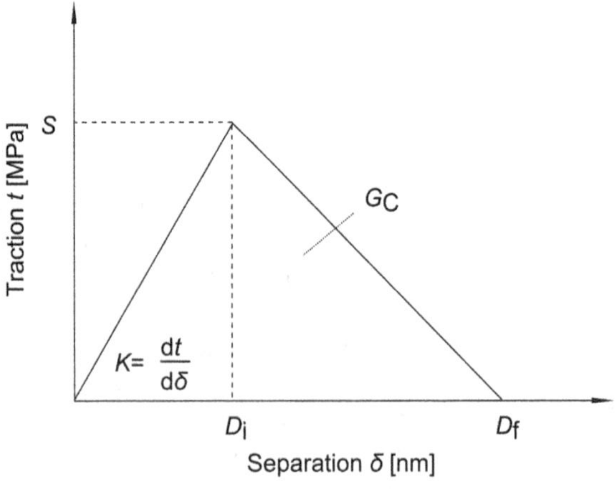

Table 1 Cohesive zone parameters for the interphases (mode II)

Interphase	Shear strength S [MPa]	Shear stiffness K [MPa/ nm]	sp^3-bond density* f [%]	Energy release rate G_{IIC}[MPa nm]	Damage initiation displacement D_i [nm]	Failure displacement D_f [nm]
A	30	20	–	50	1.5	3.333
B	30	100	–	50	0.3	3.333
C	30	100	–	150	0.3	10
D	30	100	–	100	0.3	6.667
E	90	100	–	150	0.9	3.333
F	120	100	–	150	1.2	2.5
G	30	180	–	50	0.167	3.333
O	5	10	0.00019	2.5	0.5	1
P	25	50	0.00097	12.5	0.5	1
Q	250	500	0.00966	125	0.5	1
R	2500	5000	0.09659	1250	0.5	1

* In this work sp^3-bond density f was calculated from $G = G_0 f$ for the given shear stiffness K (cases O-R) and the interwall thickness 0.34 nm, where $G_0 = 1760$ GPa (see [6]) – values of f in Table 1 correspond approximately to 1, 6, 58 and 580 sp^3-bonds respectively

be separated. Moreover, shear strength, shear stiffness and energy release rate in mode I were assumed to be twice as much of their counterparts in mode II, while damage initiation displacement D_i and failure displacement D_f in mode I and mode II were assumed to be of same value.

Parameters of the cohesive law were defined by considering results of molecular and experimental studies reported in the literature, and a short summary is given here. Ganesan et al. [10] estimated interfacial fracture energies between 50 and 250 MPa nm from pull-out experiments. A discrepancy can be found between interfacial shear strength from molecular predictions and experiments. Molecular simulation results predicted the interfacial shear strength for non-functionalised CNTs within polymer to be around 2.7 MPa [9], while from pull-out experiments shear strengths of ∼30 MPa for non-functionalised and ∼150 MPa for functionalised MWCNTs were found [3]. Therefore, the shear strength value was chosen to be based on the experimental investigation of Barber et al. [3]. Various combinations of CNT–epoxy interphase are denoted by letters from A to G (see Table 1). They are used to capture the shear strength range (∼30–150 MPa) and fracture energy range (∼50–250 MPa nm), as described above.

In addition, short summary of properties for the CNT–CNT interwall bonding reported in the literature is given as follows. Cumings and Zettl [8] estimated through controlled and reversible telescopic extension of MWNTs using transmission electron microscopy the static friction force per surface area of contact smaller than 6.6×10^{-15} N/Å^2 and the dynamic friction force per surface area of contact smaller than 4.3×10^{-15} N/Å^2. These values are negligible small and correspond to an interwall shear strengths of around 0.5 MPa. In another experimental work, it was shown that the interwall shear resistance can be enhanced through irradiation of MWCNTs, which leads to sp^3-bonds [17]. The grade of irradiation may be reflected by the interwall sp^3-bond density. It is also possible that standard CVD provide MWCNTs with covalent sp^3-bonding [1, 2]. Hence, the range of shear stiffnesses reflect variations in the sp^3 interwall bond density, which is defined as the number of sp^3-bonds divided by the number of carbon atom-pairs in the DWCNT system [6] (see the footnote of Table 1). It is noteworthy to mention that those sp^3-bonds can damage the mechanical performance of CNTs. However, no effect of the interwall bond density on the mechanical properties (i.e. axial Young's modulus) of the CNT walls was assumed in this work. Various combinations of CNT–CNT interwall bonding are denoted by letters from O to R (see Table 1).

The virtual crack closure technique (VCCT), which is based on assumptions of linear elastic fracture mechanics (LEFM), was used to determine the total energy release rate (*TERR*). According to this technique the work necessary to extend the crack from $a + \Delta a$ to $a + 2\Delta a$ is the same as the work required to close the crack from $a + \Delta a$ to a (see Fig. 3). It is assumed that crack extension from $a + \Delta a$ to $a + 2\Delta a$ does not significantly alter the total energy release rate [14]. Applied tensile loads and model geometry imply that the total energy release rate is equal to the mode I energy release rate G_I, i.e. no mode II was assumed to be present here. The computation was carried out at a post-processing stage, by calculating the reaction force F_2^j at node j, the displacement g_2^k of node k in x_2 direction and *TERR* at each increment of applied displacement:

$$TERR = \frac{F_2^j g_2^k}{\pi \Delta a (2R_{\mathrm{RVE}} - 2a - \Delta a)}. \tag{1}$$

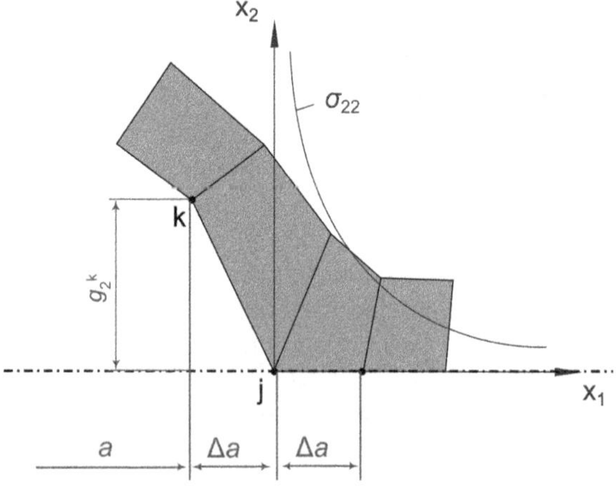

Fig. 3 Scheme of computation of *TERR*

2.2 Results and Discussion

The effect of the epoxy/CNT interphase shear stiffness on *TERR* for different combinations of interphase properties is shown in following Fig. 4. The first letter (A-G) refers to properties (Table 1) of the CNT–epoxy interphase and the second letter (O-R) corresponds to properties of the interwall phase. The composite *TERR* is normalised against *TERR* of the same axisymmetric model made of unfilled epoxy. The normalised TERR is then plotted as a function of applied strain to track the evolution of the normalised *TERR* with the applied load.

The CNT–epoxy interphase shear stiffness significantly affected normalised *TERR* at the beginning of strain until approximately 2 % of strain. In this range, a higher value of shear stiffness produced a lower value of normalised *TERR*, which is directly related to a lower intensity of the stress field around the matrix crack tip—all this is caused by the increased stress transfer from the matrix to the nanofiller. Hence, if the matrix crack is in the vicinity of functionalised CNT–epoxy and sp^3-bonded CNT–CNT interphases, the fracture parameter is reduced.

Then, all interphase combinations that possess the weakest interwall phase (denoted by O) revealed a fracture parameter jump around 1 % of strain. Damage initiated in the interwall phase at the start of the jump, and subsequently the first cohesive zone element failed completely. Fracture propagation in the interwall

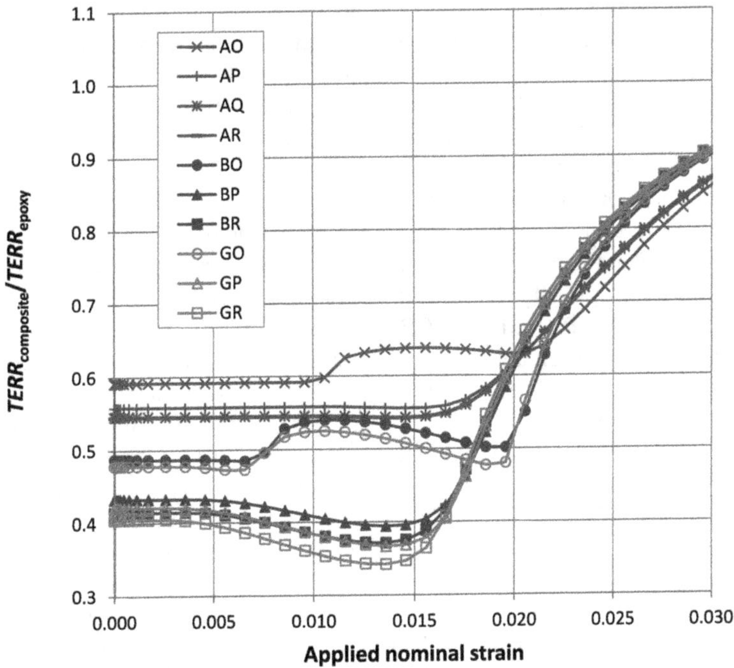

Fig. 4 Effect of CNT-epoxy interphase shear stiffness on *TERR*

phase continued (final failure of 60–70 % of the interwall phase) until the onset of failure in a first cohesive zone element in the CNT–epoxy interphase. At that point of around 0.02 applied nominal strain a local minimum of *TERR* was observed. Then, as damage and failure progressed in the CNT–epoxy interphase, the slope of *TERR* increased with increasing applied nominal strain. At larger applied strains the normalised *TERR* tended to uniformity slightly above 1, because of full debonding of the CNT–epoxy interphase, where the model behaves as an unfilled epoxy. A value of 1 was passed at around 0.045 of applied strain, which lies out of the investigated range in Fig. 4. At this level of strain around 80 % of the CNT–epoxy interphase has failed.

It is noteworthy to mention that the graph reveals a nonlinear relationship between the shear stiffness of CNT–epoxy interphase and fracture parameter *TERR*. In particular, a significant decrease of energy release rate was achieved from interphase A to B (AR and BR), no further significant decrease was found from B to G (BR and GR). A similar saturation governs the effect of sp^3-bond density—a decrease of the fracture parameter was observed when increasing the sp^3-bond density from 0.00019 to 0.00097 %. Further increase did not provide significant improvements. This is shown to be dominant at smaller magnitudes of applied nominal strain.

The decrease in *TERR*, especially pronounced for BP, BR, GP and GR between applied nominal strain values of 0.005 and 0.013, is probably caused by a stress redistribution within the matrix at the CNT end. The constraints on the matrix at the CNT end were relaxed due to the localized stiffness degradation of cohesive elements. As a result further applied nominal strain primarily strained this matrix region, while the rest of the model was hardly affected. Temporarily, this led to a less pronounced increase of *TERR* for the composite. Normalisation with *TERR* of pure epoxy resulted then in a local decrease of *TERR*. The less dominant reduction of TERR with interphases of lower shear stiffnesses (AP, AQ, AR) would support this explanation, because the matrix is less constrained in those cases.

Before fracture in the CNT–epoxy interphase occured, the sp^3-bond density of the interwall phase have had a larger effect on the energy release rate when combined with stiff CNT–epoxy interphases. This combined effect was further pronounced for BP, BR, GP and GR by a decrease of normalised *TERR* with increasing strain between applied nominal strain magnitudes of 0.005 and 0.013.

The effect of the CNT–epoxy interphase shear strength on *TERR* is shown in Fig. 5a. The interphase shear strength (considering only the mode II traction–separation law) increased from 30 to 90 to 120 MPa for C, E and F respectively. Again, according to the pull-out experiments of [3] 30 MPa corresponds to non-functionalised CNTs and 120 MPa may correspond to functionalised CNTs.

It was found that the shear strength of the CNT–epoxy interphase affected damage in the interwall phase—whereas for C, interwall damage/fracture occurred only in combination with interwall phase O, the CNT–epoxy interphase E exhibited interwall damage in combination with O and P. However, for visualisation purposes curves that possess damage in the interwall phase are excluded in Fig. 5.

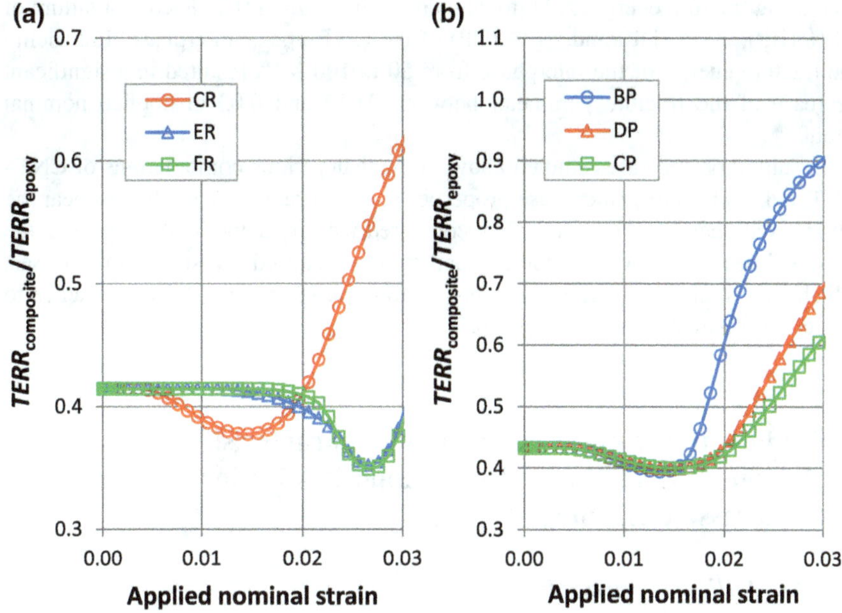

Fig. 5 Effect of **a** CNT–epoxy interphase shear strength and **b** CNT–epoxy fracture energy on TERR

The graph shows, how the CNT–epoxy interphase shear strength influenced the position and shape of the minimum. When increasing the CNT–epoxy interphase shear strength, the minimum of *TERR* was compressed and shifted to larger strain values. Significant improvements are visible from 30 to 90 MPa, but the difference between 90 and 120 MPa is negligibly small, which suggests that there will be no significant difference in the stress field near the crack tip. The larger applied strain value localising the minimum was caused by the increase in interphase shear strength that allowed the interphase to last longer. The smaller magnitude of the minimum for larger interphase shear strength values (90, 120 MPa) can be attributed to the larger stress that was carried by the CNT.

The effect of the CNT–epoxy interphase fracture energy on *TERR* is shown in Fig. 5b. From interphase B to D to C the mode II interphase fracture energy changed from 50 to 100 to 150 MPa nm respectively. The graph shows how the CNT–epoxy interphase mode II fracture energy affected the normalized *TERR* at later stages of strain due to the delay of failure in the CNT–epoxy interphase—smaller values of the interphase fracture energy (associated with the smaller separation at failure) resulted in an accelerated interphase damage propagation and failure, thus causing an increase of stresses at the crack tip, and hence the increase of *TERR*. For all interphase combinations, damage in the CNT–epoxy interphase started at around 0.005 of applied strain. The start of failure was delayed from 1.8 % (BP) to 3.0 % (DP) and to 4.3 % (CP) of applied nominal strain with increasing fracture energy. Additional increase in strain of 0.5 % led to a failure of 38 % (BP), 20 % (DP) and 14 % (CP) of the CNT–epoxy interphase. Enhancing the fracture energy of the interphase from 50 to 100 MPa resulted in a significant decrease of the fracture parameter between 0.015 and 0.03 of applied nominal strain.

In summary, the cases studied above show that various combinations of CNT–CNT and CNT–epoxy interphase properties lead to a reduction of stresses near the tip of a stationary matrix crack when compared with pure epoxy. As expected, the most efficient reduction of fracture parameter is caused by strong interphases, which lead to an improved stress transfer from the matrix to CNTs, and hence to reduction of stresses at the crack tip.

3 Prediction of Energy Absorption Characteristics of Aligned CNT/Epoxy Nanocomposites Under Compressive Loading

3.1 Modelling Approach

The ductility of epoxy in compression suggests that the energy absorption characteristics of CNT/epoxy can be improved through a CNT-enhanced nonlinear deformation of the matrix in zones undergoing compression. Hence, this study was focused on the prediction of the compressive macroscopic stress–strain response of the CNT/epoxy nanocomposite at strain rates ranging from quasi-static to impact rates. In particular, the emphasis was on the effect of CNT aspect ratio and volume fraction on energy absorption characteristics. An idealised aligned morphology of CNTs was investigated in this work to obtain an initial insight into the rate-dependent response of aligned CNT/epoxy nanocomposites. A rate-dependent constitutive model for epoxy was implemented into a finite element (FE) framework, and combined with the representative volume element (RVE) and numerical homogenisation. Predictions of nanocomposite behaviour for different rates were compared with pure epoxy, to demonstrate the effect of CNT aspect ratio and CNT volume fraction on energy absorption characteristics of CNT/epoxy nanocomposites.

In order to capture the macroscopic response of the nanocomposite it was assumed that the CNT morphology is globally periodic and that the macroscopic deformation is uniform. This enabled the application of the RVE concept. A simple axisymmetric RVE, including a single-walled CNT embedded in an epoxy matrix, was proposed because of its simplicity. The single-walled CNT was modelled as an effective continuum fibre in this work (see Fig. 6). The elastic properties of the effective continuum CNTs were obtained by homogenizing discrete versions of CNTs. The effective Young's modulus $E_{\mathrm{CNT}}^{\mathrm{eff}}$ was derived based on the axial CNT direction [21], as model and loading conditions used in this study suggest that this is a good approximation, and it is given by:

$$E_{\mathrm{CNT}}^{\mathit{eff}} = E_{\mathrm{CNT}} \frac{2R_{\mathrm{CNT}}T_{\mathrm{CNT}} - T_{\mathrm{CNT}}^2}{R_{\mathrm{CNT}}^2}, \tag{2}$$

where R_{CNT} is the CNT radius, T_{CNT} is the CNT wall thickness and E_{CNT} the Young's modulus of the CNT—based on a tubular cross section. A Young's modulus of $E_{\mathrm{CNT}} = 3.36$ TPa for a discrete CNT was used to obtain the effective modulus of the homogenised CNT. This value is higher than commonly used, but it was calculated using elastic shell theory for a smaller (than usual) wall thickness i.e. $T_{\mathrm{CNT}} = 0.1$ nm and Poisson's ratio of $v = 0.2$ [19]. The effective Young's modulus of homogenised CNTs was calculated as 133.056 GPa.

It is noteworthy to mention that different geometries of CNTs were compared (discrete and effective, capped and uncapped) to investigate shape of CNT ending. As a result, negligible differences were found for the stress–strain response of RVEs simulated with (1) solid, homogenised isotropic CNT cylinder and (2) discrete, tubular isotropic CNT cylinder exhibiting a circular cap at the CNT ending (see Fig. 6).

Perfect bonding between CNTs and the matrix was assumed in this work. Hence, a possible debonding and pull-out of CNTs from the matrix was excluded as possible energy dissipation/absorption mechanisms. As a result, nonlinear deformation of the epoxy matrix was expected to be the major cause of energy absorption for the nanocomposite.

The non-linear and rate-dependent matrix behaviour was represented using a physically-based constitutive model proposed and validated by Buckley et al. [5] for thermosetting resins. The model accounts for the strain softening and adiabatic

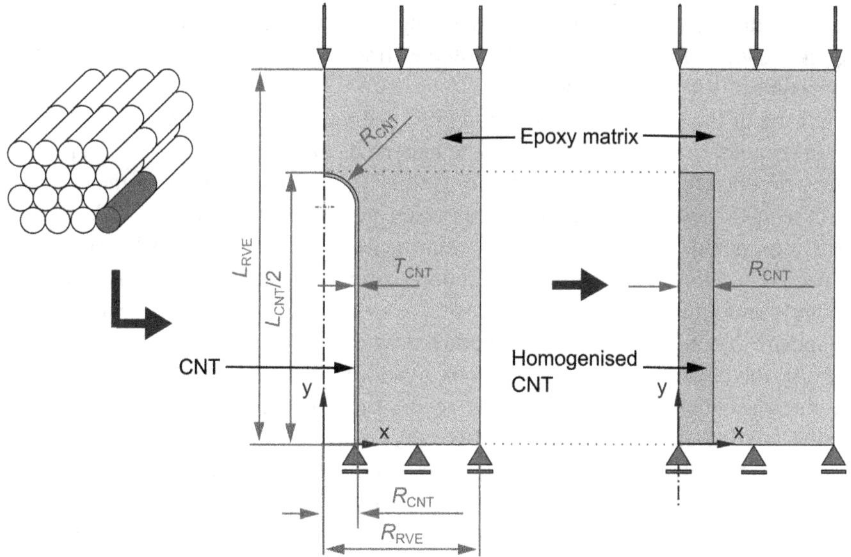

Fig. 6 2D Axisymmetric RVEs for aligned CNT/epoxy nanocomposites

heating effect (active upon applied impact loads), and enabled capturing basic phenomena of epoxy matrices subjected to varying strain rates. The mechanical constitutive behaviour was implemented as a user subroutine UMAT (ABAQUS/ Standard), with the model parameters obtained by [4] for the standard bisphenol A resin.

Matrix and CNTs were discretized using up to 3000 CAX8 (8-node biquadratic axisymmetric quadrilateral) elements. Uniform displacement boundary conditions were applied to the axisymmetric RVE in the y-direction, i.e. a uniform deformation of the right edge in the x-direction was enforced by setting the corresponding nodal displacements (right edge) as equal to the top right corner node of the RVE (using the ABAQUS' '*Equation' command). The vertical movement of the RVE was constrained at the bottom as shown in Fig. 6.

Then, the macroscopic true (Cauchy) stresses in the longitudinal direction were predicted based on the sum of reaction forces and the current cross-sectional area of the RVE. The normalized energy absorption was calculated as the ratio of areas under the stress–strain curves for the nanocomposite and for pure epoxy up to 13 % of applied nominal strain.

3.2 Results and Discussion

Predicted compressive stress–strain curves and energy absorption characteristics of CNT/epoxy nanocomposites subjected to different strain rates are shown and discussed in this section. The CNT aspect ratio was changed by varying the CNT length L_{CNT} of the axisymmetric RVE. Variations in CNT volume fraction in the axisymmetric CNT were obtained by changing the dimensions of the RVE with $L_{RVE}\text{-}L_{CNT}/2 = R_{RVE}\text{-}R_{CNT}$. CNT volume fractions refer to a nanocomposite composed of a hexagonal packing.

Effects of the CNT volume fraction (VF) and varying strain rates on the stress–strain response are shown in Fig. 7. As expected, the incorporation of CNTs into epoxy produced a considerable stiffening effect in the linear portion of the curve— the Young's modulus of the nanocomposite increased by around 15, 30, 46 and 77 % compared to pure epoxy for volume fractions of 0.5, 1, 1.5 and 2.5 % respectively. Hence, CNT volume fraction exhibited almost a linear effect on the Young's modulus of the nanocomposite. As expected, no change in the nanocomposite Young modulus was predicted with increasing strain rate. This is because the constitutive model for epoxy used in this work does not account for the rate-dependent stiffness. However, experimental data across a wide range of compressive strain rates for an RTM-6 epoxy system reported in [11] shows that there is no significant change in Young's modulus up to true strain rates $\sim 10^3\ \text{s}^{-1}$. The latter is greater than the maximum magnitude of nominal strain rates considered in this work, as the true strain rate $\dot{\varepsilon}_M$ is related to the applied (nominal) strain $\dot{\varepsilon}_{M(app)}$ through $\dot{\varepsilon}_M = \dot{\varepsilon}_{M(app)}/\left(1 + \varepsilon_{M(app)}\right)$

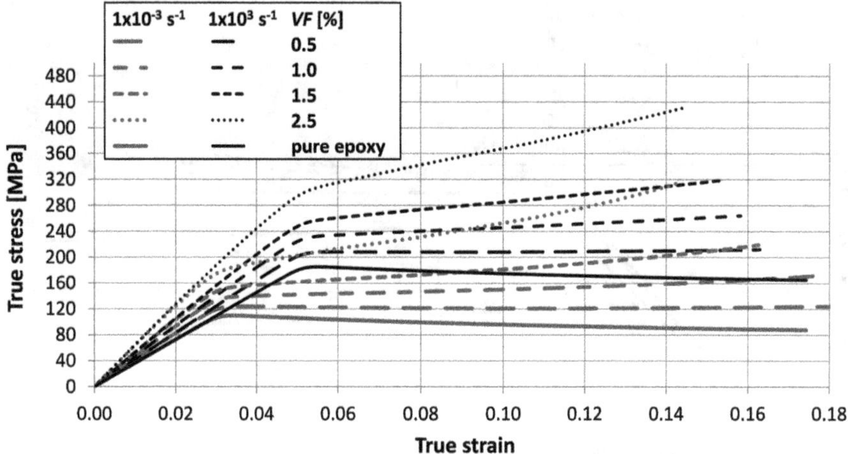

Fig. 7 Predicted stress–strain curves for pure epoxy and nanocomposites for different CNT volume fractions *VF* and applied strain-rates; CNT aspect ratio $AR = 50$

The stiffening effect on the stress–strain response caused by the CNT volume fraction continued beyond the elastic limit. In particular, for the given magnitude of applied strain, a gradual increase in the macroscopic true stress with CNT volume fraction was predicted, while the onset of nonlinear deformation was shifted to a slightly lower value of applied strain with increasing CNT volume fraction. In all cases a higher yield stress, defined here as the stress of the transition zone between the elastic and nonlinear behaviour, was predicted, when compared with the pure epoxy. The post-yield strain softening, which have been predicted and validated experimentally for the pure epoxy [4], was no longer present macroscopically for volume fractions equal to or greater than 1 %. Furthermore, an increase in the CNT volume fraction resulted in an increased post-yield strain-hardening effect.

Effects of CNT aspect ratio (*AR*) and varying strain rates on the stress–strain response were also investigated (see Fig. 8). As expected, the predicted trends were found to be in a qualitative agreement with the effects of varying CNT volume fraction. However, quantitatively, only a slight increase of the nanocomposite modulus was predicted with increasing CNT aspect ratio within the investigated range from 50 to 250. As indicated by different curves, the effect of CNT aspect ratio on the stress–strain response tended to saturate with increasing aspect ratio—less significant improvement of the overall stress–strain response was obtained for aspect ratios greater than 100 within the simulated strain range. However, in general there was an enhanced post-yield behaviour (increasing strain stiffening) with increasing CNT aspect ratio, which suggests that energy absorption characteristic can be enhanced only at large deformations. Again, the major contribution to that comes from the enhanced nonlinear deformation (strain stiffening) of the matrix caused by the presence of CNTs.

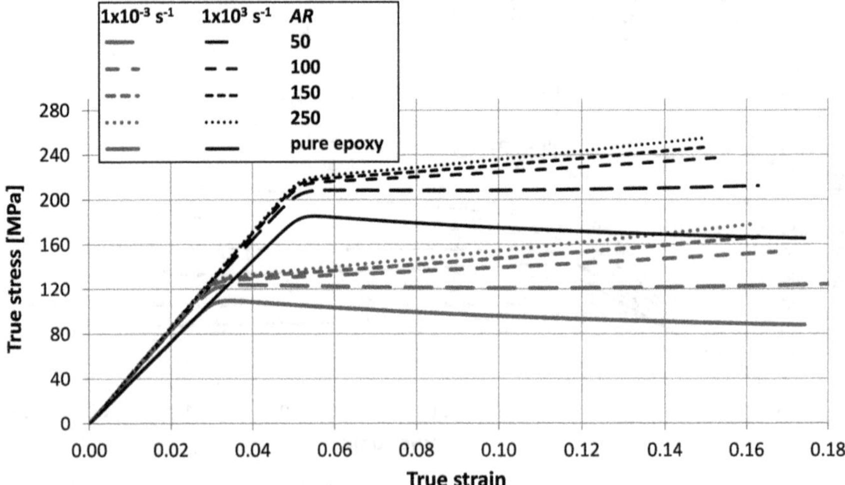

Fig. 8 Predicted stress–strain curves for pure epoxy and nanocomposites (axisymmetric RVEs) for different CNT aspect ratios *AR* and strain-rates; CNT volume fraction *VF* = 0.5 %

Hence, assuming that the energy absorbed by the material system can be represented by the area under the stress–strain curves, it is postulated that the improved post-yield behaviour of the nanocomposite can largely contribute to the improved energy absorption of the CNT/epoxy nanocomposite in compression. This increase in energy absorption characteristics results from an enhanced nonlinear deformation (pronounced by strain stiffening) of the matrix, caused by the presence of CNTs.

Therefore, normalised energy absorption characteristics were then evaluated based on the compressive stress–strain curves. Areas under those curves were evaluated up to strains of ~ 13 %, both for the nanocomposites and matrix, at two strain rates ($1 \times 10^{-3}\text{s}^{-1}$ and $1 \times 10^{3}\text{s}^{-1}$), and with different volume fractions and aspect ratios. Our predictions are shown in Fig. 9, where generally, the normalised energy absorption was enhanced due to the presence of CNTs at both strain rates.

A significant correlation between CNT volume fraction and energy absorption characteristics was found. An increase in CNT volume fraction contributed to a considerable increase in normalised energy absorption under quasi-static loading ($1 \times 10^{-3}\text{s}^{-1}$). This was especially the case for CNT aspect ratios greater than 150. There exists a significant combined effect of CNT volume fraction and aspect ratio. In comparison, predictions for a volume fraction of 0.5 % did not show such a significant increase in the normalised energy absorption with increasing aspect ratio than predictions for a volume fraction of 1.5 %.

In summary, our results suggest that CNTs can be used to enhance the nonlinear response of epoxies under compression to improve their energy absorption characteristics. The main contribution comes from the strain stiffening effect occurring

Fig. 9 Predicted normalised energy absorption characteristics as a function of CNT aspect ratio and CNT volume fraction for quasi-static 1×10^{-3} s^{-1} (**a**) and high strain rates 1×10^{3} s^{-1} (**b**). The normalised energy absorption characteristics at the data points (crosses) were used to determine the coefficients of the nonlinear regression functions via the least squares method

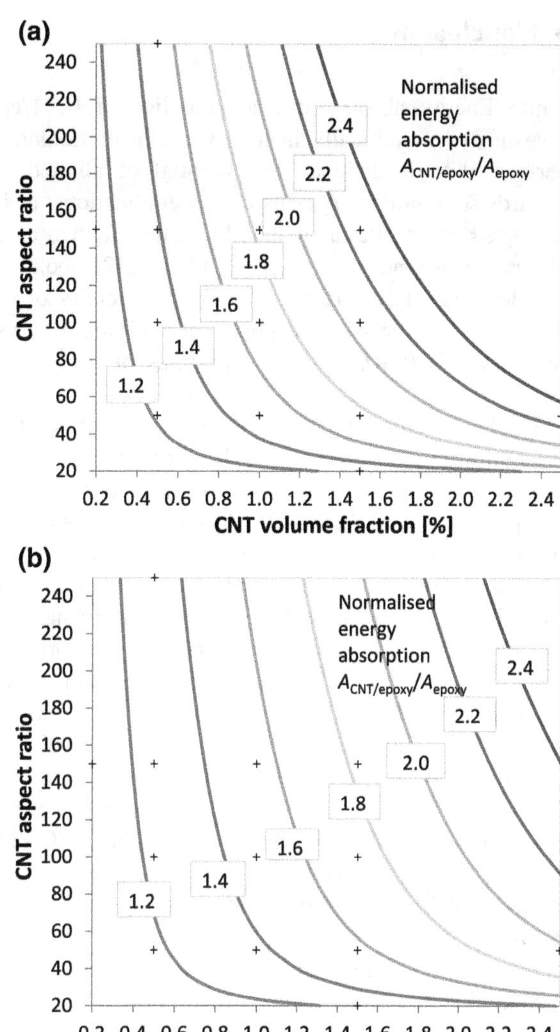

in the post-yield regime. The effect is improved with the increasing CNT volume fraction, which should be used for a carefully selected average CNT aspect ratio to maximize the energy absorption of CNT/epoxy nanocomposites. Results of our studies hold under the assumption that the nonlinear matrix deformation is the primary energy absorption mechanism. However, future studies will include other possible damage mechanisms such as debonding [20] and pull-out of CNTs from the matrix, which will enhance our predictions of energy absorption.

4 Conclusions

Finite Energy absorption characteristics of CNT/epoxy nanocomposites were investigated numerically in this work. In particular, computational studies were performed to investigate the potential of aligned CNT/epoxy nanocomposites towards structural energy absorption applications, as impact-resistant coatings for advanced composite laminates. In particular, two case studies were investigated: (1) crack resistance characteristics of the CNT/epoxy nanocomposites with aligned double-walled CNTs (DWCNTs) in the vicinity of a matrix crack, and (2) rate-dependent compressive response of CNT/epoxy systems with aligned single-walled CNTs (SWCNTs). The findings of these studies can be summarized as follows:

(1) The fracture parameter (total energy release rate) of a stationary matrix can be significantly reduced due to the presence of a CNT. Furthermore, the value of that fracture parameter depends on the properties of CNT–epoxy and CNT–CNT interphases such as shear strength, shear stiffness and fracture energy. Whereas the shear stiffness significantly reduced the fracture parameter at early stages of applied strain, the interphase shear strength and fracture energy contributed to its reduction at larger applied strains. Premature nanocomposite failure due to the existence of matrix cracks may therefore be prevented through relevant functionalisations. In particular, the reinforcement of epoxy with CNTs can reduce the crack driving force and promote larger strains to failure while progressive damage at the CNT–epoxy interphase takes place. This might be achieved through enhancements of shear stiffness, shear strength and mode II fracture energy of CNT–epoxy interphases via CNT functionalisation, and minor increases of low sp^3-bond densities in the interwall phase of DWCNTs.

(2) The compressive behaviour of epoxies can be significantly affected by the presence of CNTs at different loading rates (from static to impact). In particular, the initial stiffness and post-yield behaviour of nanocomposites were shown to be influenced considerably by CNT volume fraction and aspect ratio. In both cases, increased volume fraction and aspect ratio caused increased strain stiffening. This led to increased area under the stress–strain curves, which suggests that energy absorption in compression can be improved by addition of CNTs. The energy absorption characteristics evaluated from the stress–strain curves were improved with increasing CNT volume fraction and CNT aspect ratio, both for static and impact rates. Additionally, a combined effect of CNT aspect ratio and CNT volume fraction on energy absorption characteristics was found. This suggests that the average aspect ratio of CNTs should be carefully selected in order to maximise the energy absorption for the given CNT volume fraction. Our ongoing work on inclusion of interphase properties into the model will further improve our description of energy absorption of CNT/epoxy nanocomposites, and give some new insight into the effect of CNT functionalisation on the compressive response of the nanocomposites across different strain rates.

Acknowledgments The project is supported by the Irish Research Council (IRC). Computational facilities and support are provided by the SFI/HEA Irish Centre for High-End Computing (ICHEC).

References

1. Barber, A.H., Andrews, R., Schadler, L.S., Wagner, H.D.: On the tensile strength distribution of multiwalled carbon nanotubes. Appl. Phys. Lett. **87**, 203106 (2005)
2. Barber, A.H., Kaplan-Ashiri, I., Cohen, S.R., Tenne, R., Wagner, H.D.: Stochastic strength of nanotubes: an appraisal of available data. Compos. Sci. Technol. **65**, 2380–2384 (2005)
3. Barber, A.H., Cohen, S.R., Eitan, A., Schadler, L.S., Wagner, H.D.: Fracture transitions at a carbon-nanotube/polymer interface. Adv. Mater. **18**, 83–87 (2006)
4. Buckley, C.P., Harding, J., Hou, J.P., Ruiz, C., Trojanowski, A.: Deformation of thermosetting resins at impact rates of strain. Part I: Experimental study. J. Mech. Phys. Solids **49**, 1517–1538 (2001)
5. Buckley, C.P., Dooling, P.J., Harding, J., Ruiz, C.: Deformation of thermosetting resins at impact rates of strain. Part 2: Constitutive model with rejuvenation. J. Mech. Phys. Solids **52**, 2355–2377 (2004)
6. Byrne, E.M., Letertre, A., McCarthy, M.A., Curtin, W.A., Xia, Z.: Optimizing load transfer in multiwall nanotubes through interwall coupling: theory and simulation. Acta Mater. **58**, 6324–6333 (2010)
7. Cooper, C.A., Young, R.J., Halsall, M.: Investigation into the deformation of carbon nanotubes and their composites through the use of Raman spectroscopy. Compos. Part A-Appl. S. **32**, 401–411 (2001)
8. Cumings, J., Zettl, A.: Low-friction nanoscale linear bearing realized from multiwall carbon nanotubes. Science **289**, 602–604 (2000)
9. Frankland, S.J.V., Caglar, A., Brenner, D.W., Griebel, M.J.: Molecular simulation of the influence of chemical cross-links on the shear strength of carbon nanotube-polymer interfaces. J. Phys. Chem. B **106**, 3046–3048 (2002)
10. Ganesan, Y., Peng, C., Lu, Y., Loya, P.E., Moloney, P., Barrera, E., Yakobson, B.I., Tour, J.M., Ballarini, R., Lou, J.: Interfacial toughness of carbon nanotube reinforced epoxy composites. ACS Appl. Mater. Int. **3**, 129–134 (2011)
11. Gerlach, R., Siviour, C.R., Petrinic, N., Wiegand, J.: Experimental characterisation and constitutive modelling of RTM-6 resin under impact loading. Polymer **49**, 2728–2737 (2008)
12. Gojny, F.H., Wichmann, M., Fiedler, B., Schulte, K.: Influence of different carbon nanotubes on the mechanical properties of epoxy matrix composites—A comparative study. Compos. Sci. Technol. **65**, 2300–2313 (2005)
13. Hu, N., Li, Y., Nakamura, T., Katsumata, T., Koshikawa, T., Ara, M.: Reinforcement effects of MWCNT and VGCF in bulk composites and interlayer of CFRP laminates. Compos. Part B-Eng. **43**, 3–9 (2012)
14. Krueger, R.: Virtual crack closure technique: History, approach, and applications. Appl. Mech. Rev. **57**, 109–143 (2004)
15. Lourie, O., Wagner, H.D.: Evidence of stress transfer and formation of fracture clusters in carbon nanotube-based composites. Compos. Sci. Technol. **59**, 975–977 (1999)
16. Peigney, A., Laurent, C., Flahaut, E., Bacsa, R.R., Rousset, A.: Specific surface area of carbon nanotubes and bundles of carbon nanotubes. Carbon **39**, 507–514 (2001)
17. Pregler, S.K., Sinnott, S.B.: Molecular dynamics simulations of electron and ion beam irradiation of multiwalled carbon nanotubes: The effects on failure by inner tube sliding. Phys. Rev. B **73**, 224106 (2006)
18. Treacy, M.M., Ebbesen, T.W., Gibson, J.M.: Exceptionally high Young's modulus observed for individual carbon nanotubes. Nature **38**, 678–680 (1996)

19. Wang, C.Y., Zhang, L.C.: A critical assessment of the elastic properties and effective wall thickness of single-walled carbon nanotubes. Nanotechnology **19**, 195704 (2008)
20. Weidt, D., Figiel, Ł., Buggy, M.: Preparation testing and modelling of nanocomposite surface coatings to improve the impact behaviour of advanced composite laminates. Mater. Sci. Forum **714**, 3–1 (2012)
21. Weidt, D., Figiel, Ł., Buggy, M.: Prediction of energy absorption characteristics of aligned carbon nanotube/epoxy nanocomposites. IOP Conf. Ser: Mater. Sci. Eng. **40**, 012028 (2012)
22. Wernik, J.M., Meguid, S.A.: Recent developments in multifunctional nanocomposites using carbon nanotubes. Appl. Mech. Rev. **63**, 050801 (2010)
23. Wicks, S.S., de Villoris, G.R., Wardle, B.L.: Interlaminar and intralaminar reinforcement of composite laminates with aligned carbon nanotubes. Compos. Sci. Technol. **70**, 20–28 (2010)
24. Yokozeki, T., Iwahori, Y., Ishiwata, S., Enomoto, K.: Mechanical properties of CFRP laminates manufactured from unidirectional prepregs using CSCNT-dispersed epoxy. Compos. Part A-Appl. S. **38**, 449–460 (2007)
25. Zhang, W., Picu, R.C., Koratkar, N.: Suppression of fatigue crack growth in carbon nanotube composites. Appl. Phys. Lett. **91**, 193109 (2007)

Silver Nanocluster/Silica Composite Coatings Obtained by Sputtering for Antibacterial Applications

Cristina Balagna, Sara Ferraris, Sergio Perero, Marta Miola, Francesco Baino, Andrea Coggiola, Daniela Dolcino, Alfio Battiato, Chiara Manfredotti, Ettore Vittone, Enrica Vernè and Monica Ferraris

Abstract Bacterial contamination is a critical issue which concerns different fields related to people and everyday life products and goods. Authors at Politecnico di Torino developed a new silver nanocluster/silica composite coating obtained by sputtering able to confer antibacterial properties to several materials. Silver nanocluster/silica composite coatings were deposited by radio frequency co-sputtering technique on glasses, ceramics, metals and polymers. The sputtering method is extremely versatile and suitable for most of substrates, because it does not require high temperatures which could decrease the mechanical properties of coated materials (e.g. polymers). The main results will be discussed for each coated substrate, in terms of characterization techniques, morphology, composition, antibacterial effect and adhesion to the substrate.

1 Introduction

1.1 Microbial Contamination in Different Fields

Materials with antibacterial properties are more and more requested in several fields where the risk of microbial contamination is considered a relevant issue, such as biomedical implants, agricultural/food industry, facilities in crowded

C. Balagna (✉) · S. Ferraris · S. Perero · M. Miola · F. Baino · E. Vernè · M. Ferraris
Department of Applied Science and Technology, Institute of Materials Physics
and Engineering, Politecnico di Torino, Corso Duca degli Abruzzi 24 10129 Turin, Italy
e-mail: cristina.balagna@polito.it

A. Battiato · C. Manfredotti · E. Vittone
Experimental Physics Department/Centre of Excellence Nanostructured Interfaces and
Surfaces, University of Torino and INFN sez, Torino, Via P. Giuria 1 10125 Turin, Italy

A. Coggiola · D. Dolcino
Azienda Ospedaliera SS. Antonio e Biagio e Cesare Arrigo, Via Venezia 16,
Alessandria, Italy

J. Njuguna (ed.), *Structural Nanocomposites*, Engineering Materials,
DOI: 10.1007/978-3-642-40322-4_10, © Springer-Verlag Berlin Heidelberg 2013

places (hospitals, public transportation systems), cell phones, personnel protective systems as well as aerospace structures [1–3].

Bacterial contamination of surfaces is mainly related to the formation of biofilm, a complex 3D structure of microorganisms and extracellular materials which adheres to solid surfaces and makes bacteria more resistant to antimicrobial agents. Biofilm formation includes the absorption of macromolecules on the surface, bacterial cell adsorption and adhesion on the conditioned surface, bacterial production of biopolymeric substances and cell signaling molecules, cell replication and growth and bacteria production of polysaccharide matrix [4, 5]. Biofilm represents a problem from different points of view, i.e. human health (infection development), industrial processes and equipments (products contamination and material damage). A brief review of bacterial contamination in different fields is reported below.

1.1.1 Medical Implants

Implant-associated infection is a widely discussed and studied topic in the medical field. Despite of systemic antibiotic prophylaxis and special antiseptic operative procedures, the problem is not completely solved. The development of an infection can lead to implant failure and need for its removal. The presence of an artificial material within the body increases the susceptibility to infections. The patient immune system results compromised and the implant surface constitutes a preferential site for bacterial adhesion. Moreover bacterial susceptibility to antimicrobials is reduced in presence of an artificial material [6–10]. The physical–chemical properties of the implant surface (e.g. roughness, charge, chemical composition) affect the initial proteins absorption and consequently bacterial adhesion as well as the biofilm formation [6, 7, 9]. Microbial contamination and biofilm formation can affect different materials for medical implants, mainly metals for orthopedic and dental applications [6–8] and polymers for catheters and abdominal wall repair prostheses [9, 10]. Bacterial contamination issues in ophthalmic applications, such as eye surgery and implants, can give significant clinical problems together with the need for post-surgical additional treatments, both expensive and stressful for patients.

1.1.2 Home

Microbial contamination of the household environment has recently been reviewed [11]. *Staphylococcus aureus* and other staphylococci have been found in different household areas and surfaces: the kitchen (working surfaces, sink, refrigerator and sponges), the toilet (cabinet top or shelves, hand towels and taps) the bedroom (pillows and bedding) but also on toys, television, remote controls, telephones, door knobs and carpets [11]. The main sources for bacterial contamination have been individuated as people, animals and environment.

1.1.3 Aerospace

Microbial contamination in space missions is generally limited and maintained below strict, high quality standard thanks to specific prevention and monitoring strategies. However it constitutes an extremely tricky topic because of the hampered immune system of the astronauts (due to high working pressure, defined diet, restricted hygienic practices, microgravity and radiation) [12]. The main bacterial sources are humans and the contamination can interest air, surfaces, water and food [13, 14]. As far as surfaces are concerned, microorganisms can adhere onto materials and then form a structured and resistant biofilm. In space structures surface bacterial contamination is minimized before the flight start by means of different surface treatments (e.g. heat, radiation, chemicals) depending on the material resistance characteristics. In flight contaminations are reduced by using disinfectant wipes based on quaternary ammonium compounds or a mixture of quaternary ammonium compounds and hydrogen peroxide. The repeated failure of the disinfection process may cause the replacement of the contaminated surfaces/components [12].

1.1.4 Food

Bacterial contamination of food from processing and handling is an extremely critical issue. In particular for products that cannot undergo any antibacterial treatment before consumption, such as cheese, they can be the cause of poison, as reported in the literature [15]. In these case, bacteria can come from the raw milk, milking machines, farm environment, cheese processing plants and also from operators [15–17]. The basic rule for prevention of contamination is based on an effective cleaning and disinfection program (including removal of all residuals from the surface, application of detergents and disinfectants, removal of all traces of cleaning products) and an equipment design easy for cleaning (free of dead ends, corners or cracks and preferably made of stainless steel) [5].

1.2 Silver as Antibacterial Agent

Silver is the most known and documented antimicrobial agent and its powerful action can be expressed in several forms as metallic film/layer, nanoparticles and ions [18, 19].

Silver ions (Ag^+) can bind to different specific chemical sites (characterized by thiol groups) causing inactivation and cell death. For example Ag^+ can bind to proteins altering their structure and cause the rupture of cell walls, or to enzymes preventing their function, or to DNA interfering with cell division and replication [18, 20].

The antimicrobial properties of silver nanoparticles can be related to their extremely high surface area that provide silver release. Moreover nanoparticles can directly attach to the cell surface or enter the cell causing membrane rupture and cell damage [18]. The formation of pits on the cell wall and the alteration of cellular structures and function have been documented both for Gram negative [21] and Gram positive [22] bacteria. It has been observed that the antibacterial activity of silver nanoparticles depends both on their dimensions and shape [18].

Thanks to the multiplicity of action routes, a reduced resistance development has been documented for silver in comparison to common antibiotics.

The bactericidal effect of silver has been widely employed in the treatment of burns, wounds and infection and also in the water purification [18]. Different forms of silver (e.g. silver nitrate, sulfadiazine, metallic silver) have been considered for these purposes.

Recently, an increasing interest in silver nanoparticles has been observed [18] and numerous antibacterial products based on nano-silver can be found [23]. They include medical devices (masks, tubes, catheters, fillers, diagnosis tools, coatings...), personal care and cosmetics products, textiles and shoes, electronic devices and domestic appliances (refrigerators, washing machines, mobile phones...), household products (antibacterial surfaces, paints and cleaning products), filtration devices, food handling and production tools [23]. This widespread use of nano-sized silver can lead to different routes of exposure to nano-Ag (mainly inhalation, dermal and oral contact) with possible long-term toxic effects, due to the exposition to free ion and silver nanoparticles [23].

1.3 Techniques to Obtain Silver-based Coatings

Silver-related antibacterial properties can be conferred to glasses, ceramics, metals and polymers by means of several techniques such as: absorption on zeolites or in silica microsphere [24–26], Ag embedding in polymers [27], ion-exchange on Na^+/K^+ based glasses [28–36], chemical vapor deposition (CVD) [37, 38], sol gel [27, 40–45], sputtering [46, 47] etc.

Advantages and disadvantages of each technique are listed in Table 1.

Sputtering is one of the most versatile coating methods, suitable for most substrates, because it does not need high temperatures during process, which could decrease the mechanical and thermal properties of materials to be coated (e.g. polymers) [47].

The sputtering technique consists in extracting from a bulk material (target) some atoms/molecules or clusters of atoms/molecules and deposit them on the substrate. All the process is performed in vacuum.

Sputtering is one of the most common coating technique and there are several literature references that describe in details technology and equipments [55]. Nowadays sputtering is widely diffused in industrial processes also because it can be merged in the industrial process flow as an in-line production technology.

Table 1 Comparison among different techniques suitable to provide antibacterial coatings

Method	Process	PROs	CONs	Ref.
Ag doped zeolites or silica microspheres	Ag^+ absorption on zeolites or silica microsphere	Low cost, industrial products available, suitable to coat large surfaces	Ag doped zeolites and silica microspheres must be embedded in a matrix to obtain a coating Low thermo-mechanical properties of the polymer matrix	[24–26]
Ag embedded in polymers	Ag particles embedded in polymers or synthesis of polymer containing Ag ions	Low cost, industrial products available, suitable to coat large surfaces and to large scale production	Limited mechanical and thermal Stability due to polymers	[27]
Silver Ionic-exchange on glasses	Ag^+/Na^+ or Ag^+/K^+ Ion-exchange on glasses containing Na^+ or K^+	Low cost, industrial products available, suitable to coat large surfaces and to large scale production	Suitable only for glasses containing K^+ or Na^+	[28–36]
Chemical Vapour Deposition (CVD) of silver or silver doped coatings	Reaction of volatile precursors on the substrate in a reaction chamber, in vacuum	No chemical solution manipulation and waste solution production No needing of special and expensive precursor gas and related industrial equipments	Limited number of suitable precursors. Expensive technique	[37, 38]

(continued)

Table 1 (continued)

Method	Process	PROs	CONs	Ref.
Sol gel deposition of silver doped coatings	Mixing, stirring, gelation and drying of organic and inorganic precursors containing silver	Cheap process, suitable for large scale production	Thermal treatment needed for drying and stabilization. Time consuming process. Possible cracks on the coating after drying	[27, 40–45]
Sputtering of silver or silver doped coatings	Atomic clusters deposition from a bulk target to the desired substrate in a vacuum chamber	Suitable for large scale production. Suitable for almost every substrate. Quick, one step process	High initial costs for the sputtering equipment. Substrates must resist under vacuum	[46, 47]
STOBER method	Mixing, stirring, filtering and drying of silica and silver nanoparticle precursors, which are then dispersed in a suitable carrier (polymeric) to be used as coating	Coating production in one step process. Easy scaling up process. Suitable to coat large surfaces	Limited mechanical and thermal stability due to the polymer matrix	[48–50]
MELTING of Ag doped glasses	Mixing, melting and pouring of raw materials. Glass bulk/frit that can be used to coat surfaces by a second process	Coating in one step process. High thermal resistance. Suitable for large scale production	Unsuitable to coat polymers	[28, 39]
PAINTING with silver doped paints	Silver doped polymeric-based paints	Suitable to every substrate	Suitable only for low temperature applications. Low thermal/chemical/mechanical stability of polymeric paint	[51–54]

In order to obtain antibacterial layers with high thermal, chemical and mechanical stability ideally on every class of materials, co–sputtering is the preferred technique [56–63]. For the preparation of silver nanocluster/silica coatings, two different targets are used at the same time to obtain a composite material with Ag nanoparticles embedded in a matrix. Modifying the power density applied to the targets and/or the deposition time, it is possible to tailor the Ag content in the silica matrix, therefore the coating antibacterial properties, characteristic that is mandatory for most applications.

2 A New Approach to Confer Antibacterial Properties to Materials: Sputtering Deposition of Silver Nanocluster/Silica Composite Coatings onto Several Substrates

A new approach was developed by authors from Politecnico di Torino [56] to confer antibacterial power to glasses, ceramics, metals and polymers for different applications [64, 65]. Silver nanocluster/silica composite coatings were deposited onto the substrates by means of radio frequency (RF) co-sputtering technique using two targets, a silica one and a silver one, both placed in the sputtering chamber [56–63]. The antibacterial silver nanoclusters are well embedded into the silica matrix which provides good mechanical and thermal resistance together with a suitable release of silver ions. These coatings allow the reduction of the silver amount and thereby the risk of toxicity, if compared with pure silver coatings [23, 66]. As discussed in the introduction, the sputtering technique is a method easy-fitting to all the classes of materials as the process parameters (deposition time, power on the targets, pressure, pre-treatment of the samples) can be properly set as a function of the substrate and the final application. In particular some properties of the coatings, such as the thickness and the total silver content, could be increased with the deposition time. The optimization of the silver amount is of primary importance, especially for materials whose applications are in direct contact with human organs and tissues, where the risk of a cytotoxic effects is enhanced.

Generally, the temperature reached in sputtering chamber is quite low (maximum 80 °C, even lower with systems for cooling the substrate). This temperature, or in some case the exposures to high temperatures during sputtering can damage thermo-sensitive substrates, such as polymers. For these materials, short sputtering treatments have been optimized. Figure 1 summarizes the influence of the sputtering deposition time on the coating characteristics.

In the following sections, an overview about the main techniques for the characterization and the deposition of silver nanocluster/silica composite coatings on several substrates will be presented. The materials (silica and sodalime glasses, different polymers and steel) were chosen because of their suitability for different applications, from the food industry to biomedical implants, aerospace structures and personal protective systems.

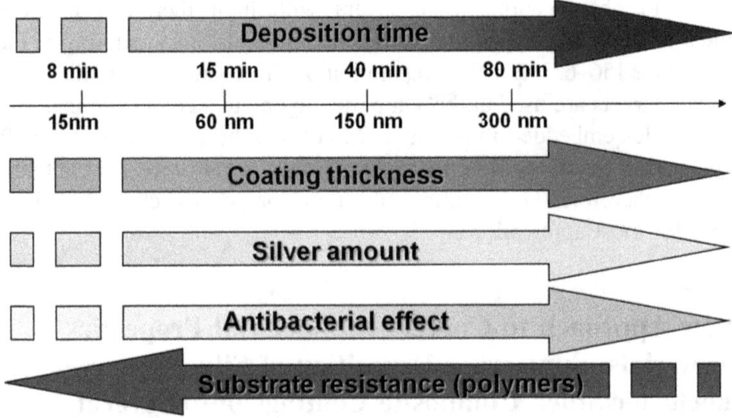

Fig. 1 Schematic representation of the influence of deposition time on silver amount in the coating and its effects

2.1 Characterizations and Analysis

There are many different techniques that can be used for characterizing thin films; the most commonly used in this research are shortly presented below.

Spectroscopy techniques, such as electron dispersive spectroscopy (EDS) and X-ray photoelectron spectroscopy (XPS), allowed the identification of the coating composition by measuring the Ag/Si ratio, whereas the size of silver nanoclusters can be directly observed by transmission electron microscope (TEM) or, indirectly, by UV–visible absorption spectroscopy (UV–Vis), by measuring the Localized Surface Plasmon Resonance (LSPR). The presence of metallic silver nanoclusters and/or other crystalline phases was detected by means of X-ray diffraction (XRD).

The morphology of the coating was studied and analyzed by means of field emission scanning electron microscopy (FESEM).

The antibacterial behavior of the silver nanocluster/silica composite coating was verified through the inhibition zone evaluation towards *Staphylococcus aureus* strain, a Gram-positive bacteria in accordance with the National Committee for Clinical Laboratory (NCCLS) [67]. In other specific cases, different bacteria strains as Gram positive (*Bacillus cereus*) and Gram negative (*Morganella, Klebsiella pneumonite, Escherichia coli*) were tested.

Finally, the mechanical behavior was evaluated in terms of adhesion of the coating to the substrate through tape test according to the ASTM D3359–97 standard [68] and nanohardness by means of nanoindentation test.

Some thermal treatments were performed in order to control the effect of the temperature on the coating properties.

2.2 Silver Nanocluster/Silica Composite Coatings on Silica and Sodalime

Silver nanocluster/silica composite coatings were deposited on silica (SiO_2) and sodalime (SL) glass substrates [57, 58, 61]. The coated substrates have been heated from 150 to 600 °C in order to control the effect of heat treatments on their properties.

Figure 2 shows the as deposited (AS DEP) and thermal treated (450 °C) (TT 450 °C) coatings on silica (SiO_2) and sodalime (SL) substrates. The coating thickness ranges from 50 to 300 nm.

The typical brown color of the coating is due to the silver nanoclusters embedded into the silica matrix [57, 58]. The higher is the coating thickness, and so the silver amount, the darker is the deposited layer (as in the case of SiO_2, coated by a 300 nm layer). On the other hand, the thermal treatment tends to bleach the coating. This modification is correlated with an enlargement of nanoclusters size [57, 58] as confirmed by FESEM and XRD in Fig. 3 (with permission of [58]) and by UV–visible absorption spectra (Fig. 4).

FESEM in Fig. 3 shows the morphology of the coating surfaces, as deposited (a) and after thermal treatment at 150 °C (c), 300 °C (e), 450 °C (g), and 600 °C (i), on silica substrate [58].

The typical porous structure of the sputtered layers is observable in the as deposited coating (Fig. 3a). The coating becomes gradually more compact and the silver nanoclusters, visible as bright dots, decrease in number but enlarge in dimensions with the increase of thermal treatment temperatures. In fact, the nanocluster size ranges from less than 7 nm (as deposited coating) to about 20–60 nm (thermally treated coating). It was observed that a treatment from 500 to 800 °C decreases the porosity of silica containing silver nanoparticles [69]. In addition, the XRD patterns in Fig. 3b, d, f, h, l relative to the same samples show the presence of a broad diffraction band due to the silica substrate and two peaks due to metallic silver. A gradual increase of silver peaks intensity is directly influenced by the increment of the treatment temperature.

Figure 4 shows the comparison of UV–visible spectra of coatings on SiO_2 and SL substrates, as-deposited and thermally treated at 450 °C. An absorption peak due to the well-known Localized Surface Plasmon Resonance (LSPR) absorption of metal silver nanoclusters [61, 70].

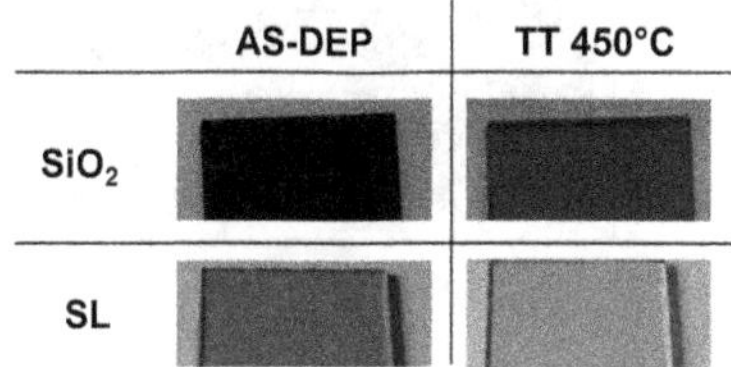

Fig. 2 Silver nanocluster/ silica composite layer on silica (SiO_2) and sodalime *SL* substrates, as deposited (AS DEP) and after a thermal treatment at 450 °C (TT 450 °C)

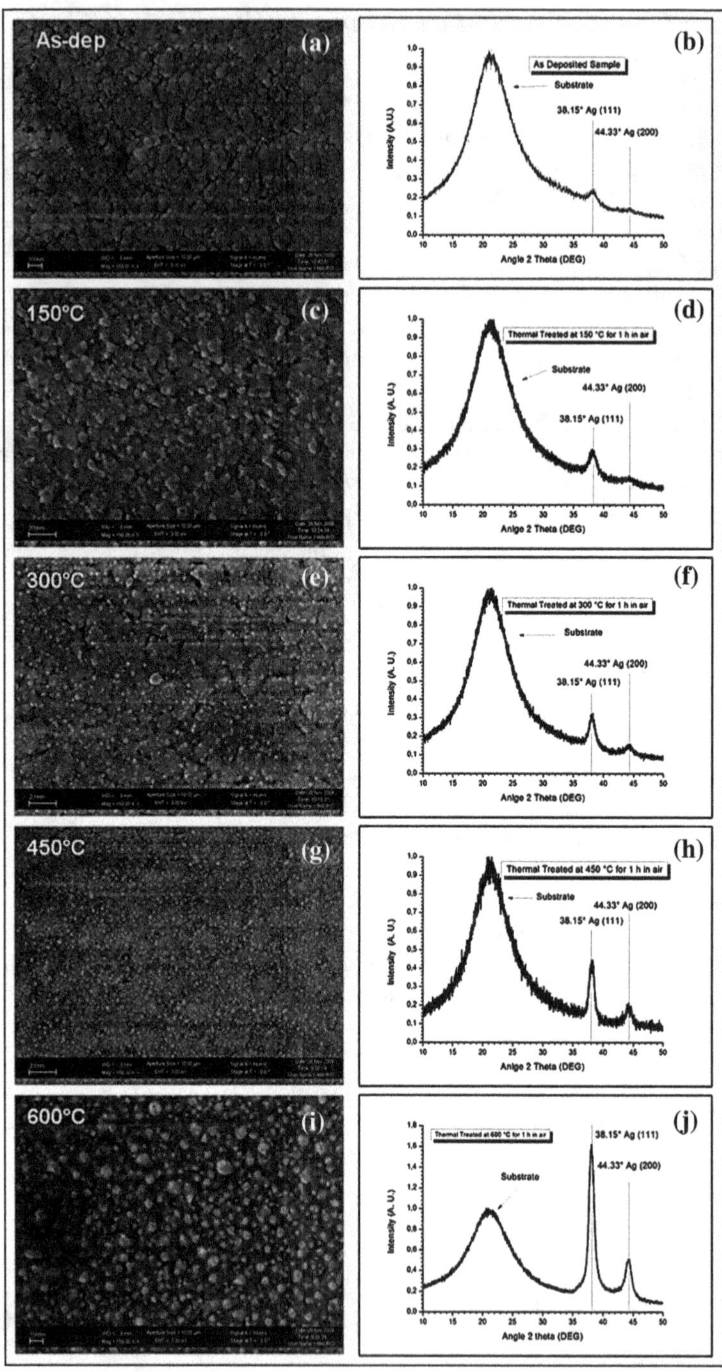

Fig. 3 FESEM images and XRD spectra of as deposited silver nanocluster/silica composite coatings on silica (**a**, **b**) and after thermal treatment at: 150 °C (**c**, **d**), 300 °C (**e**, **f**), 450 °C (**g**, **h**), and 600 °C (**i**, **l**) (with permission of [58])

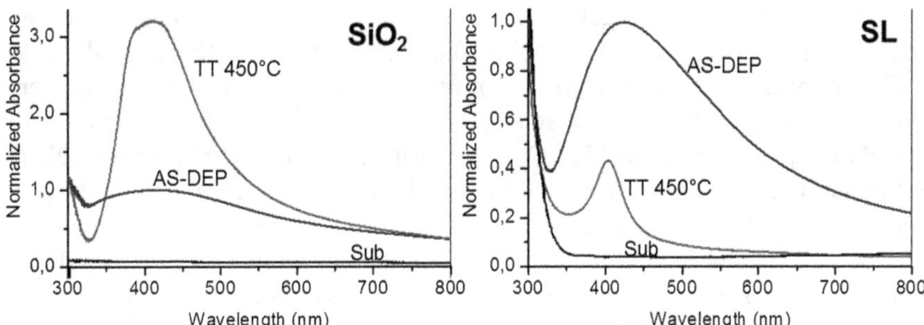

Fig. 4 Comparison of UV–Vis absorption spectra of silver nanocluster/silica composite coatings, as-deposited and treated at 450 °C, on silica (SiO_2) and soda lime *SL* substrates. The uncoated substrates are reported for comparison purposes

In particular, the absorption peak at 414 nm obtained from the silver nanocluster/silica composites sputtered on silica substrates, increases from the as deposited coating to the one treated at 450 °C [57]. Both scattering and SPR effects were produced from the silver nanoclusters embedded into a matrix. The prevalence of the first or second effects depends on the nanoclusters dimension. The dark color brown color of the coated samples, noticed in Fig. 2, is a consequence of the scattering in the case of small nanoclusters. When the nanocluster dimension size enlarges, the LSPR shifts to at 420 nm of wavelength and increases the absorption, with a generating bleaching effect resulting in a yellow/orange color.

On the contrary, the curves of coatings deposited on soda-lime glass show a shift of the absorption peak from 403 to 423 nm, and a decrement in intensity from the as-deposited coating to the treated one. Elemental depth profiles obtained by ERDA (Elastic Recoil Detection Analysis) have evidenced, in the case of SL an effective ionic exchange between Na^+ of the substrate and Ag^+ of the silver nanocluster/silica composite coating during thermal treatment [61].

The antibacterial activity of the silver nanocluster/silica composite coating, tested towards *Staphylococcus aureus* strain, is observable in Fig. 5. A well visible inhibition halo of about 4–5 mm is formed around the as deposited coating on silica substrate. The halo results slightly reduced after the thermal treatment at 450 °C. The silver nanoparticles with smaller size have a higher antibacterial

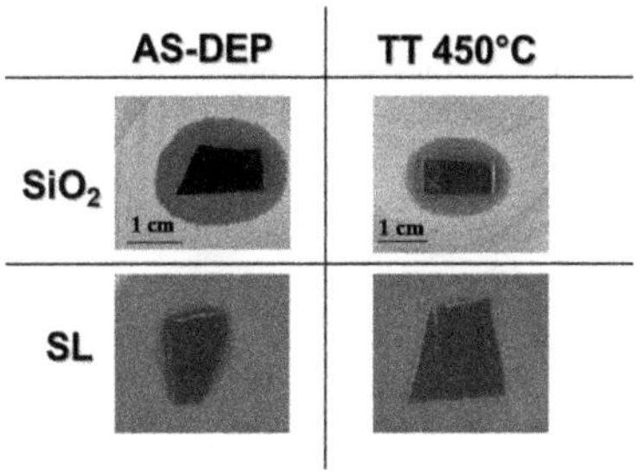

Fig. 5 Antibacterial halo against *S. aureus* for the silver nanocluster/silica composite coatings, as-deposited *AS DEP* and treated at 450 °C (TT 450 °C), on silica (SiO_2) and sodalime *SL* substrates

behavior because of their larger surface area. When the nanocluster size increases with the temperature, the antimicrobial efficiency is consequently reduced.

The antibacterial coating deposited on the SL substrate is able to form a very thin halo of about 1 mm in the case of the as deposited layer. The inhibition zone around the sample treated at 450 °C, is not visible, but the bacteria are not able to proliferate under it. This different behavior of the two substrates is a direct consequence of the ionic exchange process which occurs between Ag and Na ions of the sputtered coatings and the soda-lime substrate. The amount of silver is reduced after ionic-exchange on the coating surface and consequently the antibacterial power decreases, as discussed previously and in [61, 71].

Another relevant feature to be considered is the adhesion of the coating to the substrate. The silver nanocluster/silica composite coatings, as deposited and after the treatment at 450 °C, show good cohesion and adhesion to both the considered substrates (silica [57, 58] and soda-lime [61]). As reported in Fig. 6, no sign of macroscopic damage or detachment is noticed on the coatings. All the samples can be classified as 5B (0 % damage) according to ASTM D3359 standard [68] .

The nano-mechanical properties of the coatings before and after thermal treatment in terms of hardness (Fig. 7 with permission of [61]) and reduced modulus were tested by nano-indentation tests and compared with those of the substrates (silica or sodalime glasses). The indentation depth and the maximum applied load were chosen at 30 or 50 nm and 1 or 5 mN, respectively. The thermal treatment performed at 450 °C significantly improves the hardness of the coating deposited on the silica substrate (Fig. 7a). Consistently, an opposite effect occurs with the heated coating deposited on the sodalime substrate (Fig. 7b) in agreement with the above discussed Na/Ag ionic exchange [61]. The ionic exchange increases the amount of Na^+ which acts as a modifier for the silica matrix of the coating, thereby reducing the coating mechanical properties [61].

Similar effect has been observed for the reduced modulus which is related to the coatings Young modulus [61, 72, 73].

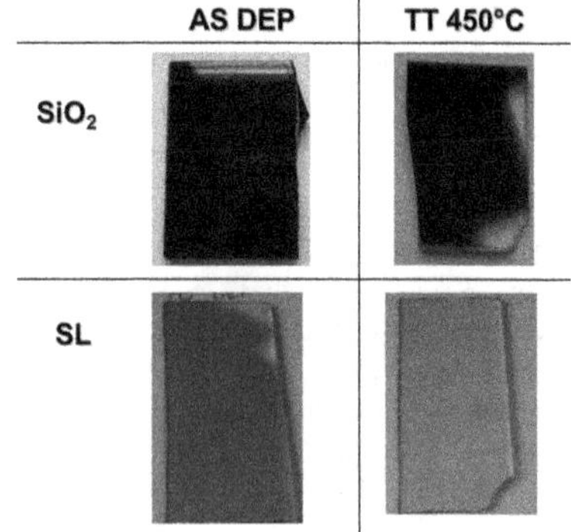

Fig. 6 Silver nanocluster/silica composite coating as-deposited (AS DEP) and treated at 450 °C (TT 450 °C) on silica (SiO₂) and sodalime *SL* after tape test (ASTM D3359): the coating is still well adhered on the substrates after removal of the tape

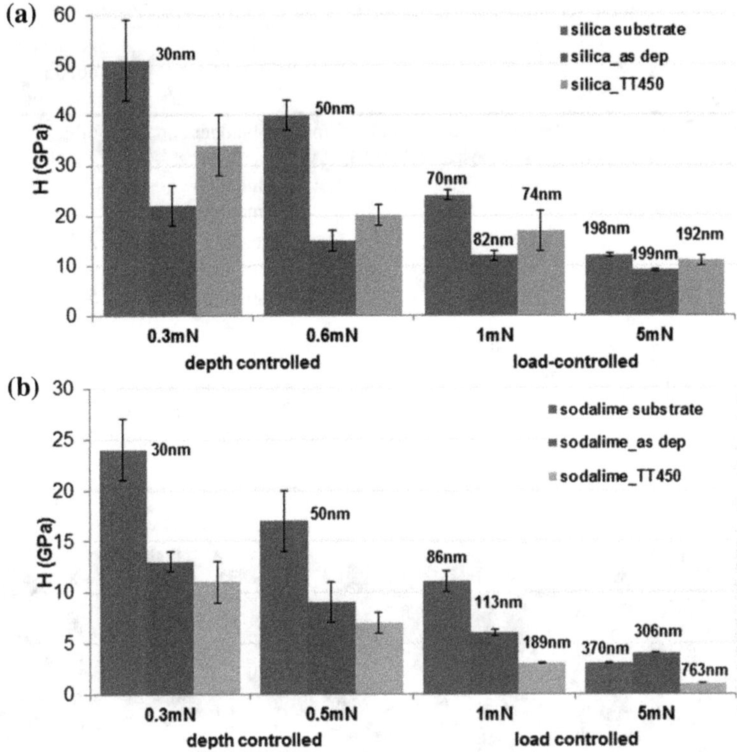

Fig. 7 Hardness of silver nanocluster/silica composite layers on silica **a** and soda-lime **b** substrates, as deposited and after thermal treatment at 450 °C (with permission of [61])

2.3 Silver Nanocluster/Silica Composite Coatings on Polymers

The same co-sputtering technique have been applied to several polymeric substrates, taking into account the limited thermal resistance of some of them (Table 2).

The thickness of the silver nanocluster/silica composite coating, as previously discussed, is optimized for each substrate after several preliminary tests according to the polymer thermo-mechanical resistance and final application. As an example, an antibacterial coating for a polymer to be used for biomedical implants such as PP or PMMA, has to be carefully controlled in terms of silver amount in order to avoid eventual toxicity due to silver release.

The surface of sputtered PMMA (Fig. 8a) was investigated by FESEM, revealing a porous morphology typical of sputtered silica with Ag nanoclusters visible as bright spots (Fig. 8b). EDS (Fig. 8c) confirms the composite nature of the sputtered coating, showing the presence of Ag, Si and oxygen with Ag/Si atomic ratio of about 0.3.

Table 2 Silver nanocluster/silica composite coatings on several polymeric substrates

Polymer	Composition	Application	Antibacterial coating thickness (nm)
Combitherm®	Multilayer co-extruded film composed of EVOH, PA and PE	Air bladder for aerospace inflatable modulus	60
Kevlar fabric	Aramidic fibers	Aerospace inflatable modulus	300
		Personal protective systems	50
Nylon film	Polyamide	Parachute	50
Polypropylene film	PP	Biomedical implant	60
Polyurethane film	PU	Biomedical implant	150
Poly(methylmethacrylate) (flat substrate)	PMMA	Biomedical implant	50

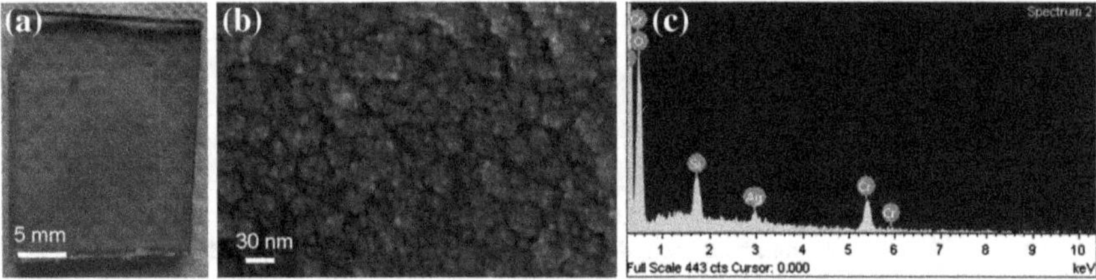

Fig. 8 Silver nanocluster/silica composite coatings sputtered onto PMMA substrates: **a** appearance of the coated sample, **b** FESEM micrograph of the deposited layer (magnification 900000 ×), **c** EDS of the layer

Combitherm ® film is currently used for food packaging, but also for the air bladder layer in the aerospace inflatable modulus of the International Space Station [74, 75].

Figure 9 shows the visual appearance of the Combitherm ® film before (a) and after (b) the deposition of silver nanocluster/silica composite coating with a thickness of about 60 nm. Since the substrate has a low thermal resistance compared to inorganic substrates, a deposition time of only 15 min was chosen to avoid any damage [60].

The typical brown color of the coating is present also on this substrate, as discussed before, and reported in [60] where the LSPR typical of the silver nanoclusters was detected at about 414 nm.

As discussed in the introduction, different kinds of bacteria were discovered into the International Space Station after long-term missions. For this purpose, several bacterial species, both Gram positive (*Staphylococcus aureus, Bacillus cereus*) and Gram negative (*Morganella, Klebsiella pneumoniae and Escherichia coli*), were used for evaluating the antibacterial activity of the silver nanocluster/

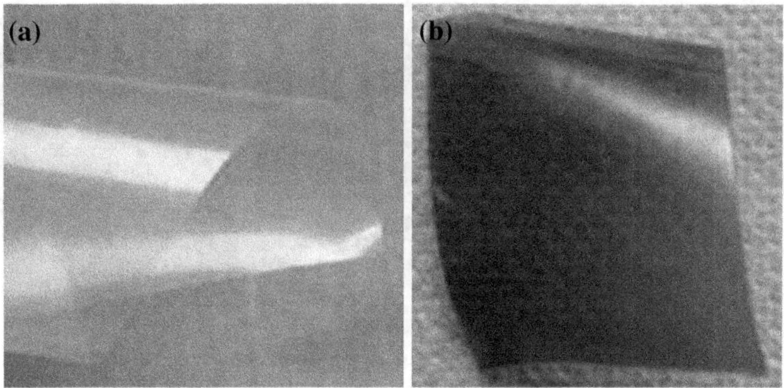

Fig. 9 Combitherm ® film **a** before and **b** after the silver nanocluster/silica composite coating deposition

silica composite coating deposited on the Combitherm ® substrate. The formed inhibition halos towards all the chosen bacteria are reported in Fig. 10. The dimension and morphology of the halo is strongly affected by the sensitivity of the bacteria strain to silver ions and the structure of their cell wall [76]. The effectiveness of silver ions on *E. coli* and *B. cereus* is more evident and a halo of about 2–3 mm is formed around these samples.

As discussed in [60], the most relevant features, i.e. mechanical properties and air permeability, of the Combitherm ® for aerospace applications are not significantly affected by the antibacterial coating. The transmission rate and permeability to air remained unchanged and equal to about 14 ml/m^2*day and 2 ml*mm/m^2*day*atm, respectively, for the uncoated and coated substrate [60]. Some tested mechanical properties such as tensile strength and perforation were improved by the presence of the coating; the tear resistance was not altered after coating [60]. However, the coating was gradually removed during an abrasion test against an aramidic fabric until the complete elimination after 3,000–5,000 cycles.

Some textiles, such as mimetic aramidic fibers, polyamide and activated carbon fiber textiles (Fig. 11), have been coated by the antibacterial silver nanocluster/silica composite layer. The brownish color typical of the antibacterial coating is not visible on these substrates because of their original color.

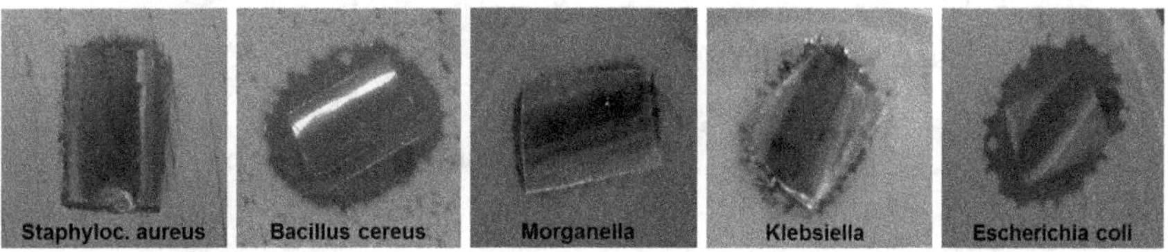

Fig. 10 Inhibition halo test against different bacteria of the silver nanocluster/silica composite coating deposited on Combitherm ® substrates

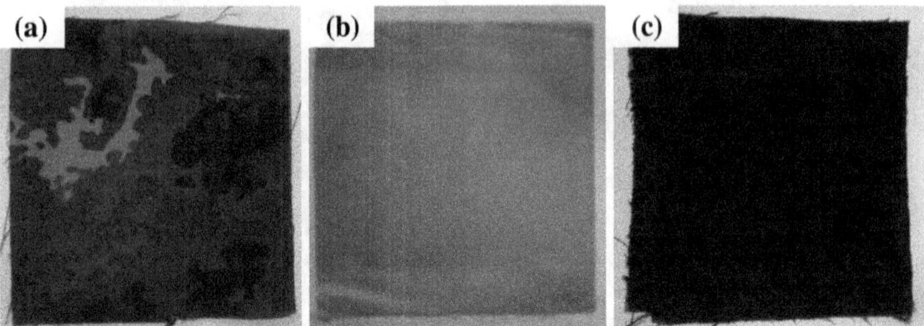

Fig. 11 Some textiles coated by silver nanocluster/silica composite coating: **a** mimetic aramidic fibers, **b** polyamide and **c** activated carbon fiber textiles

However, a pale yellow aramidic fabric becomes darker after the coating deposition as reported in Fig. 12, for the same reasons discussed in Sect. 2.2

The antibacterial activity and the unchanged mechanical properties of these textiles after coating deposition have been demonstrated.

2.4 Silver Nanocluster/Silica Composite Coatings on Metals

The sputtering technique has been successfully used to obtain silver nanocluster/silica composite coatings on metals (aluminum and steel) for different applications such as food handling industry and space applications.

Figure 13 shows the antibacterial coating deposited on stainless steel before (a) and after thermal treatment at 450 °C (b). The morphology of the sputtered silver nanocluster/silica coating is the same as reported in [57, 58] with silver

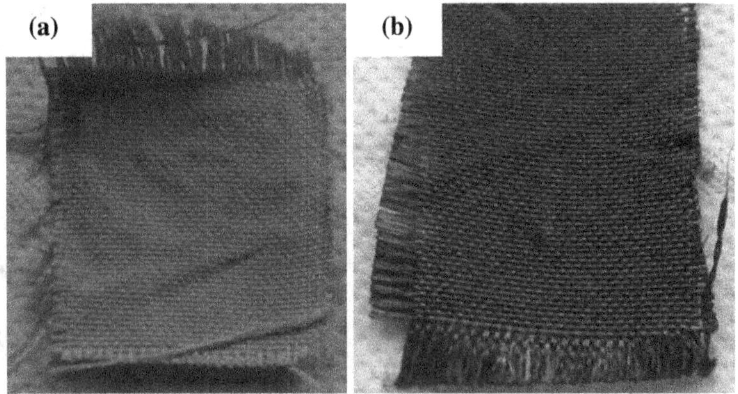

Fig. 12 Aramidic fabric **a** before and **b** after antibacterial coating deposition

Fig. 13 Silver nanocluster/silica composite coatings as deposited **a** and after thermal treatment at 450 °C **b** on a steel substrate

nanoclusters well embedded in the sputtered silica matrix. The effect of the thermal treatment, with the increase of nanoclusters size and the consequent chromatic change, occurs also for these samples, similarly to what observed for the silica substrates. Moreover a significant improvement in the adhesion and cohesion of the coating have been measured after the heating, due to a further densification of the composite coating, as reported in [57] for thermal treated coatings on silica substrates.

Figure 14 reports inhibition halos against *S. aureus* for as deposited and thermally treated samples. The antibacterial behavior is evident for the as deposited samples (at about 5 mm halo) and still present after heating at 450 °C (about 1 mm halo).

These results are in accordance with those obtained for coated silica substrate and confirm the ability of this coating to retain its antibacterial effect after being heated at 450 °C, also on steel substrates.

Silver nanocluster/silica composite coating depositions on aluminum alloys for aerospace application have also been considered in order to prevent bacterial contamination. Sputtered aluminum tiles have been tested in two international experiments. The first one, MARS 500, is an international experiment in which crewmembers stayed 520 days in a confined environment, simulating the travel to Mars [77]. The second one (VIABLE) is an activity on the International Space

Fig. 14 Inhibition halo against *S. aureus* for as deposited **a** and heated antibacterial silver nanocluster/silica composite coatings **b** (450 °C) on stainless steel

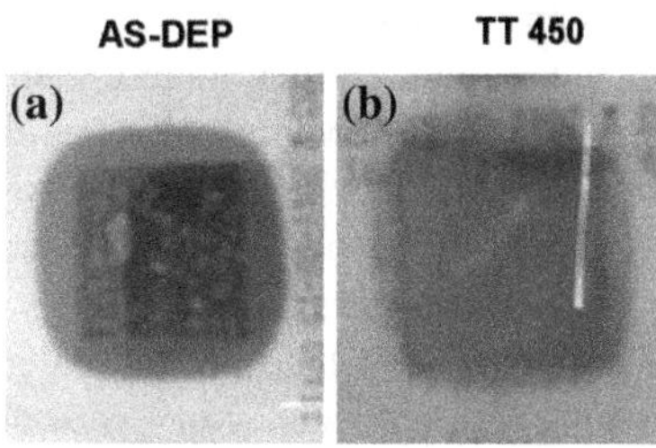

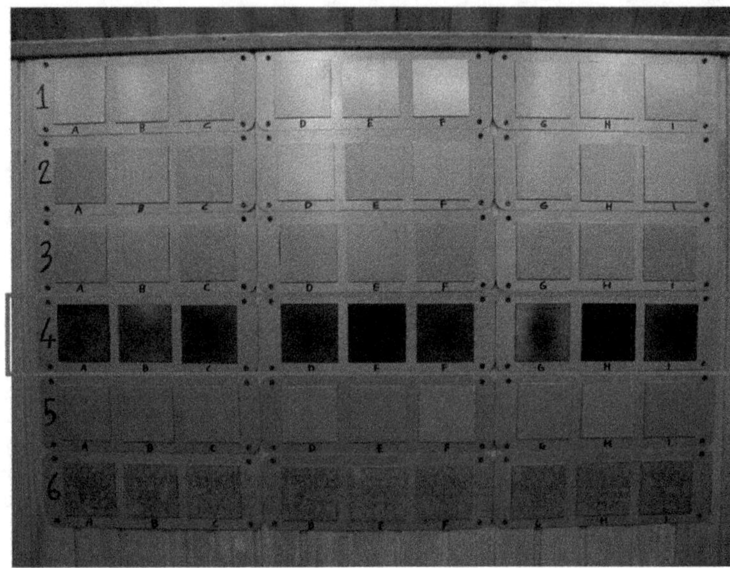

Fig. 15 Silver nanocluster/silica composite coating on aluminum alloys for space application in the MARS 500 experiment

Station, actually ongoing: it will last 4 years and, together with other scientific activity, will compare the behavior of different surface treatments and coatings against bacterial contamination [78]. In both cases the bacterial biofilm formation on the different surfaces is under investigation.

Figure 15 reports the different tiles under testing. In the fourth row (evidenced in red) the silver nanocluster/silica composite coating on aluminum tiles. Results of both tests are under evaluation.

3 Summary and Future Trends

A new antibacterial silver nanocluster/silica composite coating was successfully deposited on glasses, metals and polymers by co-sputtering technique. The coating thickness and the amount of silver was easily modulated by sputtering deposition parameters, according to the substrate properties and final application. The coating has a brownish color due to the presence of silver nanoclusters which resulted well embedded in the silica matrix with a typical nano-structured morphology. Nanocluster size and coating color could be modified by means of appropriate thermal treatments. Coating bleaching and increase of the clusters size occur with the increment of heating treatment. The antibacterial activity against *S. aureus* and other bacterial strains was widely demonstrated with the formation of a well-visible inhibition halo on all coated substrates. The adhesion of the antibacterial coating resulted satisfactory with all the substrates.

The new silver nanocluster/silica composite coating exhibit an intrinsic advantage with respect to other antibacterial coatings:

-being silver antibacterial agent different from antibiotics, no development of bacterial resistance that could compromise the efficiency of the antibacterial effect, are expected to occur.

-it is possible to tailor the silver content by varying the co-sputtering parameters. In such a way it is possible to design the antibacterial effect in terms of both efficiency (amount of released silver) and life-time (release kinetic of silver).

-thermal and mechanical resistance of the coating is higher than those of polymer-based alternatives: the antibacterial effect of this coating it has been demonstrated up to 450 °C.

-RF sputtering technique is potentially applicable to substrates with irregular or curved geometries

A crucial issue to be investigated is the durability of the composite coating after prolonged contact with biological fluids; in order to overcome this problem, materials with higher stability in wet environment than silica in the biological environment will be tested in the future, as suggested in [62] by the authors.

Another aspect deserving accurate investigation is the assessment of the form and amount of released Ag. Early studies seem to suggest that Ag is released from the nanoclusters in ionic form rather than as nanoparticles: this is a very significant characteristic of these coatings, since several in vitro and in vivo studies demonstrated the toxicity of metal nanoparticles [79, 80].

Acknowledgments This activity was funded by Regione Piemonte, Italy (NABLA, Nanostructured Antibacterial Layers) and by REA (EU Project-NASLA-FP7-SME-2010-1—Project 262209).

The authors kindly acknowledge Dr Giacomo Fucale (Traumatology Orthopaedics and Occupational Medicine Department, University of Turin, Italy) for antibacterial tests facilities, and all industrial partners of the projects (Thales Alenia Space—IT, Dipromed, Alce Calidad, Easreth, Aero Sekur, Reply, KTH Royal Institute of Technology, Bactiguard).

References

1. Tunc, K., Olgun, U.: Microbiology of public telephones. J. Infect. **53**(2), 140–143 (2006)
2. Shia, X., Zhua, X.: Biofilm formation and food safety in food industries. Trends Food Sci. Technol. **20**, 407–413 (2009)
3. Gu, J.: Microbial colonization of polymeric materials for space applications and mechanisms of biodeterioration: a review. Int Biodeterior Biodegrad **59**, 170–179 (2007)
4. Hori, K., Matsumoto, S.: Bacterial adhesion: from mechanism to control. Biochem. Eng. J. **48**, 424–434 (2010)
5. Simoes, M., Simoes, L.C., Vieira, M.J.: A review of current and emerging biofilm control strategies. LWT—Food. Sci Technol. **43**, 573–583 (2010)
6. Wagner, C., Aytac, S., Hansch, G.N.: Biofilm growth on implants: bacteria prefer plasma coats. Int. J. Artif. Organs. **34**(9), 811–817 (2011)

7. Harris, L.G., Richards, R.G.: Staphylococci and implant surfaces: a review. Injury, Int. J. Care Injured **37**, S3–S14 (2006)
8. Subramani, K., Jung, R.E., Molenberg, A., Hammerle, C.H.F.: Biofilm on dental implants: a review of the literature. Int. J. Oral Maxillofac. Implants **24**, 616–626 (2009)
9. Engelsman, A.F., Van der Mei, H.C., Ploeg, R.J., Busscher, H.J.: The phenomenon of infection with abdominal wall reconstruction. Biomaterials **28**, 2314–2327 (2007)
10. Eggimann, P., Sax, H., Pittet, D.: Catheter-related infections. Microbes Infect. **6**, 1033–1042 (2004)
11. Davis, M.F., Iverson, S.A., Baron, P., Vasse, A., Silbergeld, E.K., Lautenbach, E., Morris, D.O.: Household transmission of meticillin-resistant Staphylococcus aureus and other staphylococci. Lancet Infect Dis **12**, 703–716 (2012)
12. Van Houdt R, Mijnendonckx K, Leys N (2012) Microbial contamination monitoring and control during human space mission Planet Space Sci 60:115–120
13. Ilyin, V.K.: Microbiological status of cosmonauts during orbital spaceflights on Salyut and Mir orbital stations. Acta Astronaut. **56**, 839–850 (2005)
14. Novikova, N., De Boever, P., Poddubko, S., Deshevaya, E., Polikarpov, N., Rakova, N., Coninx, I., Mergeay, M.: Survey of environmental biocontamination on board the International Space Station. Res. Microbiol. **157**, 5–12 (2006)
15. Kousta, M., Mataragas, M., Skandamis, P., Drosinos, E.H.: Prevalence and sources of cheese contamination with pathogens at farm and processing levels. Food Control **21**, 805–815 (2010)
16. Callon, C., Gilbert, F.B., De Cremoux, R., Montel, M.C.: Application of variable number of tandem repeat analysis to determine the origin of S. aureus contamination from milk to cheese in goat cheese farms. Food Control **19**, 143–150 (2008)
17. Temelli, S., Anar, S., Sen, C., Akyuva, P.: Determination of microbiological contamination sources during Turkish white cheese production. Food Control **17**, 856–861 (2006)
18. Rai, M., Yadav, A., Gade, A.: Silver nanoparticles as a new generation of antimicrobials. Biotechnol. Adv. **27**, 76–83 (2009)
19. Guzman, M., Dille, J., Godet, S.: Synthesis and antibacterial activity of silver nanoparticles against gram-positive and gram-negative bacteria. Nanomed. Nanotechnol. Biol. Med. **8**, 37–45 (2012)
20. Ovington LG (2004) The truth about silver. Ostomy Wound Manage 50 (9A): 1 s–10 s
21. Sondi, I., Salopek-Sondi, B.: Silver nanoparticles as antimicrobial agent: a case study on E. coli as a model for Gram negative bacteria. J. Colloid Interf. Sci. **275**, 177–182 (2004)
22. Mirzajani, F., Ghassempour, A., Aliahmadi, A., Esmaeili, M.A.: Antibacterail effect of silver nanoparticles on Staphylococcus aureus. Res. Microbiol. **162**, 542–549 (2011)
23. Wijnhoven, S.W.P., et al.: Nano-silver—a review of available data and knowledge gaps in human and environmental risk assessment. Nanotoxicology **3**(2), 109–138 (2009)
24. Wang, J.X., Wen, L.X., Wang, Z.H., Chen, J.F.: Immobilization of silver on hollow silica nanospheres and nanotubes and their antimicrobial effects. Mater. Chem. Phys. **96**, 90–97 (2006)
25. Bugla-Płoskońska, G., et al.: Bactericidal properties of silica particles with silver islands located on the surface. Int. J. Antimicrob. Ag **29**, 738–748 (2007)
26. Lee, J.M., Lee, Y.G., Kim, D.W., Oh, C., Koo, S.M., Oh, S.G.: Facile and novel route for preparation of silica/silver heterogeneous composite particles with hollow structure. Colloid Surface A **301**(1–3), 48–54 (2007)
27. Masuda, N., Kawashita, M., Kokubo, T.: Antibacterial activity of silver-doped silica glass microspheres prepared by a sol-gel method. J. Biomed. Mat. Res. B **83B**(1), 114–120 (2007)
28. Verne, E., Di Nunzio, S., Bosetti, M., Appendino, P., Vitale Brovarone, C., Maina, G., Cannas, M.: Surface characterization of silver-doped bioactive glass. Biomaterals **26**, 5111–5119 (2005)

29. Di Nunzio, S., Verne', E.: Process for the production of silver-containing prosthetic devices WO 2006/058906 (2006)
30. Di Nunzio, S., Vitale Brovarone, C., Spriano, S., Milanese, D., Verne, E., Bergo, V., Maina, G., Spinelli, P.: Silver containing bioactive glasses prepared by molten salt ion exchange. J. Europ. Ceram Soc. **24**, 2935–2942 (2004)
31. Carturan, S., et al.: Formation of silver nanoclusters in transparent polyimides by Ag-K ion-exchange process. Eur. Physic. J D **42**, 243–251 (2007)
32. Kamiiya, Y., Kanamaru, S.: Antibacterial glass and method for producing antibacterial glass WO/2005/087675 (2005)
33. http://www.yourglass.it/agc-flatglass-europe/news/it/04-09-2007_AB.html
34. Verné, E., et al.: Surface silver-doping of biocompatible glass to induce antibacterial properties. Part I: massive glass. J. Mater. Sci. Mater. Med. **20**, 733–740 (2009)
35. Borrelli, N F., Morse, D L., Senaratne, W., Verrier, F., Wei, Y.: Coated, Antimicrobial, chemically strengthened glass and method of making, US 2012/0034435 (2012)
36. http://www.agc-flatglass.sg/products-detail.aspx?id=32
37. Brook, L.A., et al.: Highly bioactive silver and silver/titania composite films grown by chemical vapour deposition. J. Photoch. Photobio. **A187**, 53–63 (2007)
38. Brook, L.A., et al.: Novel multifunctional film. Surf. Coat. Technol. **201**, 9373–9377 (2007)
39. Baia, L., Muresan, D., Baia, M., Popp, J., Simon, S.: Structural properties of silver nanoclusters–phosphate glass composites. Vib. Spectrosc. **43**, 313–318 (2007)
40. Kawashita, M., Tsuneyama, S., Miyaji, F., Kokubo, T., Kozuka, H., Yamamoto, K.: Antibacterial silver-containing silica glass prepared by sol–gel method. Biomaterials **21**, 393–398 (2000)
41. Tarimala, S., Kothari, N., Abidi, N., Hequet, E., Fralick, J., Dai, L.L.: New approach to antibacterial treatment of cotton fabric with silver nanoparticle–doped silica using sol–gel process. J. App. Polym. Sci. **101**, 2938–2943 (2006)
42. Jeon, H.J., Yi, S.C., Oh, S.G.: Preparation and antibacterial effects of Ag–SiO2 thin films by sol–gel method. Biomaterials **24**, 4921–4928 (2003)
43. Tatar, P., Kiraz, N., Asiltürk, M., Sayılkan, F., Sayılkan, H., Arpaç, E.: Antibacterial thin films on glass substrate by sol–gel process. J. Inorg. Organomet. Polym. Mater. **17**, 525–533 (2007)
44. Jürgens, R., Schwindt, S.: Coating compound made of an agent which generates SiO2 with at least two antibacterial agents EP1825752 A1 (2011)
45. Lei, Z., Wang, D.: Antibacterial sol-gel coating solution, method for preparing antibacterial sol-gel coating solution, antibacterial articles, and method and equipments for preparing antibacterial articles EP1975132A1 (2008)
46. Sant, S.B., Gill, K.S., Burrell, R.E.: Nanostructure, dissolution and morphology characteristics of microcidal silver films deposited by magnetron sputtering. Acta Biomater. **3**, 341–350 (2007)
47. Dowling, D.P., et al.: Deposition of anti-bacterial silver coatings on polymeric substrates. Thin Solid Films **398–399**, 602–606 (2011)
48. Kim, Y.H., et al.: Synthesis and characterization of antibacterial Ag-SiO2 nanocomposite. J. Phys. Chem. C **111**, 3629–3635 (2007)
49. Wang, J.X., Wen, L.X., Wang, Z.H., Chen, J.F.: Immobilization of silver on hollow silica nanospheres and nanotubes and their antibacterial effects. Mater. Chem. Phys. **96**, 90–97 (2006)
50. Kim, Y.H., Lee, D.K., Cha, H.G., Kim, C.W., Kang, Y.C., Kang, Y.S.: Preparation and characterization of the antibacterial Cu nanoparticle formed on the surface of SiO2 nanoparticles. J. Phys. Chem. B **110**, 24923–24928 (2006)
51. Wakamura M (2008) Antibacterial paint for materials and materials coated therewith EP1985675 (A1)

52. Wakamura, M.: Antibacterial and anti-staining paint for building material and building material coated therewith EP1522564 (A1) (2005)
53. Imai, T., Takahashi, T., Hibino, Y.: Coating material, building material and method for coating building material EP2067759 (A2) (2009)
54. Nakagawa, Y., Iwasaki, Y., Takenaka, K., Shano, M.: Antibacterial laminate EP2018960 (A1) (2009)
55. Rauscher, H., Perucca, M., Buyle, G.: Plasma technology for hyperfunctional surfaces—food, biomedical, and textile applications Wiley-VCH (2009)
56. Ferraris, M., Chiaretta, D., Fokine, M., Miola, M.,Verne', E.: Pellicole antibatteriche ottenute da sputtering e procedimento per conferire proprietà antibatteriche ad un substrato TO2008A000098 (2008)
57. Ferraris, M., Perero, S., Miola, M., Ferraris, S., Verné, E., Morgiel, J.: Silver nanocluster-silica composite coatings with antibacterial properties. Mater. Chem. Phys. **120**, 123–126 (2010)
58. Ferraris, M., et al.: Chemical, mechanical, and antibacterial properties of silver nanocluster-silica composite coatings obtained by sputtering. Adv. Eng. Mater. **12**, B276–B282 (2010)
59. Ferraris, S., Perero, S., Vernè, E., Battistella, E., Rimondini, L., Ferraris, M.: Surface functionalization of Ag-nanoclusters–silica composite films for biosensing. Mater. Chem. Phys. **130**, 1307–1316 (2011)
60. Balagna, C., et al.: Antibacterial coating on polymer for space application. Mater. Chem. Phy. **135**, 714–722 (2012)
61. Ferraris, M., et al.: Effect of thermal treatments on sputtered silver nanocluster/silica composite coatings on sodalime glasses: ionic exchange and antibacterial activity. J. Nanopart. Res. **14**, 1287–1306 (2012)
62. Baino, F., Pererom S., Miola, M et al.: Rivestimenti e trattamenti superficiali per impartire proprietà antibatteriche a dispositivi per oftalmoplastica. TO2012A000512 (2012)
63. Ferraris, M., et al.: Silver nanocluster/silica composite coatings obtained by sputtering for antibacterial applications Iop Science Iop Conf. Ser. Mater. Sci. Eng. **40**, 012037 (2012)
64. http://www.composites.polito.it/?p=projects_european_nasla
65. http://en.wikipedia.org/wiki/NASLA
66. Beer, C., et al.: Toxicity of silver nanoparticles—Nanoparticle or silver ion? Toxicol. Lett. **208**, 286–292 (2012)
67. NCCLS M2-A9. Performance standards for antimicrobial disk susceptibility tests, approved standard, 9th Edn, NCCLS, Villanova, PA, USA (2003)
68. ASTM D3359–97 Standard test methods for measuring adhesion by tape test
69. Cai, W., Zhang, L., Zhong, H., He, G.: Annealing of mesoporous silica loaded with silver nanoparticles within its pores from isothermal sorption. J. Mater. Res. **10**, 2888–2895 (1998)
70. Hoa, X.D., Kirk, A.G., Tabrizian, M.: Towards integrated and sensitive surface plasmon resonance biosensors: A review of recent progress. Biosens. Bioelectron. **23**, 151–160 (2007)
71. Durucan, C., Akkopru, B.: Effect of calcination on microstructure and antibacterial activity of silver-containing silica coatings. J. Biomed. Mater. Res. B Appl. Biomater. **93**(2), 448–458 (2010)
72. Almasri, A.H., Voyiadjis, G.Z.: Nanoindentation in FCC metals: experimental study. Acta Mech. **209**, 1–9 (2010)
73. Lucca, D.A., Herrmann, K., Klopfstein, M.J.: Nanoindentation: measuring methods and applications. CIRP Annals- Manuf. Technol. **59**, 803–819 (2010)
74. Cadogan, D., Stein, J., Grahne, M.: Inflatable composite habitat structures for lunar and mars exploration. Acta Astronaut. **44**, 399–406 (1999)
75. Brandon, E.J., et al.: Structural health management technologies for inflatable/deployable structures: Integrating sensing and self-healing. Acta Astronaut. **68**, 883–903 (2011)
76. Monteiro, D.R., et al.: The growing importance of materials that prevent microbial adhesion: antimicrobial effect of medical devices containing silver. Int. J. Antimicrob. Agent **34**, 103–110 (2009)

77. Canganella, F. et al.; Microbial ecology of space confined habitats and biofilm development on space materials: the project MARS500–MICHA. 63rd International Astronautical Congress (2012)
78. Canganella, F et al.: VIABLE: a current flight experiment on ISS to investigate biocontamination and human life support in space. 63rd International Astronautical Congress (2012)
79. Kawata, K., Osawa, M., Okabe, S.: In vitro toxicity of silver nanoparticles at non cytotoxic doses to HepG2 human hepatoma cells. Envir. Sci. Technol. **43**, 6046–6051 (2009)
80. Haase, A., Tentschert, J., Jungnickel, H., et al.: Toxicity of silver nanoparticles in human macrophages: uptake, intracellular distribution and cellular response. J. Phys. Conf. Series **304**, 012030 (2011)

Recent Advances on the Utilization of Nanoclays and Organophosphorus Compounds in Polyurethane Foams for Increasing Flame Retardancy

Javier C. Quagliano, Victor M. Wittemberg
and Irma C. Gavilán García

Abstract Recent advances on fire retardants utilized in polyurethane foams were reviewed, mainly focusing on organophosphorus compounds. The application of nanotechnology for developing novel fire retarded nanocomposites is discussed as well, focusing in the addition of nanoclays to polyurethanes. The review is focused on flame retardants known or believed by the authors to be in actual use at the time of writing, without expanding on recent proprietary or patented information. Although fire retarded nanocomposite is an increasing area of research, still market utilization is to be developed.

1 Introduction

All organic polymers are combustible. They decompose when exposed to heat, decomposition products burn, smoke is generated, and the products of combustion are highly toxic, even if only CO and CO_2 [24]. The use of flame retardants to

J. C. Quagliano (✉)
Organic Synthesis Division, Applied Chemistry Department, Instituto de Investigaciones
Científicas y Técnicas para la Defensa (CITEDEF), Avenue Juan Bautista de Lasalle 4397
B1603ALO Buenos Aires, Argentina
e-mail: jquagliano@citedef.gob.ar

V. M. Wittemberg
Analytical Chemistry Division, Applied Chemistry Department, Instituto de Investigaciones
Científicas y Técnicas para la Defensa (CITEDEF), Avenue Juan Bautista de Lasalle 4397
B1603ALO Buenos Aires, Argentina
e-mail: vwittemberg@citedef.gob.ar

I. C. Gavilán García
Faculty of Chemistry, National Autonomous University of Mexico (UNAM), Mexico,
Mexico
e-mail: irmac@unam.mx

J. Njuguna (ed.), *Structural Nanocomposites*, Engineering Materials,
DOI: 10.1007/978-3-642-40322-4_11, © Springer-Verlag Berlin Heidelberg 2013

reduce the flammability of polymers and production of smoke or toxic products during their combustion has become an important aspect of the research, development, and application of new materials. Among polymers, cellular materials like polyurethanes possess a high surface area per unit mass and this resulted in almost complete pyrolysis of combustible matter [20]. Polyurethanes, in the absence of flame retardants, are extremely combustible.

There are four main chemical groupings of flame retardants. These are inorganics (e.g. aluminum trioxide and magnesium hydroxide), nitrogen-based organic, organophosphorus (e.g. phosphate esters), and halogenated flame retardants [3]. Considering a broader range of applications as with plastics and textiles, fire retardants fall into several distinct classes (1) alumina trihydrate (2) halogenated compounds usually used in combination with antimony oxide (3) borax and boric acid and (4) the phosphorus nitrogen and phosphorus halogen compounds [25]. Respect to their reactivity with polymers, fire retardants are classified in reactive and additive. Polyurethane foam is one area where phosphorus flame retardants are often used due to their effectiveness [44].

Recently, nanotechnology has been described as the next great frontier of materials research because nanocomposite formation brings about improved material performance, including enhanced mechanical, thermal, optical, dimensional, and barrier performance properties [14]. This technology has also been developed for polyurethanes. Regarding fire retardancy, polyurethane nanocomposite preparation methods were reviewed, including characterization of the composites, thermal stability and combustion behaviour [9, 22, 43].

2 Fire Retardants in Polyurethanes

Because of the banning of pentabromodiphenyl ether (PBDE) in Europe and voluntary withdrawal of this product from the market in the US, the polyurethane (PU) industry is searching for a more environmentally acceptable low-scorch alternative. For polyurethane foams, the most commonly used flame retardants include chlorinated phosphate esters that provide cost-effective performance at a significant disadvantage in thermal resistance (scorch resistance) [26]. Also a mixture of an alkyl tetrabromobenzoate and a triaryl phosphate has been used commercially for the purpose of avoiding scorch. Both halogenated and halogen-free solutions are being considered but the PU industry now seems to have a preference for the halogen-free products, generally containing phosphorus [18]. Progress in flame retardancy of polyurethane and polyisocyanurate foams was also reviewed by Levchik and Weil [17, 19]. Major progress in the area of flame-retardant PUs in recent years is found in the field of phosphorus- or silicon-containing products, especially reactive ones. Inorganic additives remain of great interest, especially in PU-based intumescent coatings [40].

Organophosphorus compounds have a long history of use as fire retardant in polymers in general and particularly polyurethanes. Since the advent of polyvinyl chloride (PVC) in the 30, tricresyl phosphate was known as a fire retardant for this plastic. Later it was shown that the o-isomer is neurotoxic, and tricresyl phosphates made from *m,p*-cresol continued to be used. Several other alkyl phenyl and dialkyl phenyl phosphates were developed [41]. Many other organophosphorus fire retardants include halogens, such as halo alkyl phosphates. Tris(1,3-dichloro-2-propyl) phosphate (TDCPP) or tris(chloroisopropyl) phosphate (TCPP) have been, and are currently, extensively used. This later has substantially reduced volatility and is more stable than TDCPP. There was some earlier use of tris(2,3-dibromo-1-propyl) phosphate but that compound was removed from the market when it was found to be mutagenic and carcinogenic. The chlorine analogs were found not to share that problem. Also, chlorinated alkyl phosphate, available as AMGARD V6 is utilized in polyurethanes. Tetrabromophthalic anhydride and its derivative diol are reactive flame retardants mainly used in polyurethanes. Also tribromoneopentyl alcohol is used as a reactive flame retardant for polyurethanes. Its high solubility in urethanes allows reaction of the single hydroxy functionality to form pendant urethane groups [26]. Generally phosphorus appears to act as an acid precursor in the solid phase to induce decomposition pathways that result in a reduced combustion and an increase in charring.

A huge amount of research was done in the last two decades to develop low-toxicity fire retardants for polyurethanes. Aromatic phosphates are another type of flame retardants which are currently used in polyurethane foam found in home furnishings. However, there is also the concern of exposure to aromatic phosphates [3].

The leading method for flame retarding rigid foam is to use additives, although reactive diols are occasionally employed where there is some special requirement. The leading additives are tris(2-chloroethyl) phosphate (now less used) and tris (1-chloro-2-propyl) phosphate [40]. Altogether, several types of fire retardants have been utilized in polyurethanes: both organic, as for example melamine [28], melamine polyphosphate and melamine cyanurate [26, 34] and inorganic, such as alumina trihydrate [27]; zinc borate [42]; zinc stearate (not used as a flame retardant but as a miscibility adjuvant) [23] and expanded graphite [13]. Among organics, organophosphorus compounds in polyurethanes, we found oligomer phosphates [1]; phosphonates like dimethyl methylphosphonate (DMMP) and oligomer phosphonates; phosphine oxides and other less common such as pyrophosphoric lactone as fire retardant coating [11] and 5,5,5′,5′,5″,5″-hexamethyltris (1,3,2,-dioxaphosphorinanemethan)amine 2,2′,2″-trioxide (XPM-1000 from Monsanto), advocated to be used in flame-retardant polyurethanes formulations [45]. Oligomeric cyclic phosphonates are utilized in rigid polyurethanes foams, and have been known since more than 50 years. They have excellent thermal and hydrolytic stability and are extensively used. On the other side, oligomeric chloroalkyl phosphonates such as Antiblaze 78 (Mobil) and Phosgards, once used in rigid PU foams, are not utilized in the market because of environmental and toxicological reasons [36]. Phosphazenes, both cyclic and polymeric, have been examined for their application as flame-retardant materials. A blend of poly [bis(carboxylatophenoxy)-phosphazene] with polyurethane precursors resulted in a urethane foam which exhibited increased thermal stability relative to the pure polyurethane [2]. An aromatic-substituted cyclic phosphazene, 2,4,6-triphenoxy-2,4,6-tri-(hydyoxyethoxy) cyclotriphosphazene (TPTHCP) was synthesized and used as a flame retardant to improve the flame retardancy of PU foams. Self extinguishing foam could be obtained when the phosphorus content was 1.5–2.0 wt% [10]. However, they are not utilized in practical applications. Sivriev et al. [31] synthesized a phosphorus- and nitrogen-containing polyol by condensation of tetrakis(hydroxymethyl)phosphonium chloride with diethanolamine. The effects of the structure and the content of the phosphorus- and nitrogen-containing polyol on properties of the polyurethanes, especially resistance to combustion, have been investigated. They demonstrated that the new phosphorus and nitrogen-containing polyol is an efficient flame retardant for rigid polyurethane foams. However, more recent research on these materials was not continued. There is abundant literature in this area regarding fire retardancy, and utilization is driven by a balance of cost and performance.

Phosphine oxides have been found highly effective in polypropylene [38]. Research on these compounds has continued. A phosphorus and nitrogen-containing polyol has been synthesized by condensation of tetrakis(hydroxymethyl) phosphonium chloride with diethanolamine. The newly synthesized polyol-tris[N,N-bis-(2-hydroxyethyl)aminomethyl] phosphine oxide (TPO) has been used in the preparation of rigid polyurethane foams with reduced flammability [31]. Phosphine oxides are expensive and this significantly limits their application [26].

It was shown early in fire retardant development that for a given series of related structures, either aliphatic or aromatic, the phosphates and phosphonates were superior to the phosphites (which are often hydrolytically sensitive) [6]. Novel organophosphorus fire retardant methyl-DOPO: 9,10-dihydro-9-oxa-methylphosphaphenthrene-10-oxide, which is a phosphinate, was considered for flexible polyurethane foam although more usually used in epoxy laminates. At temperatures where urethanes depolymerize, methyl-DOPO releases low molecular weight species like HPO, CH_3PO, or PO_2. These species are able to scavenge the H- and OH-radicals in the radical chain reactions of the flame [16].

Some researchers synthesized inherent flame retardant polyurethanes containing phosphorus that can react with isocyanate. Phosphorus-containing polyurethanes were synthesized by the reaction of phosphorus-containing diisocyanates and diols. Some newer phosphorus containing polyurethanes have highly flame retardant properties; the LOI values of these polyurethanes are around 29–33 [10]. A phosphorus- and nitrogen-containing polyol [polyol-bis(hydroxymethyl)-N, N-bis(2-hydroxyethyl) aminomethylphosphine oxide, AMPO] was used in the preparation of rigid polyurethane foams and were compared to polyurethane foams prepared on the basis of the commercially available flame retardant diethyl-N, N-bis(2-hydroxyethyl)aminomethylphosphonate (Fyrol 6). The polyurethane foams with AMPO exhibit a slight increase in the resistance to combustion (oxygen index) and a noticeable improvement in the thermal and mechanical properties [30].

The influence of several fire retardants on compressive strength and fire behaviour of rigid polyurethane foams was studied with ammonium polyphosphate, melamine cyanurate, alumina trihydrate, borax, and expanded graphite. In general, the cell size decreased, and compressive strength increased, as filler percentage was increased. One exception to this trend was borax, which led to a significant loss in compression strength of PU when it was added at the 15 % level [5].

3 Polymer-Layered Silicate Polyurethane Nanocomposites

Polymer-layered silicate nanocomposites are a new type of material, based on smectite clays usually rendered hydrophobic through ionic exchange of the sodium interlayer cation with an onium cation. They may be prepared via various synthetic routes comprising exfoliation adsorption, in situ intercalative polymerization and melt intercalation [12].

The use of nanoclays in polymers to impart fire retardancy to polyurethanes has been subject to numerous investigations, including preparation, properties and uses of this new class of materials. Nanoclays are interesting materials that are nano-metric in only one dimension: they are formed by sheets of few nanometeres thick to hundred thousands nanometers long. There are many investigations about the use of clays both alone or with other compounds to flame retard polyurethanes. These compounds usually contain phosphorus, nitrogen or both, and are added to the polyol component of PU. Although the increasing research on nanocomposites for its fire retardant properties, very often polymer nanocomposites exhibit low flammability in terms of peak of heat release but fail other tests such as UL-94 rating and limiting oxygen index (LOI). This can be overcome combining nano-particles with conventional fire retardants [7].

Wang and Pinnavaia [37] have synthesized intercalated nanocomposites based on elastomeric PUs. They used an organomontmorrillonite modified with dode-cylamine or octadecylamine swollen in a polyol and then cross-linked with an isocyanate prepolymer. PU elastomeric matrices exhibit a sizeable increase in tensile stress at break upon the addition of small quantities of nanofillers. Pro-cessing, structure and properties of polyurethane/clay nanocomposites were studied by Cao and coworkers [8]. They found that the morphology and properties of PU nanocomposites and foams greatly depend on the functional groups of the organic modifiers, synthesis procedure and molecular weight of the polyols because of the chemical reactions and physical interactions involved. A novel intumescent flame retardant, namely montmorillonite (MMT) modified with a compound containing phosphorus–nitrogen structure (called c-MMT), was pre-pared by ion exchanging of the nanometer Na-montmorillonite (Na-MMT) with a phosphorus-nitrogen compound. The results showed that the addition of flame retardant c-MMT enhanced the thermal stability and flame retardancy of poly-urethanes significantly. The authors concluded that exfoliated montmorillonite incorporated into polyurethane systems by in situ polymerization enhanced the thermal stability and flame retardancy of PU significantly [15]. Also, the nanoscale morphology of segmented polyurethane (SPU) nanocomposites containing various proportions of organomodified montmorillonite (MMT) was reported [21]. Nanoparticles such as organoclay MMT and carbon nanotubes in thermoplastic PU were reported to act mainly in a physical way (no chemical interactions) reinforcing intumescent char [7]. Mechanical properties of nanoclay-PU composites were studied. Data obtained from the compressive stress–strain curves revealed that the strength and modulus of polyurethane foam increase by addition of organoclay up to 1 wt% and then decrease [35]. Thirumala et al. [33] found that the compressive strength of organoclay/organically modified nanoclay (OMC) filled PU foam (PUF) increased up to 3 parts per hundred of polyol (php) by weight of OMC loading and then decreased. Dynamic rheological studies were performed on thermoplastic PU (TPU) with the incorporation of organoclay, showing increased dynamic viscosity and storage modulus, which was attributed to the formation of an interphase between soft and hard segments of the TPU matrix and effective dispersion of the organoclay [4].

Incorporation of commercial and laboratory-prepared nanoclays in a TDI-based polyurethane was done in our laboratory. Organically-modified clays interchanged with cetyl trimethyl ammonium bromide (CTAB) were incorporated in the polyol component of the PU at a high shear mixing rate (in situ intercalative polymerization), resulting in an exfoliated clay, which is nanometric in one dimension. This nanoclay, when dispersed in the polyurethane, resulted in the same fire retardancy rating (UL-94) than when the polyurethane was treated with a commercial nanoclay [29]. These results are in agreement with those obtained by Huang et al. [15].

The influence of surface-modified nanoclay on the self-organized nanostructure of segmented polyurethane composites was investigated [21]. Local researchers have studied the mechanical properties of montmorillonite-filled polyurethane foams and found an enhancement in the mechanical properties for the PUF sample at 5 wt% of MMT [32].

4 Commercial Applications

Despite the huge amount of literature existent in this field, which we have only reviewed in a condensed way, it is important to stress that fire retardant additives should not jeopardize polyurethane performance in general uses. The choice of suitable polymeric flame-retardants has to be restricted to species that allow retention of the advantageous mechanical properties of the polyurethane. Commercial applications were thoroughly reviewed by Levchik and Weil [18]. Tris(2-chloroisopropyl) phosphate and tris(dichloroisopropyl) phosphate are adequately stable in many formulations and have been used widely in polyurethanes. A non-halogenated phosphorus additive, which has had usage in rigid polyurethane foam for a long time, is dimethyl methylphosphonate (DMMP). It contains 25 wt% phosphorus (near the maximum possible for a phosphorus ester) and it is therefore highly efficient on a weight basis as a flame retardant. Triaryl phosphates, specifically triphenyl phosphate, isopropylphenyl diphenyl phosphate, tricresyl phosphate and trixylenyl phosphate all find some use in rigid foam formulations. Although reactive phosphorus- or halogen-containing fire retardants are on the market, it is believed that the use of additives is dominant in both rigid and flexible polyurethanes. From an environmental point of view, the reactive or reacted-in flame retardant has less likely to become an air and water pollutant in contrast with additive flame retardants. A good compromise is an oligomeric additive that has no vapor pressure or water-solubility, thus also meeting the environmental requirements.

Among reactive fire retardants, the Exolit OP 5xx series for polyurethanes in manufactured by Clariant. They are reactive flame retardants which integrate into the polyurethane matrix like a polyol. They show no migration, are more effective than the usual additive flame retardants and have only a minimal impact on the material properties. The three largest markets are transportation, furniture and carpet backing. Carpet backing uses mostly alumina trihydrate (ATH) as the flame-retardant additive, the other markets use an ever-increasing variety of approaches. A major fraction of the flexible polyurethane foams used in furniture is flame retarded. Additives are the dominant means, although much research has been expended on reactives.

When slightly volatile additives such as tris(1,3-dichloro-2-propyl) phosphate are used as flame retardants in automobile seating foams, there is often a detectable fogging of the inside of the windshield if the passenger compartment of the vehicle is warm. Other components of the foam, such as the catalysts and surfactants can also contribute to the fogging. There are various industry tests for windshield fogging. Where the monophosphate fails this test, it is usual to use diphosphates or oligomeric phosphates or phosphonates. The most widely used non-fogging flame retardant is tetrakis(2-chloroethyl) 2, 2-bis(chloromethyl)-1,3-propanediyl diphosphate.

The commercial development of several phosphorus-containing diols occurred in response to the need to flame-retard rigid urethane foams used for insulation in the transportation and construction industry. The earliest produced was a diol obtained from propylene oxide and dibutyl acid pyrophosphate [38]. A commercially significant phosphorus diol is diethyl bis(2-hydroxyethyl)aminomethylphosphonate (a trade mark for this chemical is Fyrol 6), especially useful in spray foams. It is synthesized by reaction of diethanolamine, formaldehyde and diethyl phosphite. It is stable in polyol-catalyst mixtures and imparts humid-aging resistance to foams. This is a reactive flame retardant, as the diol structure permits permanent incorporation of the phosphonate group into the urethane foam. The phosphonate linkage is on a side chain rather than in the backbone of the polyurethane, which increases the hydrolytic stability [39]. Most recent non-halogenated polyols containing phosphorus are utilized, as Clariant's hydroxy-terminated ethyl phosphate oligomer, as well as brominated diols.

5 Conclusions

Organophosphorus flame retardants are in an active state of development driven by their good performance and also by environmental and regulatory problems with the brominated flame retardants. Literature sources were mostly taken from the publications from the last decade, with reference to previous ground older publications. Nanoclay incorporation into polyurethane foams for increasing flame retardancy appears to be a new area for development of commercial future products, the same as with other industrial polymers. Current publications show that nanoclays exert improved flame retardant properties mainly when acting synergistically with conventional fire retardants.

References

1. Aaronson, A., Bright, D.: Oligomeric phosphate esters as flame retardants. Phosphorus, Sulfur, and Silicon and the Related Elements **109**, 83–86 1,4 (1996)
2. Allcock, H.: Chemistry and applications of polyphosphazenes. Wiley, New York (2003)
3. Andrae, N.: Durable and environmentally friendly flame retardants for synthetics. North Carolina State University, 2007. Cited: U.S. Consumer Product Safety Commission (CPSC) Staff. In Nomination of FR Chemicals for NTP Testing (2007)
4. Barick, A.K., Tripathy, D.K.: Effect of organically modified layered silicate nanoclay on the dynamic viscoelastic properties of thermoplastic polyurethane nanocomposites. 8th International Conference on Composite Science and Technology, ICCST8, Kuala Lumpur (2011)
5. Barikani, M., Askari, F., Barmar, M.: A comparison of the effect of different flame retardants on the compressive strength and fire behaviour of rigid polyurethane foams. Cell. Polym. **29**(6), 343–358 (2010)
6. Batorewicz, W., Hughes, K.: Some structure—activity correlations for halogen and phosphorus flame retardants in rigid urethane foam. Fire Retardants, International Symposium on Flammability and Fire Retardants, Proceedings, Cornwall, 1 May 1974 through 2 May 1974 (1974)
7. Bourbigot, S., Samyn, F., Turf, T., Duquesne, S.: Nanomorphology and reaction to fire of polyurethane and polyamide nanocomposites containing flame retardants. Polym. Degrad. Stab. **95**(3), 320–326 (2010)
8. Cao, X., Lee, J., Widya, T., Mocosko, C.: Polyurethane/clay nanocomposites foams: processing, structure and properties. Polymer **46**, 775–783 (2005)
9. Chattopadhyay, D., Webster, D.: Thermal stability and flame retardancy of polyurethanes. Prog. Polym. Sci. **34**, 1068–1133 (2009)
10. Chen, L., Wang, Y.: A review on flame retardant technology in China. Part I: development of flame retardants. Polym. Adv. Technol. **21**, 1–26 (2010)
11. Choi Y.: Synthesis of phosphorus-containing two-component polyurethane coatings and evaluation of their flame retardancy fire and polymers, Chap. 9, pp. 110–122. ACS Symposium Series, vol. 797, Washington DC (2001)
12. Dubois, P., Alexandre, M.: Polymer-layered silicate nanocomposites: preparation, properties and uses of a new class of materials. Mater. Sci. Eng. **28**, 1–63 (2000)
13. Duquesne, S., Le Bras, M., Bourbigot, S., Delobel, R., Vezin, H., Camino, G., Eling, B., Lindsay, C., Roels, T.: Expandable graphite: a fire retardant additive for polyurethane coatings. Fire Mater. **27**(3), 103–117 (2003)

14. Hu, Y., Song, L.: Nanocomposites with halogen and non intumescent phophorus flame retardant additives. In: Morgan, A.B., Wilkie, C.A. (eds.) Flame Retardant Polymer Nanocomposites. Wiley, New York (2007)
15. Huang, G., Jao, J., Li, Y., Han, L., Wang, X.: Functionalizing nano-montmorillonites by modification with intumescent flame retardant: preparation and application in polyurethane. Polym. Degrad. Stab. **95**, 245–253 (2010)
16. König, A., Kroke, E.: Methyl-DOPO-a new flame retardant for flexible polyurethane foam. Polym. Adv. Technol. **22**(1), 5–13 (2011)
17. Levchik, S., Weil, E.: Thermal decomposition, combustion and fire-retardancy of polyurethanes—a review of the recent literature. Polym. Int. **53**(11), 1585–1610 (2004)
18. Levchik, S., Weil, E.: A review of recent progress in phosphorus-based flame retardants. J. Fire Sci. **24**(5), 345–364 (2006)
19. Levchik, S., Weil, E.: Recent progress in flame retardancy of polyurethane and polyisocyanurate foams. Fire and Polymers IV, Chap. 22, pp. 280–290. ACS Symposium Series, vol. 922 (2006)
20. Modesti, M., Lorenzetti, A.: FR design for foam materials. In: Wilkie, A.C., Morgan, A.B. (eds.) Fire Retardancy of Polymeric Materials, 2nd edn, pp. 763–781. CRC Press, Boca Raton (2009)
21. Mondal, M., Chattopadhyay, P., Setua, D., Chas, N., Chattopadhyay, S.: Influence of surface-modified nanoclay on the self-organized nanostructure of segmented polyurethane composites. J. Fire Sci. **24**(5), 345–364 (2006)
22. Mouritz, A., Gibson, A.: Fire properties of polymer composite materials. Solid Mechanics and its Applications. Chapter 2. Thermal Decomposition of Composites on Fire. Springer, New York
23. Najafi-Mohajeri, N., Jayakody, C., Nelson, C.G.: Calorimetric analysis of modified polyurethane elastomers and foams with flame-retardant additives fire and polymers, Chap. 7, pp. 79–89. ACS Symposium Series, vol. 797 (2001)
24. Nelson, G. L., Wilkie, C. A.: Fire Retardancy in 2001, Chap. 1, pp. 1–6. Fire and Polymers—ACS Symposium Series, vol. 797 (2001)
25. NPCS Board of Consultants & Engineers: Handbook on Textile Auxiliaries, Dyes and Dye Intermediates Technology. Asia Pacific Business Press Inc. (2009)
26. Papazoglou, E.S.: Handbook of building materials for fire protection. In: Harper, C.A. (ed.) Chapter 4. McGraw-Hill, New York (2004)
27. Pinto, U., Yuan Visconte, L., Gallo, J., Reis Nunes, R.: Flame retardancy in thermoplastic polyurethane elastomers (TPU) with mica and aluminum trihydrate (ATH). Polym. Degrad. Stab. **69**(3), 1, 257–260 (2000)
28. Price, D., Liu, Y., Milnes, G. J., Hull, R., Kandola, B. K. and Horrocks, A. R.: An investigation into the mechanism of flame retardancy and smoke suppression by melamine in flexible polyurethane foam. Fire Mater., **26**, 201–206. (2002)
29. Quagliano, J.: Studies on incorporation of exfoliated bentonitic clays in polyurethane foams for increasing flame retardancy. International Conference on Structural Nano Composites (NANOSTRUC 2012) IOP Publishing IOP Conference Series: Materials Science and Engineering 40 (2012)
30. Sivriev, H., Borissov, G., Zabski, L., Walczyk, W., Jedlinski, Z.: Synthesis and studies of phosphorus-containing polyurethane foams based on tetrakis(hydroxymethyl) phosphonium chloride derivatives. J. Appl. Polym. Sci. **27**(11), 4137–4147 (1982)
31. Sivriev, H., Kaleva, V., Borissov, G., Zabski, L., Jedlinski, Z.: Rigid polyurethane foams with reduced flammability, modified with phosphorus- and nitrogen-containing polyol, obtained from tetrakis(hydroxymethyl)phosphonium chloride. Eur. Polym. J. **24**(4), 365–370 (1988)
32. Stefani, P., Esposito, L., Manfredi, L., Vazquez, A.: Mechanical properties of montmorillonite filled polyurethane foams. Conference on Science and Technology of Composite Materials COMAT, 169–170, Buenos Aires. ISBN: 987-544-162-7 (2005)

33. Thirumala, M., Khastgira, D., Singhaa, N., Manjunathb, B., Naik, Y.: Effect of a nanoclay on the mechanical, thermal and flame retardant properties of rigid polyurethane foam. J. Macromol. Sci., Part A 704–712 **46**(7) (2009)
34. Thirumala, M., Khastgira, D., Nando, G., Naik, Y., Singha, N.: Halogen-free flame retardant PUF: effect of melamine compounds on mechanical, thermal and flame retardant properties. Polym. Degrad. Stab. **95**, 1138–1145 (2010)
35. Valizadeh, M., Rezaei, M., Eyvazadeh, A.: Effect of nanoclay on the mechanical and thermal properties of rigid polyurethane/organoclay nanocomposite foams blown with cyclo and normal pentane mixture. Key Eng. Mater. **471–472**, 584 (2011)
36. Van der Veen, I., De Boaer, J.: Phosphorus flame retardants: properties, production, environmental occurrence, toxicity and analysis. Chemosphere **88**, 1119–1153 (2012)
37. Wang, Z., Pinnavaia, T.: Nanolayer reinforcement of elastomeric polyurethane. Chem. Mater. **10**, 3969–3971 (1998)
38. Weil, E.: Flame retardants (Phosphorus Compounds). In: Grayson, M. (ed.) Kirk-Othmer Encyclopaedia of Chemical Technology, vol. 10, 3rd edn. Wiley Interscience, New York (1980)
39. Weil, E.: Phosphorus-containing polymers. In: Kroschwitz, J. (ed.) Encyclopaedia of Polymer Science and Engineering, vol. 11, pp. 96–126. Wiley, New York (1988)
40. Weil, E., Levchik, S.: Commercial flame retardancy of polyurethanes. J. Fire Sci. **22**, 183–209 (2004)
41. Weil, E., Levchik, S.: Flame Retardants for Plastics and Textiles: Practical Applications, p. 71. Carl Hansen, Munich (2009)
42. Yildiz, B., Seydibeyoglu, M., Guner, F.: Polyurethane–zinc borate composites with high oxidative stability and flame retardancy. Polym. Degrad. Stab. **94**(7), 1072–1075 (2009)
43. Zammarano, M.: Thermoset fire retardant nanocomposites. In: Morgan, A.B., Wilkie, C.A. (eds.) Flame Retardant Polymer Nanocomposites. Wiley, New York (2007)
44. Zhang, S., Horrocks, A.R.: A review of flame retardant polypropylene fibres. Prog. Polym. Sci. **11**(28), 1517–1538 (2003)
45. Zhu, W., Weil, E., Mukhopadhyay, S.: Intumescent Flame-Retardant System of Phosphates and 5,5,5′,5′,5″,5″-hexamethyltris(1,3,2,-dioxaphosphorinanemethan)amine 2,2′,2″-trioxide for Polyolefins. J. Appl. Polym. Sci. **62**, 2267–2280 (1996)

Ecological Assessment of Nano Materials for the Production of Electrostatic/Electrochemical Energy Storage Systems

M. Weil, H. Dura, B. Simon, M. Baumann, B. Zimmermann,
S. Ziemann, C. Lei, F. Markoulidis, T. Lekakou and M. Decker

Abstract Electrochemical double layer capacitors, also known as supercapacitors are considered as a promising option for stationary or mobile electric energy storage. At present lithium ion and nickel metal hydride batteries are used for automotive applications. In comparison to this type of batteries supercapacitors possess a high specific power, but a relatively low specific energy. Therefore, the goal of ongoing research is to develop a new generation of supercapacitors with high specific power and high specific energy. To reach this development goal particularly nano materials are under investigation on cell level. In the presented study the ecological implications (regarding known environmental effects) of carbon based nano materials are analysed using Life Cycle Assessment (LCA). Major attention is paid to efficiency gains of nano material production due to scaling up of such processes from laboratory to industrial production scales. Furthermore, a developed approach will be displayed, how to assess the environmental impact of nano materials on an automotive system level over the whole life cycle.

M. Weil (✉) · H. Dura · B. Simon · M. Baumann · B. Zimmermann ·
S.Ziemann · M. Decker
Karlsruher Institute of Technology (KIT), Institute of Technology Assessment and Systems
Analysis (ITAS), POB 3640 76021 Karlsruhe, Germany
e-mail: marcel.weil@kit.edu

M. Weil · B. Simon
Helmholtz Institute Ulm for Electrochemical Energy Storage, Hermann-von-Helmholtz-
Platz 1 76344 Eggenstein-Leopoldshafen, Germany

C. Lei · F. Markoulidis · T. Lekakou
Department of Mechanical Engineering Sciences, University of Surrey,
Guildford GU2 7XH, UK

J. Njuguna (ed.), *Structural Nanocomposites*, Engineering Materials,
DOI: 10.1007/978-3-642-40322-4_12, © Springer-Verlag Berlin Heidelberg 2013

1 Introduction

Electro mobility is considered as one of the most promising technologies for replacing fossil fuelled mobility. Depending on the electricity generation mixture of a region, electric cars can have already today an environmental advantage against petrol cars [1]. With a growing share in regenerative production of electric energy this advantage will substantially increase, particularly when synergy effects of the progressively merging energy and transport networks will appear. The key element of an electric vehicle is the energy storage system. At present for hybrid and full electric vehicles commonly nickel metal hydride and lithium based batteries are used. If electro mobility is to succeed there are a number of obstacles to be overcome including great improvements in energy density, peak power, life time (calendar and cycle stability), operation temperature and costs. Furthermore, the energy storage systems should rely on available resources [2], be easily recyclable and should have an overall good environmental performance. None of the present energy storage systems fulfil the requirements sufficiently, that means all electrochemical and electrostatic storage systems needs significant further improvement. Nevertheless supercapacitors belong to a group of very promising energy storage systems, which could enable the breakthrough for full electric or hybrid power trains.

2 Double Layer Capacitor

Within the group of capacitors the electrochemical and electrostatic (dry) capacitors can be basically distinguished (Fig. 1). The electrostatic capacitor with no electrolyte has no importance for energy intensive mobile or stationary

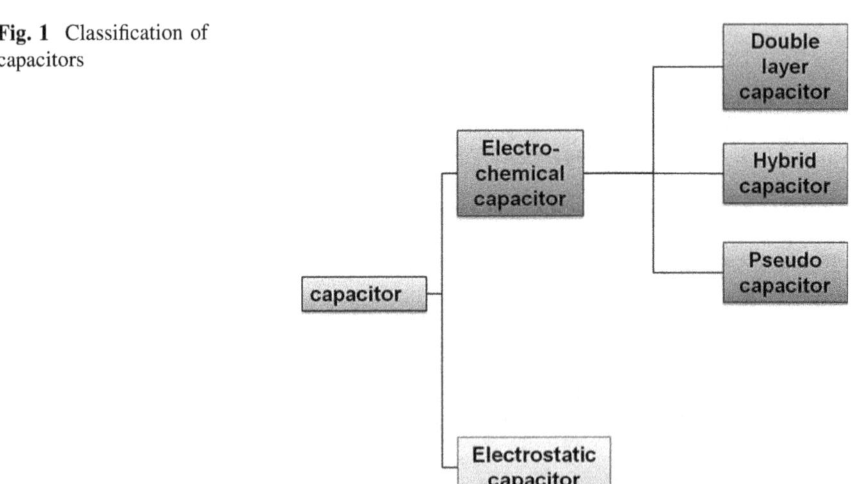

Fig. 1 Classification of capacitors

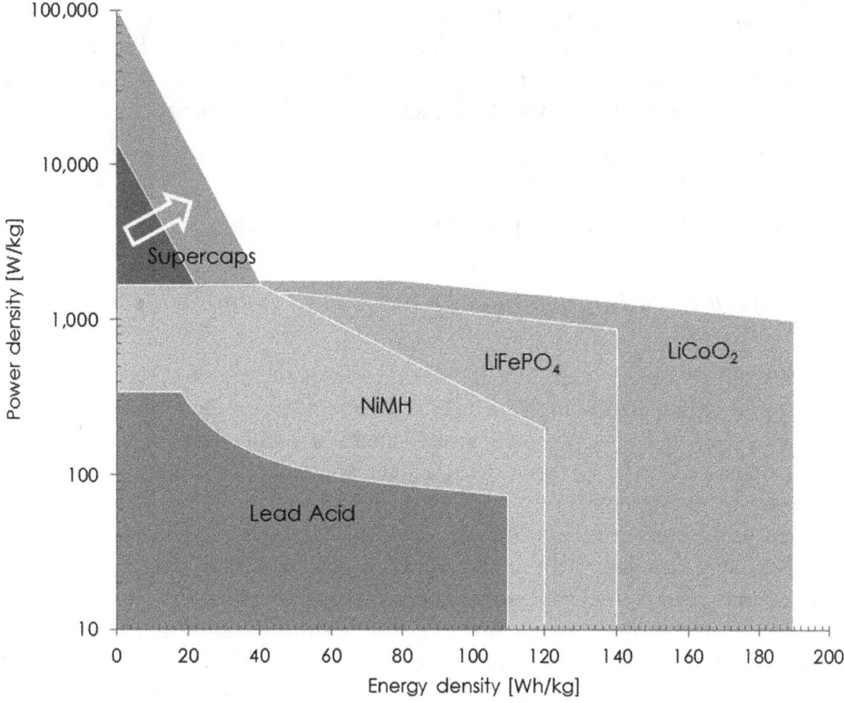

Fig. 2 Specific energy and specific power of various energy storage systems

applications, due to its relatively low capacity, which is in the range of a few microfarads or lower. In contrast, the capacity of electrochemical double layer capacitors is thousands of times higher.

Thus double layer capacitors are a very promising candidate as an alternative energy storage system in automotive applications.

They have the advantage of high power density, high cycle stability, high calendar life time, a broad temperature range for operation and rely only on good available raw materials like carbon. However, they are facing a number of problems including low energy density and high cost.

In Fig. 2 the performance regarding power and energy density of capacitors, supercapacitors and batteries are shown (c.f. [3]). In this way, supercapacitors can be considered as a link between electrostatic capacitors and electrochemical batteries. Also regarding the cycle life time supercapacitors take a centre position between electrostatic capacitors and electrochemical batteries.

Despite of the high price and the low energy density, there are some first applications of supercapacitors in vehicles. Buses for public transport in the USA, Russia and China are already in use with no traction batteries, but with supercapacitors as the only electric energy storage system [4].

But to enable the mobility with supercapacitors (with low energy density) all the bus stops are equipped with an induction charge stations. The induction charging avoids is the physical need of a wire and plug connector. In this way the buses can be recharged very simple at each stop through induction.

3 Improved Supercapacitors by Nano Materials

3.1 Performance Implications by the Use of Nano Materials

The structure of a super capacitor is related to the structure of a battery encompassing current collectors, high surface area electrodes, an electrolyte and separator. However, the energy is not stored chemically but electrostatically in an electric double layer (EDL) forming at the interface between the electrodes and the electrolyte as depicted in Fig. 3 [5]. Thus, the amount of energy stored is proportional to the surface area of the electrode [6], the higher the surface area, the greater the specific energy. Since carbonaceous nano materials offer great surface areas at low specific weight and volume, their application for supercapacitor electrodes is increasingly being investigated. As already shown in Fig. 3 supercapacitors have a very good power density, but need great improvement in energy density in order to increase their applicability for automotive energy storage.

To overcome the stated problem a second generation of electrochemical double layer capacitors is under development.

Often research largely focuses on the development of high area electrodes. This may be achieved using carbonaceous nano materials, as these offer high surface to volume and weight ratio.

The EU project "Autosupercap" focuses on the development of supercapacitors with high power and high energy density while reducing weight and costs. To achieve these ambitious goals different kinds of carbon based nano objects for the production of the electrodes are being investigated and tested [7].

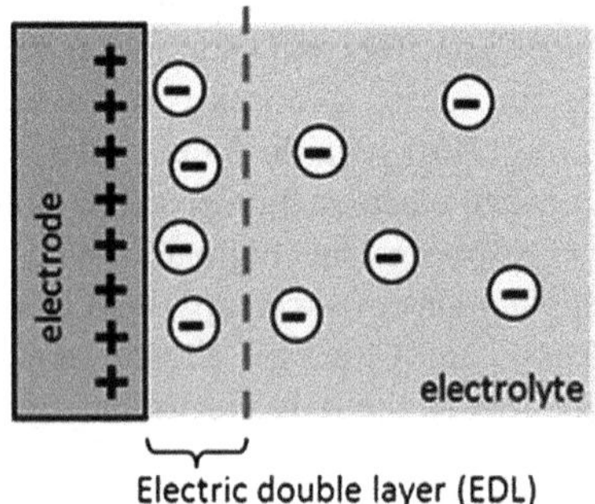

Fig. 3 Simplified drawing of the electric double layer [18]

The energy and power performance of the capacitor as well as the lifetime are strongly influenced by the internal resistance. The internal resistance is the product of the carbon material itself, the carbon particle–particle contact, and the carbon-current collector contact [7].

Normally carbon based aerogels and powders are used for the synthesis of electrodes. A major improvement of the electrode resistance can be achieved by using conducting additives, such as black carbon or carbon nanotubes [8].

In both ways the intelligent use of nano materials may improve the techno-logical performance of the supercapacitor.

At present there a several investigations which focus on the use of graphene or graphene derivate for the development of high performance electrochemical double layer capacitors [9, 10]. A research group at the University of Alberta and the National Research Council of Canada have synthesized a new graphene material with a sponge-like structure. They could show, that this material, which has a 3D macroporous structure, can be used for the manufacturing of double layer capacitors electrodes. The produced capacitors with these new electrodes on a laboratory scale have an incredible ultrahigh power density of approximately 48,000 W/kg, and provide still a relatively high energy density of 7.1 Wh/kg [10].

3.2 Production of Nano Materials and Their Environmental Implications

High performance carbonaceous nano materials may be produced by three methods: arc discharge, laser ablation and chemical vapour deposition (CVD). The arc discharge method and laser ablation are based on vaporization of carbon feedstock. In the arc method two carbon rods are placed in inert gas or liquid nitrogen. A high temperature discharge between the two rods causes a vaporization of the surface of one of the carbon rods while deposition takes place on the other. In laser ablation vaporization of graphite rods is attained using a dual pulsed laser and a subsequent thermal treatment. Due to the underlying principal of evaporation of the carbon source, these two production approaches appear to be difficult for up scaling to an industrial level [11].

Chemical Vapour Deposition (CVD) on the other hand has the greatest potential for large scale production. In a batch process, plates containing a catalyst are placed on the bottom of a reaction chamber, which is filled with a synthetic gas mixture. Upon heating, the gas reacts with the catalyst forming nanotubes on the catalyst surface [12]. A further development of the CVD process uses a fluidized bed reactor and allows the continuous and mass production of CNTs [13].The production of nano materials is very often a energy and resource intensive process, which may partially explain the high price of these products. Using such materials increases the efforts for the production of the supercapacitor. To reach a net environmental and economic gain the efforts for the production have to be at least compensated if not significantly outweighed by a better environmental performance during the use phase and at the end of life treatment of a nano enabled supercapacitor.

In this regard an environmental assessment of the entire life cycle is a key element. But until today only few LCAs exist on the production of nano materials [14]. That means the real environmental impacts of such materials are neither well assessed nor well understood. Furthermore, there is a need for additional models that describe adequately the paths of release and effects of nano materials throughout their life cycle [15]. This problem is further enhanced by the high variability of nano materials even within one material group, i.e. a multi walled carbon nano tube (MWCNT) is not necessarily comparable to another MWCNT. The associated environmental impact depends on several factors including processes of synthesis, purification and dispersion as well as functionalization. In this study it is the goal to investigate the environmental impact of commonly used carbon based nano materials for electrode production.

4 LCA Methodology, System Boundary and Approach

Life cycle assessment (LCA) offers a methodological framework for a consistent assessment and comparison of different technological systems with respect to various types of environmental impacts. According to the framework set forth by DIN EN ISO 14040 [16] and 14044 [17] a LCA is comprised of four work packages starting with a goal and scope definition followed by a life cycle inventory (LCI) and life cycle impact assessment (LCIA). The fourth work package, the evaluation is included as an iterative process which is performed during all three prior stages. In the first step the objective of the study is to be stated and further details explained including information on how the study will be performed, the functional unit, system boundaries as well as allocation methods and impact categories used. The LCI is purely descriptive documenting all input and output flows into and out of the earlier defined system. In a subsequent step, the LCIA is performed in order to determine environmental repercussions on the basis of the flow network.

For the presented work the cumulated energy demand (CED) is displayed and used as a screening indicator for environmental burdens as it depicts the direct and indirect energy used e.g. from use of raw materials.

The considered system boundaries of the investigated nano materials are from "cradle to gate", meaning all pre chains of the production process are included. A general problem of LCA investigations for nano materials is the complexity, which is described in [12], but more often also the restricted access to detailed data regarding the production process [18]. It has to be stated that at present LCA can not consider and evaluate the potential human toxicity and ecological impact of nano objects and materials.

5 LCA Results of Carbonaceous Materials on Nano and Micro Scale

The raw materials, the production technology and energy requirements of the investigated carbon derivates are quite different, thus it was expected to identify the f'differences also in the LCA results. The calculated CED values for the different carbonaceous materials and production procedures are shown in Fig. 4.

Most striking is the extremely high energy intensity of SWCNT production by laser ablation creating an energy demand of more than 200.000 MJ/kg, more than twice as much as SWCNT produced by CVD. MWCNT cause a much smaller CED, although this also strongly depends on the production method applied. Although significant reductions may be achieved with the fluidized bed and CVD production processes, in comparison with the production of black carbon and activated carbon both types of the carbon nano materials create a relatively high CED, from 4 fold larger (MWCNT fluidized bed) up to approximately 4 orders of magnitude larger (SWCNT laser ablation).

Another problematic issue regarding the production of MWCNT by a fluidized bed reactor is the tendency to have high impurities and agglomerations of CNTs, which might cause in addition (for some application fields) further efforts for purification and dispersion. There exist some evidences, that the purification and dispersion of nano objects can cause a relatively high environmental impact, in

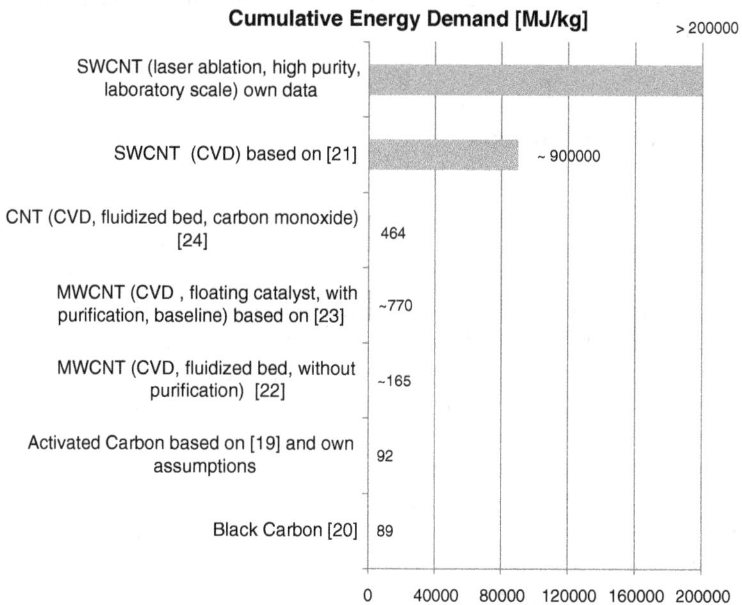

Fig. 4 Comparison of different carbon based materials for electrode production regarding the cumulative energy demand (CED)

some cases even significantly higher than the production itself, depending on the considered life cycle impact category [12].

6 Assessment of Nano Materials on Automotive System Level

The focus on the production phase is not sufficient for an ecological and economic evaluation of nano materials and to understand their inherent value.

In order to assess the economic and environmental effects of the use of nano materials in new developed supercapacitor the whole life cycle of a use case has to be considered. For this the life cycle models regarding an automotive application have to cover the production phase, use phase, and end of life phase.

In Fig. 5 the system boundary for an ecological assessment of a supercapacitor is shown. It is obvious that the production of carbon based nano materials is just one process of a long process chain. Thus there is a potential danger, that the positive and negative contribution of nano materials reaches not a high significance, due to the dominance of other processes.

But only by considering the whole life cycle the net value of nano materials with respect to economic and ecological aspects can be evaluated. Furthermore, regarding the results, general statements are not possible, due to the strong dependency of the application field. Thus, this kind of prospective evaluation has to be conducted for each single nano material and for each application field of supercapacitors.

7 Summary and Outlook

Present available electrochemical double layer capacitor technology is too expensive and the energy density is too low for a broad mass application in electric vehicles. The next generation of electrochemical double layer capacitors, which are under development, uses particularly nano materials to improve the technical performance and reduce costs. It could be shown, that the production of carboniferous materials on nano scale can have a significant environmental impact. Certainly the environmental impact of nano materials depends on several factors, but the upscale of the synthesis process from laboratory to industrial scale looks very promising in significantly reducing economic and ecological efforts and related impacts. Unfortunately the scale up process of nano particle production might have an important influence on the quality of the nano materials. The quality losses may cause a higher use of the nano materials or cause further efforts for the improvement of their quality, e.g. by an additional dispersion or purification process. To evaluate the possible environmental and economic savings due to the improved technical performance, the supercapacitor has to be investigated on a

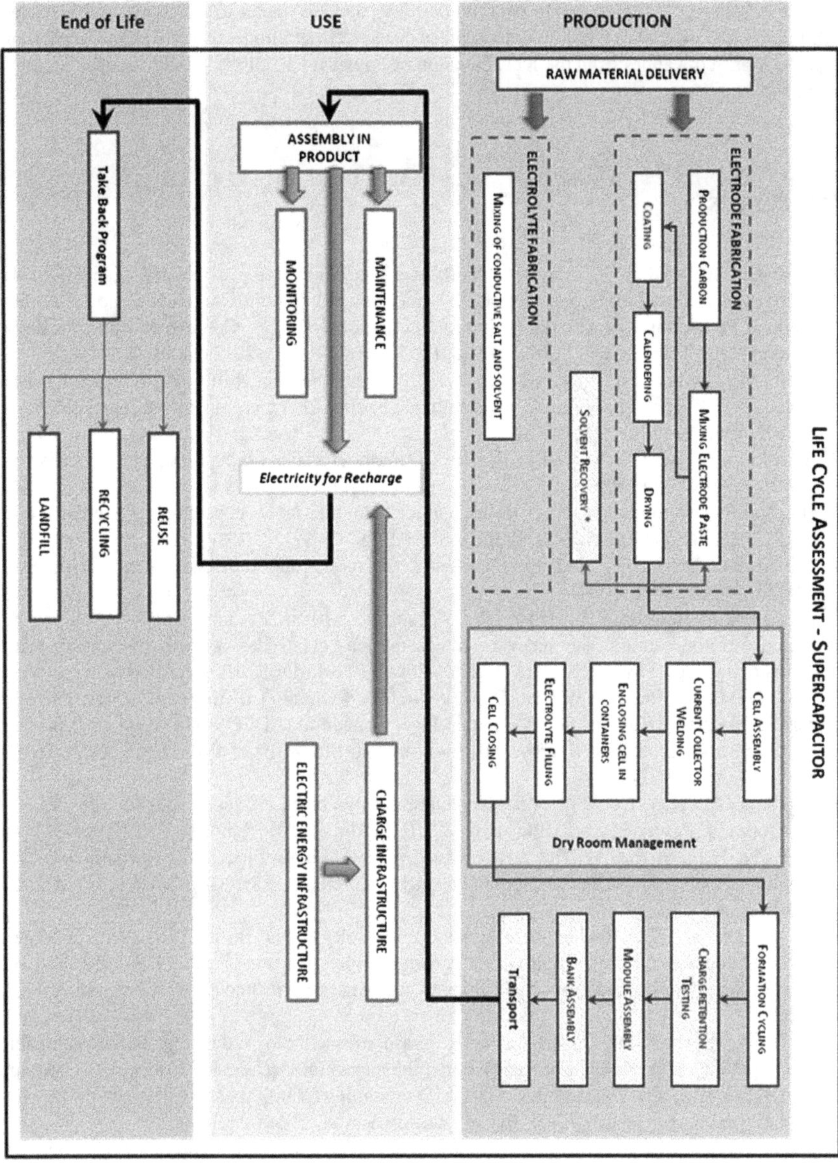

Fig. 5 System boundaries for life cycle assessment of supercapacitors

product system level. This will be conducted in the future for a simulated automotive application with measured data from a test rig. By the consideration of the whole life cycle the real value (in respects of economic and ecological savings) of nano materials applied in energy storage systems can be assessed.

Acknowledgments This research was funded by the European Commission FP7 project "AUTOSUPERCAP". We would like to express our sincere gratitude also to Regina Fischer and Prof. Dr. Manfred Kappes from the KIT, Division of Physical Chemistry of Microscopic Systems, Germany.

References

1. Baumann, M., Balint, S., Dura, H., Weil, M.: The contribution of electric vehicles to the changes of airborne emissions. In: IEEE—International Energy Conference and Exhibition (Hrsg.): Proceedings of the IEEE-Xplore International Energy Conference and Exhibition (EnergyCon), Florenz, 09.-12.09.2012. 2012, S. 1.189-1.194. (2012). doi:978-1-4673-1

2. Weil, M., Ziemann, S., Schebek, L.: How to access the availability of resources for new technologies? Case study: Lithium a strategic metal for emerging technologies. Rev. de Metall. **106**(12), 554–558 (2009)

3. Bertoldi, O., Berger, S.: WP2_5Energy_3Thermoelectricity.pdf. European commission, observatory NANO (2009)

4. Harrop, P., Zhitomirsky, V.: Electrochemical double layer capacitors: Supercapacitors 2013–2023. IDTechEx. www.IDTechEx.com/EDLC (2012)

5. Winter, M., Brodd, R.J.: What are batteries, fuel cells, and supercapacitors? Chem. Rev. **104**(10), 4245–4269 (2004)

6. Diederich, L., Barborini, E., Piseri, P., Podesta, A., Milani, P., Schneuwly, A., Gallay, R.: Supercapacitors based on nanostructured carbon electrodes grown by cluster-beam deposition. Appl. Phys. Lett. **75**(17), 2662–2664 (1999). doi:10.1063/1.125111

7. Lei, C., Markoulidis, F., Wilson, P., Lekakou C.: Reduction of the internal resistance of carbon electrodes for an electric double-layer capacitor (EDLC). The First International Conference on Materials, Energy and Environments (ICMEE), Toledo, Ohio, USA, 09–11 May 2012

8. Snook, G.A., Kao, P., Best, A.S.: Conducting-polymer-based supercapacitor devices and electrodes. J. Power Sources **196**(1), 1–12 (2011). doi:10.1016/j.jpowsour.2010.06.084

9. El-Kady, M.F., Strong, V., Dubin, S., Kaner, R.B.: Laser scribing of high-performance and flexible graphene-based electrochemical capacitors. Science. **335**(6074), 1326–1330. 16 Mar 2012

10. Xu, Z., et al.: Electrochemical supercapacitor electrodes from sponge-like graphene nanoarchitectures with ultrahigh power density. J. Phys. Chem. Lett **3**, 2928–2933 (2012)

11. Wilson, M.: Nanotechnology: Basic science and emerging technologies. Chapman & Hall/CRC

12. Weil, M.: System analysis in the early phase of technology development—responsible development and production of carbon nanotube paper. In: Decker, M., Grunwald, A., Knapp M. (Hrsg.): Der Systemblick auf Innovation—Technikfolgenabschätzung in der Technikgestaltung, pp. 301–312. Berlin, Edition Sigma (2012)

13. Howe, C.S., Harris, A.T.: A review of carbon nanotube synthesis via fluidized-bed chemical vapor deposition. Ind. Eng. Chem. Res. **46**(4), 997–1012 (2007)

14. Hischier, R., Walser, T.: Life cycle assessment of engineered nanomaterials: State of the art and strategies to overcome existing gaps. Sci. Total Environ. **425**, 271–282 (2012). doi:10.1016/j.scitotenv.2012.03.001

15. Fries, R., Greßler, S., Simkó, M.: Kohlenstoff-Nanoröhrchen (Carbon Nanotubes)—Teil ll: Risiken und Regulierungen. *nano trust dossiers* (024) (Mai)

16. DIN-ISO 14040.: Environmental management life cycle assessment—requirements and guidelines 14040 (2006)

17. DIN-ISO 14044.: Environmental management-environmental performance evaluation—requirements and guidelines 14044 (2006–10)

18. Weil, M., Dura, H., Simon, B., Baumann, M., Zimmermann, B., Ziemann, S., Lei, C., Markoulidis, F., Lekakou, T., Decker, M.: Ecological assessment of nano-enabled supercapacitors for automotive applications. IOP Conf. Ser.: Mater. Sci. Eng. **40** (2012)
19. Noijuntira, I., Kittisupakorn, P.: Life cycle assessment for the activated carbon production by coconut shells and palm-oil shells. Thailand, Jul 2009
20. Ecoinvent life cycle inventory database. swiss centre for life cycle inventories. www.ecoinvent.org
21. Healy, M.L., Dahlben, L.J., Isaacs, J.A.: Environmental assessment of single-walled carbon nanotube processes. J. Ind. Ecol. **12**(3), 376–393 (2008)
22. Steinfeld, M., von Gleich, A.: Entlastungseffekte für die Umwelt durch nanotechnische Verfahren und Produkte. Umweltbundesamt (UBA) (2008)
23. Kushnir, D., Saden, B.A.: Energy requirements of carbon nanoparticles production. J. Ind. Ecol. **12**(3), 360–375 (2008)
24. Zimmermann, B.: Integration of carbon nanotubes in lithium-ion traction batteries from an environmental perspective. Master-Thesis. Karlsruhe Institute of Technology and University of Applied Science Munich

Erratum to: Thermoplastic Nanocomposites with Carbon Nanotubes

Shyam Sathyanarayana and Christof Hübner

Erratum to:
J. Njuguna (ed.), *Structural Nanocomposites*,
DOI 10.1007/978-3-642-40322-4_2

The author affiliation for the second chapter was wrong. The correct affiliation is:

S. Sathyanarayana

Advanced Materials & Systems Research, BASF SE, B001, Carl-Bosch-Strasse 38, 67056 Ludwigshafen, Germany
e-mail: Shyam.Sathyanarayana@basf.com

C. Hübner

Polymer Engineering, Fraunhofer Institute für Chemische Technologie ICT, Joseph von Fraunhofer Strasse 7, 76327 Pfinztal, Germany
e-mail: Christof.Huebner@ict.fraunhofer.de

The online version of the original chapter can be found under DOI 10.1007/978-3-642-40322-4_2

S. Sathyanarayana (✉)
Advanced Materials & Systems Research, BASF SE, B001, Carl-Bosch-Strasse 38, 67056 Ludwigshafen, Germany
e-mail: Shyam.Sathyanarayana@basf.com

C. Hübner
Polymer Engineering, Fraunhofer Institute für Chemische Technologie ICT, Joseph von Fraunhofer Strasse 7, 76327 Pfinztal, Germany
e-mail: Christof.Huebner@ict.fraunhofer.de

J. Njuguna (ed.), *Structural Nanocomposites*, Engineering Materials,
DOI: 10.1007/978-3-642-40322-4_13, © Springer-Verlag Berlin Heidelberg 2013